A Most Amazing Scene of Wonders

THOMAS J. WILSON PRIZE

The Board of Syndics of Harvard University Press has awarded this book the thirty-sixth annual Thomas J. Wilson Prize, honoring the late director of the Press. The Prize is awarded to the book chosen by the Syndics as the best first book accepted by the Press during the calendar year.

A Most Amazing Scene of Wonders

ELECTRICITY AND
ENLIGHTENMENT
IN EARLY AMERICA

James Delbourgo

HARVARD UNIVERSITY PRESS
CAMBRIDGE, MASSACHUSETTS
LONDON, ENGLAND
2006

Copyright © 2006 by the President and Fellows
of Harvard College
All rights reserved
Printed in the United States of America

Library of Congress Cataloging-in-Publication Data

Delbourgo, James, 1972–
A most amazing scene of wonders : electricity and enlightenment
in early America / James Delbourgo.
p. cm.
Includes bibliographical references and index.
ISBN-13: 978-0-674-02299-7 (alk. paper)
ISBN-10: 0-674-02299-8 (alk. paper)
1. United States—Intellectual life—18th century.   2. Electricity—Experiments—
History—18th century.   3. Enlightenment—United States.   4. Science—
United States—History—18th century.   5. Science—Social aspects—
United States—History—18th century.   I. Title.
E162.D395 2006
303.48′3097309033—dc22   2006041146

Designed by Gwen Nefsky Frankfeldt

For my brother
David John Delbourgo
(1963–2003)

Contents

List of Illustrations ix

Introduction: Seizing the Lightning 1

1 Atlantic Circuits 14

2 Lightning Rods and the Direction of Nature 50

3 Wonderful Recreations 87

4 Electrical Politics and Political Electricity 129

5 How to Handle an Electric Eel 165

6 Electrical Humanitarianism 200

7 Electricity as Common Sense 239

Conclusion: What Is American Enlightenment? 278

Notes 285
Illustration Sources 355
Acknowledgments 357
Index 359

Illustrations

1. The Myth of Electricity and Revolution 4–5
2. Electricity as an Atlantic Science 19
3. Electrical Games 27
4. Generating Enlightenment 35
5. Illustrating Electricity 36–37
6. Heroic Self-Evidence 56–57
7. Franklin as a Bookish Natural Philosopher 61
8. The Exploding Thunder House 73
9. The Direction of Lightning 79
10. The Spectacle of Enlightenment 94–95
11. Science in the Parlor 112–113
12. Franklin as Plenipotentiary 155
13. Electricity and Revolution 160–161
14. The Conspirators 163
15. An Electric El Dorado 169
16. The Electric Eel as an Organic Machine 196
17. Perkins's Tractors: Electricity or Imagination? 241

To find the truth out about a certain thing,
I would like to find it out through a game.

        BALDASSARE CASTIGLIONE
        THE BOOK OF THE COURTIER, 1528

INTRODUCTION

Seizing the Lightning

Electricity is often called wonderful, beautiful; but . . . The beauty of electricity or of any other force is not that the power is mysterious, and unexpected, touching every sense at unawares in turn, but that it is under *law*, and that the taught intellect can govern it largely.

MICHAEL FARADAY, 1858, QUOTED BY I. BERNARD COHEN IN 1941

*I*N the summer of 1771, Loammi Baldwin of Woburn, Massachusetts, set out in search of a thunderstorm in which to fly his carefully constructed "electrical kite." Wrought from a series of metal wires, its underside covered with silk, the kite measured four and a half feet in length and two feet across, with a line made of hard cord. Baldwin, a sometime designer of lightning rods who later served as a colonel in the American Revolution and a civil engineer in the early republic, had in his youth attended lectures on natural philosophy given by John Winthrop IV (a descendant of the Puritan founder) at Harvard College. These included discussions of Benjamin Franklin's theories and experiments on electricity. "My design," Baldwin later explained, "was to make some experiments in the time of a thunder-shower." He set out, in other words, to repeat one of the most dramatic experiments of the eighteenth century: Franklin's drawing electricity from the sky, which demonstrated that lightning was a form of electricity and led to the invention of the lightning rod. One July day, a violent storm approached Baldwin's home from the northwest, convulsing the clouds with gale-force winds. Baldwin eagerly advanced into the storm and began to raise his kite. Despite "the most piercing shafts of lightning, and tremendous thunder that I had ever beheld or heard," he did not relent. But what followed was less an experiment than a bizarre episode of sensory disorientation in which his very life seemed suddenly at risk:

> I discovered a rare medium of fire between my eyes and the kite.—I cast my eyes towards the ground;—the same appearance was there.—I turned myself around;—the same appearance [of fire] still between me and every object I cast my eyes upon.—I felt myself somewhat alarmed at the appearance. I stood, however, and reasoned with myself upon the cause, for some time, but gained very little satisfaction,—the same fiery atmosphere surrounded me, only more bright and apparent. I was about to discontinue my experiments for that time; but reason accused imagination with error; and supposing it might possibly be only fancy, not knowing the cause of such an appearance, and feeling no very bad effects from it, I continued to raise the kite.

The cloud, Baldwin observed, was a mass of unbalanced electrical charges, almost completely obscuring the sky above him. The din of its "incessant rattling" filled his ears. The fiery atmosphere increased and extended. Feeling "weakness in [his] joints and limbs," a "kind of listless feeling" seized him. Finally, reluctantly, he drew in the kite and took cover in his house. What had happened? Baldwin did not know until he spoke with his family, who had been "vastly more surprised" than he at the "astonishing" scene they had witnessed. "I appeared to them . . . to be in the midst of a large bright flame of fire, attended with flashings; and [they] expected, every moment, to see me fall a sacrifice to the flame." How to explain what had occurred? His report of this "very curious Appearance of the electrical Fluid," published some years later in the inaugural issue of *Memoirs of the American Academy of Arts and Sciences,* one of the first American scientific journals, left the question unanswered. "I shall make no remarks, at this time, upon the cause," he concluded, "but leave it for the present to the consideration of the learned." Whether the experiment failed because of the direct physical effect of the electrical atmosphere, or the work of his imagination, he could not say.[1]

Since almost the moment of their performance, Benjamin Franklin's electrical experiments leapt onto an international stage. Already in the 1750s, the philosopher Immanuel Kant had christened the lightning rod's inventor a "modern Prometheus" who stole fire from the heavens to give to mankind. Two decades later, the new Prometheus turned revolutionary. "Eripuit fulmen coelo sceptrumque tirannis—he seized the lightning from the heavens and the scepter from the tyrants," ran the slogan attributed to Anne-Robert Jacques Turgot, the French comptroller-general: the stealer of fire now stole liberty for new republics. This vision was immortalized in *Au Génie de Franklin* by Jean-Honoré Fragonard in 1779, the year after the Franco-American alliance against Great Britain in the American Revolution (Figure 1). This widely reproduced etching linked the cause of American republicanism to the power of experimental science, deifying Franklin as Prometheus *and* Jupiter, a God who stole fire to aid humanity by vanquishing the enemies of freedom (figures representing Avarice and Tyranny incarnate Britain in the bottom left corner, while Mars and Minerva—France—help Franklin to protect the American republic, shown as a woman bearing a fasces). Whereas power was thought to menace liberty under the ancien régime, here was a revolutionary rendition of Francis Bacon's claim that knowledge was power, and that power over nature meant freedom and security.[2]

Franklin's virtues—innate common sense, honest empiricism, secular practical utility, and civic-minded benevolence—constitute a myth of both the origins of America and the origins of American science. Homespun yet heroic, Franklin personifies an American ideal of modest mastery and, more than any other figure, American Enlightenment as the progress of reason against superstition. Baldwin's tale, however, signals the existence of a far more shadowy epistemological landscape, unimaginable from such happy myths, where the difference between knowledge and error was decidedly unclear. "Reason accused imagination with error," Baldwin wrote. The Enlightenment was, of course, famously the "Age of Reason." But reason did not and could not operate alone in human engagements with electricity, and Baldwin unfurls a language of nonrational experience that opens up American understandings of nature in the Enlightenment: imagination, error, fancy, surprise, sublimity, astonishment, awe, wonder, terror. Because experiments with electricity used the human body as an instrument of knowledge, they engaged the passions as well as the rational faculties, sometimes overwhelmingly so. Baldwin's is an extreme but not atypical case in point. It exemplifies the problem of secularized enthusiasm that bedeviled electricity in the Enlightenment, and with which this book is largely concerned: because there was no physical separation between the experimenter and the phenomenon under observation, distinguishing external causes from ones originating only in the mind often proved difficult, if not impossible.

Today, the idea of bodily experiences of electricity conjures up unnerving visions of electroconvulsive therapy, lightning strikes, electrocution as torture and execution, and unholy experimentation epitomized by the fable of Frankenstein. The very notion of electricity coursing through the human body is illicit, even perverse, associated as it is with the bizarre, madness, and death. Electricity as we now know it is silently contained and circulated by machines to provide the power on which modern life depends, but seemingly untouchable therefore, never to be known by us firsthand and surely inimical to our

---

1. THE MYTH OF ELECTRICITY AND REVOLUTION

This much-reproduced etching links the invention of the lightning rod to American independence, portraying Franklin as a Prometheus/Jupiter figure whose godlike mastery of nature allows him to overcome British enemies with electrical lightning bolts. This link between experiment and republicanism was forged a generation after the lightning rod's invention as propaganda in the American Revolution. This French, pagan view of Franklin as a god of reason is at odds with American emphases on the limits of reason and human dependence on divine will.

soft flesh.[3] Things were different in the eighteenth century, an era in which Kant summed up enlightenment through the phrase *sapere aude:* dare to know. In electrifying their bodies, people from different walks of life—not just scientific and medical experimenters—dared to know and use electricity for themselves. Unlike in our own day, electricity was an intimate and exciting presence in the eighteenth century, known through the senses, through the very vessels and fibers of the body. Such experiences of electrification that we now shrink from were a vogue and a fascination. More than this, they opened a path to enlightenment, toward rational understanding and control, yet also to wonder and the unpredictability of strange new experience.

What were eighteenth-century Americans seeking through electricity, and what kind of enlightenment did they find in it? This book explores the origins of American science and the meaning of the American Enlightenment. Thanks to historians of science, we know a great deal about the changing character of natural knowledge in early modern Europe. We also know much about Franklin's scientific experiments, although their character and significance have often been distorted by patriotic celebration. But the history of electricity beyond Franklin's Philadelphia laboratory, and indeed the history of science as a whole in early America, remain underexplored. This neglect has occurred because, for a long time, historians of science canonized lives of heroic achievement; besides Franklin's spectacular achievements, few if any early American scientific lives seemed worthy of recounting. Early American intellectual history, which has often limited itself to retelling familiar stories of the rise of republican politics and Protestant pluralism, remains somewhat on the margins of colonial American studies in the wake of the "new social history." Furthermore, in sharp contrast to early modern European or Latin American studies, early American history all but ignores the concept of the Enlightenment. The very notion seems elitist, socially irrelevant, anachronistic, and Eurocentric. The result is that we have no cultural history of science and enlightenment in America.

This book opens up this cultural history by moving beyond the lone figure of Franklin to offer a new approach to the American Enlightenment through stories of fleshy bodies and experimental machines, rather than Protestant sermons and republican constitutions (although these bodies, machines, sermons, and constitutions will turn out to be related). The view that the history of early modern knowledge-making is essentially an asocial, elite history of ideas, somehow separate from the main currents of cultural life, is misguided. In science, understanding the universal operations of nature always requires particular instruments used by individual actors in specific settings. So I emphasize throughout the centrality of material technologies—in particular, the human body—to the practice of experiment, precisely to see how the social and intellectual aspects of culture were enmeshed in the pursuit of electricity. Through the individual stories and careers of non-elites, many told for the first time here, it becomes clear that the influence of enlightened science extended more broadly than among a colonial upper class. American science came to life as a culture of experimental performance in a commercial public sphere spanning the Atlantic Ocean, and it allowed a variety of Americans to experience electricity in ways connected to their local senses of self and society. Both electricity and enlightenment mattered in America as experience and as ideas; they offered compelling visions of society, nature, and God rooted in practical social action.[4]

But what *was* enlightenment? In introducing his edition of Franklin's *Experiments and Observations on Electricity* in 1941, the leading historian of Franklinist electricity, I. Bernard Cohen, used the words of the nineteenth-century British scientist Michael Faraday to align Franklin exclusively with the idea that enlightenment was driven by the modernizing progress of reason. Electricity, Cohen insisted through

Faraday, was beautiful because it was lawful and governable, not because it was somehow wonderful or mysterious. Again, eighteenth-century witnesses saw things differently. In 1750, the indefatigable British diarist Horace Walpole described the earthquake that shook London that year as a "wonderful commodity." In so saying, Walpole drew attention to the multiplicity of comment generated by spectacles of natural power: they could be portrayed as lawful physical operations of nature, or the moral regulation of mankind by God, or both. Electricity was also such a commodity: "a most amazing scene of wonders" was how Ebenezer Kinnersley, the leading American electrical demonstrator, described his shows. But what kind of wonder was electricity? "Wonder is the effect of ignorance, and ignorance begets credulity," lamented the gentleman-authors of *An Epitome of Electricity and Galvanism,* published in Philadelphia in 1809. "Vulgar amazement" had made "every body . . . eager to see and to feel this prodigy of nature," but "when wonder and credulity are coupled with terror and surprise, we must look for a strange and misshapen progeny. The exaggerated accounts of those who first experienced the electric shock cannot but raise a smile; especially as we may ascertain their real sensations by like experiments upon ourselves." Gentleman-philosophers were always keen to declare who was enlightened and who was not, and to distinguish those who disciplined their experience of nature through the use of reason from those uselessly awestruck by its marvels.[5]

In snootily dismissing early encounters with electricity, these gentlemen at least put their finger on what made it enthralling to eighteenth-century publics, and why it should also be interesting to us: its double status as a rational curiosity and wonderful experience. Electricity defied the logic of Cartesian dualism, according to which mind and body were separate entities, by putting mind and body into startlingly direct communication. The centrality of the body as an experimental instrument meant that in no other branch of enlightened science were the act of cognition and the experience of passion more intensely related: to know was to feel, and to feel was to know.

How to make sense of such experience, at once rational and not? The breezy rejection of electrical wonders by our Philadelphia gentlemen would seem to suggest a now familiar antipositivist narrative of science and enlightenment (inspired at its core by the work of Michel Foucault): one in which taking the wonder out of nature maintained social order and control in the turbulent eighteenth century. When commentators insisted that electricity was wonderful, whether in praise or blame, they spoke of its power as a surprising, often bewildering phenomenon that provoked passionate responses such as awe, amazement, and terror. Before the seventeenth century, wonders (as well as "marvels" and "prodigies") were often regarded as supernatural violations of the natural order, portending evil or divine displeasure. In colonial America, there was a vibrant tradition, especially strong in early New England society, of interpreting natural phenomena as providential wonders, worked by God, for the moral regulation of humanity. Enlightened natural philosophers, however, insisted that nature was so orderly and intelligible that true wonder could only be experienced upon perceiving its rational structure. This change was not simply a historical development in epistemology, but also a response to social fear: fear of the power of wonders to dazzle the masses and move them to radical action. Wonder was socially powerful, and eighteenth-century elites now set their rational wonder against the "gawking wonder" of crowds. Empirical demonstration of cause and effect connoted mental self-possession and stable order; nonrational response signaled enthusiasm, superstition, and the pathological imagination, whose effects ranged from benighted belief in miracles to disastrous religious wars. Stripping nature of its wonders—reducing legitimate wonder to a form of reason—was thus a strategy for maintaining social order through docile bodies and tranquil minds.[6]

I draw on these rich discussions of natural wonder and social power to explore a neglected phase in their history. Exhibiting wonders traditionally conferred authority on the church, the court, and social elites. But what happened when wonder began to move out of the churches and courts and into the parlors of middling sorts of people like merchants and artisans? Our view of the Enlightenment depends on which figures we examine. A history that relies solely on elite metropolitan perspectives will likely lead us to their own dichotomous view of learned reason versus the vulgar astonishment of "the masses." To such thinkers, many of the figures in this book might be vulgar, for reasons of social status, education, or style of religious belief. But why should histories of enlightenment reproduce this dichotomous and self-interested perspective as their own? This book is set in the space between highly learned practitioners of natural philosophy and "the masses," in a capacious middle ground where experimental science and medicine were customized by bourgeois consumers, neither strictly learned nor vulgar, who spoke and wrote about electricity with utmost rational intent, yet also a persistent sense of mystery, one often explicitly religious.

This middle ground connected "high" practitioners like Franklin with a range of lesser-known figures who adopted and refashioned his methods and principles to their own ends. This is where the social history of the Enlightenment really is: in no one place, but in the movement and mutability of practices and ideas between different actors in the public sphere. And what American practitioners and consumers found in electricity—or thought they found—may surprise us: the body enlightened by electricity was not docile, but ecstatic. Was the Enlightenment, then, about freedom or control, knowledge or power? The question presents a false dichotomy, but it is unavoidable nonetheless. Electricity did signal the rise of bourgeois cultural power in eighteenth-century America, with the drives for social distinction and rational control this implies. But the enlightenment it brought was ecstatic and antinomian: it embraced nonrational experience as a positive good in an entrepreneurial and anticentralist culture awash in a sea of personal meanings.

Chapter 1 describes Franklin's Philadelphia experiments of the 1740s and places them in the context of the traffic in natural knowledge around the early modern Atlantic world, and the rise of a class of middling and leisured gentlemen in British America. It describes the use of the body in experiments—to explain the behavior of electricity in terms of a rational economy of positive and negative charge, while maintaining the mystique surrounding its powers and properties. Chapter 2 examines the crowning achievement of the Franklinist program: the lightning rod. After describing the experiments by which Franklin established that lightning was electricity (including the kite experiment), we explore theological debates about the science and "superstition" of electricity, as well as technological controversies about their effectiveness (because, despite their heroic status, lightning rods often failed as protective devices). Chapter 3 follows electricity through American public culture in the experimental demonstrations of commercial performers, in particular those of Franklin's associate Ebenezer Kinnersley. Inviting audiences from New England to as far away as the West Indies to experience electricity through their own bodies, Kinnersley's shows made electricity into a rational, polite, and pious entertainment, although it was the experience of sensory disorientation that lay at the heart of their appeal. Chapter 4 examines attempts to define the political meaning of electricity, as well as the electrical meaning of politics, in the American Revolution. In the patriotic language of "electrical politics," republican virtue was said to convulse the American body like a shock from divine nature itself. But in "political electricity," this revolution by revelation was unmasked by Loyalists as a work of art, conspiracy, and enthusiasm driven by the machinations of a Franklinist cabal.

Chapter 5 takes us from North to South America, as we examine the experiments carried out on electric eels in Dutch Guiana by the itinerant Massachusetts physician Edward Bancroft. By generating their own natural supply of electricity, electric eels amazed observers, and although attempts to circulate live specimens around the Atlantic failed, Bancroft's colonial experiments helped to inspire theories of "animal electricity" and ultimately led to the invention of the first current-generating electric battery. Returning to North America, the final two chapters deal with the practice of medical electricity, which flourished at the end of the century. Chapter 6 follows the career of another itinerant doctor, T. Gale of New York State, the first American to publish a handbook on electrotherapy and who, despite the methodical nature of his practice, insisted that electricity's mysterious powers were a sign that the Christian millennium was at hand. It also describes how the frightening phenomenon of spontaneous combustion challenged early republican confidence in using electricity to restore health. Finally, Chapter 7 examines a unique electrotherapeutic device known as metallic tractors, invented by Dr. Elisha Perkins of Connecticut, which sold as well in Britain as it did in America. Advocates of the tractors on both sides of the Atlantic successfully emphasized the reliability of common sense, useful knowledge, and matters of fact in their commercial promotions, claiming that experience showed that tractors relieved nervous disorders (even though their allegedly electrical properties could not be firmly established). In the end, however, Perkinist common sense was attacked by critics as nothing more than a trick of the imagination, its cures the effects of wonder-mongering merely posing as enlightenment.

By exploring how the ways and purposes of knowing have changed, the history of science is uniquely well placed to recapture how natural phenomena can generate different cultural meanings. I aim to show how changing developments in science and society affected the use and interpretation of electricity between roughly 1750 and 1800. But the story of electricity is really one of multiplicity and simultaneity: many things working together in the same instant, moving in different, sometimes opposite directions. The key is circulation, as Franklin tells us: "The electrical fire is never visible but when in motion, and leaping from body to body." In the case of electricity, multiple meaning was a direct result of commercial circulation. Eighteenth-century Americans may have been politically subordinate to Europeans, but the movement of people, apparatus, and techniques around the Atlantic was not controlled from the center. Circulation was voluntaristic, a function of the decentralized trade networks connecting Europe, Africa, and the Americas in the early modern era. Yet from a practical point of view, electricity was a strikingly useless commodity. Why then did it travel? We shall see that the answer lies in its cultural usefulness: its ability to engage the full range of human concerns—social, political, philosophical, and religious. By its nature, then, the history of electricity is a cultural history of early American society itself.[7]

CHAPTER ONE

Atlantic Circuits

It is amazing to what a pitch the Electrical power is carried. I am well informed that In Germany they knock'd down an Ox. Several Men have been struck down at London, one was an Irish Bishop, a Lusty strong Man & yet could not surmount the shock. I presume by this Time the apparatus is got into the Colonies, for there is no describing the Electrical power unless a person feels it himself.

PETER COLLINSON TO BENJAMIN FRANKLIN, 1747

THE origins of Benjamin Franklin's experiments with electricity lie not in contemplative wonder at the order of nature, but in terrified astonishment at its violence. In January 1746, Pieter van Musschenbroek, the eminent Dutch experimental philosopher, wrote an agonized report to the naturalist René-Antoine de Réaumur at the Académie des Sciences in Paris, in which he recounted a truly shocking encounter in his Leyden laboratory:

> I would like to tell you about a new but terrible experiment, which I advise you never to try yourself, nor would I, who have experienced it and survived by the grace of God, do it again for all the kingdom of France. I was engaged in displaying the powers of electricity. An iron tube AB was suspended from blue-silk lines; a globe, rapidly spun and rubbed, was located near A, and communicated its electrical power to AB. From a point near the other end B a brass wire hung; in my right hand I held the globe D, partly filled with water, into which the fire dipped; with my left hand E I tried to draw the snapping sparks that jump from the iron tube to the finger; thereupon my right hand F was struck with such force that my whole body quivered just like someone hit by lightning . . . the arm and the body are affected so terribly I can't describe it. I thought I was done for . . . I've reached the point where I understand nothing and can explain nothing.[1]

The "explosion" of what became known as the "Leyden jar" resulted not in the destruction of the glass bottle but a powerful electric discharge that wracked van Musschenbroek's body. So frightful were the effects of such shocks—nausea, headaches, nosebleeds, convulsions, and temporary paralysis—that several experimenters besides the Dutchman swore off the jars, seeing them as objects of terror. Half a century later, it occurred to some that these jars might be useful as weapons of war. In 1801, William Caruthers of Lexington, Massachusetts, wrote to Thomas Jefferson, then president both of the United States and the American Philosophical Society, asking "what effect a Receiver highly Charged hermetically sealed and Violently projected so as to Create a Discharge itself against an object would have"? It seems that no one did the experiment to answer the question: back in the 1740s, the power of the Leyden jar seemed to signal natural limits to electrical experiment. But despite—or perhaps because of—warnings like van Musschenbroek's, the Leyden jar soon became an irresistible object of both philosophical curiosity and spectacular corporeal experience, challenging prevailing interpretations of the behavior of electricity and the nerve (and nerves) of experimenters.[2]

The story of early American electricity begins in the circulation of experimental knowledge as a spectacular performance. Electricity became one of the leading public sciences of the eighteenth century, both in Europe and colonial British America, engaging the attention of metropolitans and provincials alike. Electricity was a wonderful commodity and, throughout its eighteenth-century career, several different commodities at once. Correspondence networks created by Atlantic trade relationships connecting Europe and the Americas provided the material preconditions for the circulation of apparatus, texts, demonstrators, and techniques for knowing about electricity, all of which were embraced by members of the British-American gentry eager to participate in metropolitan cultural life. More than ideas per se, it was the circulation of practices involving the human body as a scientific instrument that made the communication of electrical knowledge possible, and it was this continuity of practice that enabled Franklin's new philosophy of electricity to emerge. The results were strikingly palindromical. From this North Atlantic zone of commercial circulation came an account of electricity in the Leyden jar as a circulating fire that behaved in regular economic fashion and sought a natural state of equilibrium. The commercial circuits of the Atlantic economy, along which electricity also traveled from Europe to America, thus generated a rational explanation of electricity's own surprising circuits. The status of this philosophy of electricity has often been mischaracterized by whig historians and hagiographers, who claim that Franklin found electricity a curiosity and left it a modern science. In fact, in the early modern eighteenth century electricity was both a science and a marvel, and Franklin's own writings reveal a tension between experimental claims to rational knowledge and the persistence of wonder at the surprising powers of the electric fire.

Trafficking in Knowledge

In colonial British America, the systematic study of nature was primarily shaped by advances in natural history and botany, genres of scientific inquiry directly useful for the organization and exploitation of colonial natural resources. Since Francis Bacon's *New Atlantis* (1627), if not earlier, Englishmen had fantasized about the knowledge and power that overseas exploration promised in their struggle against Spain (Bacon, an English chancellor, was himself an investor in the Virginia Company). Early modern colonial natural histories were heterogeneous chorographies, which combined descriptions and illustrations of plants, animals, minerals, climate, and topography with travel and navigational narratives, and not least, depictions of the customs and behavior of native and enslaved peoples. Richard Ligon's *True and Exact History of the Island of Barbadoes* (1657), to take just one example, described not only Caribbean flora and fauna, but also the author's voyage to the West Indies via the Cape Verde Islands, the Caribbean climate and soil, the logistics of sugar cultivation, and some of the physical and cultural characteristics of the enslaved African population, including their violent resistance to the nascent plantocracy. Such natural histories were not remote or disembodied acts of classification but artifacts, both utilitarian and aesthetic, of the social experience of travel enabled by early modern overseas trading systems. Increasingly in the eighteenth century, enlightened natural history meant systematic taxonomy. But the advent of the utilitarian system for classifying plants according to sexual characteristics advocated by the Swedish naturalist Carolus Linnaeus only strengthened the relationships among natural history, botany, economy, and empire in the second half of the century. By century's end, Joseph Banks, who was after 1778 president of the Royal Society, Britain's preeminent scientific organization, had established Linnaean economic botany as national policy and made Kew Gardens in London the hub of a transoceanic network of botanical gardens following French and Iberian models of centralized imperial science.[3]

This gradual turn toward centralization came after the loss of Britain's American colonies, however, and was only fully attempted later in the nineteenth century. The story of electricity in the early modern British Atlantic, by contrast, is one of uncentralized commerce, itinerancy, entrepreneurialism, and informal institutional connections. The importance of entrepreneurial energies was not unique to the British Atlantic, but these probably played a more driving role there than in other theaters in the history of science and empire. The British state was by no means absent from the Atlantic world: it shaped the flow of commercial traffic, most obviously through mercantile legislation like the Navigation Acts and the establishment of official trading monopolies like the Royal Africa Company, intended to manage the slave trade. But as the political quarrels over imperial regulation in the

American Revolution later made plain, such legislation was typically honored in the breach; and the Royal Africa Company's monopoly was famously broken in the years after 1700, the result of independent merchants lobbying Parliament in favor of deregulation. Rather than formal agents of national institutions, it was commercial entrepreneurs who developed Britain's Atlantic system based on the profitability of slavery and the "triangular trade" in agricultural and manufactured commodities among Europe, Africa, and the Americas. This entrepreneurial traffic shaped the production and movement of natural knowledge as well. Mark Catesby's travels in the American Southeast during the 1720s, for example, were paid for by private subscription to the sumptuously illustrated volumes he eventually published, and sponsored by a group of London gentlemen including Sir Hans Sloane, then president of the Royal Society, whose membership by then included a number of colonial correspondents. Such communications between Britons and British-American colonists took advantage of the accelerating trade networks, both transatlantic and intercolonial, that developed in the decades after the Peace of Utrecht (1713), an era during which there emerged regular and faster Atlantic crossings, faster news, new intercolonial roads, colonial newspapers, and a regular postal service (Figure 2).[4]

Networks that circulate knowledge are rarely collectives of equals, but are structured by social and geographical hierarchies: who and where you are shapes your ability to make claims to knowledge. In urging the pursuit of a more empirical and useful science free from the corruptions of Aristotelian wordplay, Bacon had prescribed a division of intellectual labor according to which natural historians would collect information and specimens for natural philosophers to organize and explain. This division mapped neatly onto the center-periphery hierarchy that the English (and other Europeans) projected beyond their shores. Colonial observers would send written accounts, drawings, paintings, and specimens back for metropolitans to inter-

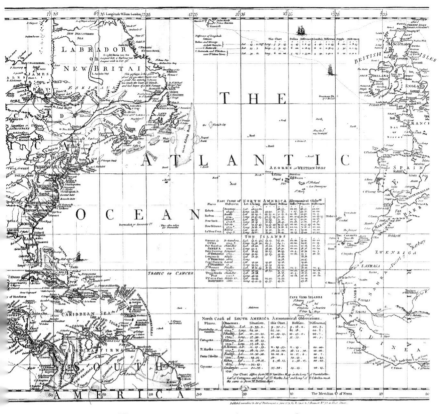

2. Electricity as an Atlantic Science

During the Enlightenment, itinerant experimenters, books, apparatus, and techniques traveled from London (as well as Edinburgh, Leyden, Paris, and Bologna) as part of an entrepreneurial traffic in goods and ideas to New England, the middle and southern colonies, and the West Indies. There was significant circulation from America back to Europe, too: lightning rods, accounts of experiments with electric eels in Dutch Guiana, and therapeutic devices such as Perkins's tractors.

pret, classify, and incorporate into generalized accounts of nature. Early modern Europeans disparaged both the intellectual capacities of American-born Creoles to write their own natural and civil histories, and the character of American nature itself—with the French naturalist the Comte de Buffon's attack on American physical "degeneracy" being only the best-known example of the genre. As has recently been shown, Spanish Creoles in South America wrote back voluminously to dispute such charges, but it was only after national independence in the late eighteenth century that North Americans like Thomas Jefferson began to mount similarly patriotic defenses. In the long colonial moment before independence, British-American gentlemen like William Byrd II, FRS (Fellow of the Royal Society) of Virginia were trapped in the frustrating dilemma of the provincial intellectual: lacking recourse to an independent national American identity, they tended grudgingly to accept subordination to metropolitans who claimed that their "creole humours" rendered them inferior in reason.[5]

This does not mean that the contributions of colonial gentlemen, as well as women, natives, and slaves, were not valued by metropolitan philosophers; as the division of labor implies, they were necessary as suppliers of information, specimens, and even techniques. And it should not imply that provincials were unwilling to participate in metropolitan knowledge projects like natural history classification—they often were. Despite the sweeping declarations of the so-called dispute of the New World, an ethos of cosmopolitanism in the imagined community of the "Republic of Letters" tempered discourses of Creole difference by accommodating provincial actors as participants in metropolitan projects, albeit in the limiting terms of colonial provider and metropolitan interpreter. The example of colonial New England demonstrates how provincial science could in fact serve metropolitan and local purposes simultaneously. Astronomers in Massachusetts Bay like Thomas Brattle of Harvard College provided observations that informed Isaac Newton's formulation of the laws of motion in his epochal work of 1687, *The Principia* (Newton did not name Brattle, but referred to him as "the observer in New England.") Cotton Mather and his father, Increase, also corresponded with the Royal Society, describing the American environment in terms of the new naturalism demanded by the Society's sense of epistemological decorum. Yet at the same time, each continued to pursue both private and local agendas in which the nature of New England remained a source of Providential wonder, a system of divine signs, and a theater of special interventions by a God who could overleap the normal course of physical laws. The philosophically literate clerical elite of the Congregational Church used Copernican astronomy to deride astrology as godless folklore, and the mechanical philosophy to denigrate Native American confusions of matter and Manitou (their Great Spirit), justifying their conversion and displacement of Algonquian-speaking Indians. To leading New Englanders, natural knowledge connected them to metropolitan intellectual life, shaped their sense of elect Christian selfhood, and advanced an internal, imperialistic frontier against "ignorant" American barbarians.[6]

By the mid-eighteenth century, then, ideologies of cosmopolitanism were tempering discourses of Creole difference, allowing provincial actors to be included as necessary participants in metropolitan projects. Cosmopolitan exchange was a two-way street: metropolitans received useful knowledge from the periphery, while colonials gained honor, distinction, and sometimes patronage. In a rare instance of state patronage, the Philadelphia Quaker and autodidact John Bartram was named royal botanist by King George III in 1765, receiving a pension to support his transmission of plant specimens to Britain. The traffic in knowledge increasingly occurred among multiple European centers and American provinces, connecting New England, New York, Pennsylvania, Virginia, Carolina, and the Caribbean islands to London, Edinburgh, Uppsala, and beyond. Cadwallader Colden and Alexander Garden, for example, expatriate Scots living in New York and Charleston, respectively, corresponded with Linnaeus himself, who honored their services by naming American plants after them. To these British Americans, the prestige of such high European connections provided powerful motivation in its own right, opening an irresistible path to provincial cosmopolitanism and participation in the universal projects of the Republic of Letters. When British Americans began to organize an American Philosophical Society, launched in 1743 by Franklin and based on the Royal Society, the aim was not a patriotic alternative to the London-based network so much as a provincial subsidiary of the transatlantic networks in which American members were already involved. As Franklin wrote in his proposal, "Many Observations occur, which if well-examined, pursued and improved, might produce Discoveries to the Advantage of some or all of the *British* Plantations, or to the Benefit of Mankind in general."[7]

By mid-century, theoretically regardless of participants' social status, commercial networks for the exchange of goods were becoming networks of knowledge exchange and, potentially for colonials, a means of metropolitan cultural participation and a path to enlightenment. Many knowledges circulated in the Atlantic beyond the exemplary colonial sciences of natural history and botany, and electricity was one of them. In contrast to electricity's later career, the earliest American conversations on the subject occurred within established institutions. Perhaps the first person to discuss electricity in the colonies was Charles Morton, a Dissenter and teacher of natural philosophy at the Warrington Academy in England (where he had taught Daniel Defoe), who moved to New England to take up a position at Harvard in the 1680s. Surviving manuscripts of Morton's course on natural philosophy, the *Compendium Physicae,* reveal a fusion, not uncommon at the time, of "old" and "new" sciences: Aristotelian, Cartesian, and experimental. His account of electricity was indebted to mechanical (or "corpuscularian") matter theory, according to which the universe was composed of atomistic bodies in perpetual motion. Morton distinguished "elatery" (electricity) from magnetism as "the power in which Jett, Amber, Red-wax, etc: have of being rubed, to draw small motes to themselves . . . by Volatile steames from themselves which stirred, and put in motion by the Action (or rubing) fly out every way in strait lines."[8]

In the early eighteenth century, again at Harvard, which was then virtually unique in North America for its collection of scientific apparatus, electricity formed part of a program of instruction in natural philosophy that emphasized experiment over textual accounts in the name of "Newtonianism." Self-styled Newtonian experimenters, with unprecedented confidence, claimed practical mastery over nature. Besides the formulation of universal laws of motion in the *Principia,* Newton's other great publication, the *Opticks* (1704) was widely embraced during this period as a model for experimental work. The first important Newtonian in America was Isaac Greenwood. Bostonian by birth and originally a divinity student of Cotton Mather's, Greenwood sailed for Britain in 1723, where he met the aged Newton, then president of the Royal Society, and studied under John T. Desaguliers, whose *Course of Experimental Philosophy* (1734) offered a well-illustrated and widely circulated account of Newtonian physics applied to practical mechanics. Returning to Massachusetts to become Harvard's inaugural professor of mathematics and natural philosophy in 1727 (a chair endowed at Greenwood's request by Thomas Hollis, a London merchant who also provided Harvard with apparatus, books, and scholarship money), Greenwood zealously promulgated "the wonderful Discoveries of the incomparable Sir *Isaac Newton*" through experimental demonstrations for college and public audiences. Besides explicating the laws of motion with balances and pulleys, Greenwood demonstrated the properties of electricity, such as attraction and repulsion, and performed "*Experiments* relating to the *Electrical Phosphorous*" and "the Application of *Electrical Light* to the Discovery of the true Cause of several *Uncommon Lights*"—displays based on recent work with electroluminescence by Francis Hauksbee at the Royal Society, and the stimulus for Newton's own speculations in the *Opticks* on the nature of electricity as an "active power" through which God animated his Creation. Greenwood insisted that unlike Aristotelian philosophy, or the supposedly dogmatic abstractions of Cartesian rationalism, Newtonian experiment dealt in "*Fact,* and *Experience,* and no other than what *Nature* herself has made Use of in the Fabrick of the *World.*" Even a few weeks of using instruments would provide a far better instruction in the "*Laws of Nature*" than "a *Years* Application to *Books,* and *Schemes.*" This was the earliest American articulation of the forceful new claims regarding experimentally derived knowledge, which was heralded as realizing the Baconian vision of reformed knowledge-making. Using scientific instruments to interrogate nature made "the Senses . . . the Judges of the Solutions that are given to *Natural Effects,* and all fanciful Suppositions, & *Hypotheses* (however plausible they may seem) [are] exploded; it being the excellency of this *Teaching* to take nothing for granted but what is shewn to be really in Nature, de facto."[9]

Experiment was to provide the practical foundations for the pursuit of electricity in the Enlightenment, involving human bodies as both performers and scientific instruments. A number of what would later be recognized as electrical effects had been known to the ancients, notably the attractive properties of amber and the numbing effects of torporific fish like the ray. But electricity's experimental history only properly began after 1600, when the Elizabethan court philosopher William Gilbert published *De Magnete.* It was Gilbert who coined the term "electric" to denote the attractive property in substances like amber, a term derived from *elektron,* the ancient Greek word for amber (the term "electricity" was not used until 1646 by Thomas Browne in his *Pseudodoxia Epidemica.*) Gilbert was the first natural philosopher to distinguish electric from magnetic attraction, and "electrics" from "non-electrics": nonconductors and conductors of electricity, respectively, as they would be known by the eighteenth century (the terms seem somewhat counterintuitive today). Electrical attraction, he maintained, was an effect produced by the action of a material electrical "effluvium" emitted by substances like amber. This mechanical account shaped discussions by an illustrious array of seventeenth-century philosophers: Kenelm Digby, Robert Boyle, Otto von Guericke, Athanasius Kircher, and Niccolò Cabeo.[10]

In the second quarter of the eighteenth century, the study of electricity moved from disembodied observation to experiments using human bodies. But how exactly was the human body perceived in this era? In the Platonic and Christian traditions to which the Anglophone eighteenth century was heir, flesh was mistrusted and the senses were considered unreliable for making knowledge and dangerous for morality. Christian doctrine emphasized the frailty and sinfulness of flesh and the immortality of the soul, even as it promised the body would ultimately be resurrected on the Day of Judgment. But in response to the excessive rationalism that late seventeenth-century moral and natural philosophers associated with the legacy of René Descartes, these philosophers, led by John Locke, placed renewed emphasis on the senses as the basis for knowledge. By the eighteenth century, many regarded reason as only one instrument through which embodied selves came to know the external world—an instrument that necessarily functioned in concert both with the senses and the faculty of imagination, each enabling but also checking the other (the imagination produced false ideas in the mind when it misrepresented the evidence of the senses). The eighteenth century was also distinctive for new medical obsessions with bodily health, such as the physical effects of diet and climate. While the eighteenth-century body was by no means entirely secularized, as the American context makes especially clear, the body nevertheless became for many a material, and sometimes materialist, preoccupation in its own right. It was also a social preoccupation: the historic progress of society toward a state of civility depended on the public self-restraint of polite conduct. While the grotesque, satirical images of crowds and society created by the likes of William Hogarth conjured a threatening alter ego, the rational, enlightened body was one whose head governed its corporeal passions. Bodily electrification was both epistemologically and culturally significant, therefore, precisely because it violated this dualism and challenged the sovereignty of mind or soul over matter (whether understood as Platonic, Christian, or Cartesian), by putting mind and matter into direct, convulsive communication.[11]

In the 1730s, experimenters began using human bodies as instruments to examine the behavior of electricity. In England, Stephen Gray (a former colleague of Desaguliers) and Granville Wheler, FRS, conducted experiments to measure both the types and lengths of matter along which electricity would conduct. They added human beings to their list of "non-electrics" (conductors) by attracting pieces of leaf metal to the face of a boy (probably a servant at the Charterhouse almshouse, of which Gray was a member) whom they had electrically "excited" by the proximity of a rubbed tube (the conventional method for generating charge), while he hung suspended from nonconductive silk cords (Figure 3). In the following decade, Charles Dufay, an academician and later administrator of the Jardin du Roi in Paris, restaged this experiment using his own body, becoming the first true auto-experimenter with electricity. Dufay recorded in particular the effects of electrification on his person: his face felt as though a spider web had spread over it and he noted the painfulness of electrical discharges from his fingers. Despite these unpleasant sensations, bodily electrification was to become the centerpiece of public electrical demonstrations in the 1740s. During this time, too, leading experimenters like the French court philosopher the Abbé Jean-Antoine Nollet began to investigate the body's response to electricity, measuring changes in weight and heart rate. The relative ease and safety of electrification

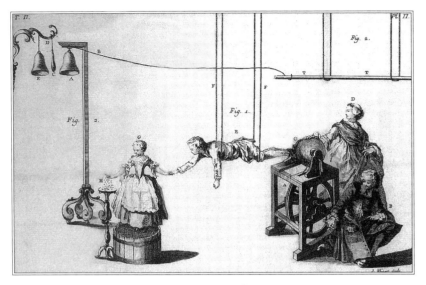

3. Electrical Games

The electrification of insulated bodies during the Enlightenment advanced experimental philosophy, provided spectacular demonstrations, and led to an interest in using electricity in medicine. The devices and techniques involved varied little, but the meanings ascribed to electrification did change over time. This image captures the social, participatory nature of electrical demonstrations after the 1730s. The gentleman at right turns the glass globe of the electric machine, producing friction against the hands of his lady companion; this charge collects in the insulated bodies of the boy suspended by silk cords in midair and the girl standing on a tub of dried pitch, allowing her to attract and repel chaff with her hands.

---

meant that electrostatics became the enlightened science of bodily experiment par excellence.[12]

It was at this juncture in the mid-1740s that electricity entered colonial America as part of the traffic in knowledge around the Atlantic world. When the Leyden jar exploded, Benjamin Franklin was almost forty years old. The son of a tallow chandler from Boston, his successful career as a printer and editor of the *Pennsylvania Gazette* in Philadelphia had allowed him to retire from full-time business and devote himself to more leisurely and genteel pursuits, including the sciences, although his time was soon to be absorbed by provincial politics, the running of the colonial post office, and ultimately the responsibilities of representing colonial interests in London, where he lived after 1757. Franklin's social transformation into a gentleman epitomized the emergence of a Creole elite in urban British America, among whom merchants, lawyers, and master artisans figured prominently. They were nowhere more in evidence than in Philadelphia, the diverse and pluralistic port city founded by Quakers and built on trade, which overtook Boston as the American hub of Atlantic commerce in the eighteenth century thanks to its proximity to an immensely fertile agricultural hinterland and the vitality of its merchant networks. To an extraordinary extent, Franklin created Philadelphia as an American exemplar of enlightenment. From the conversational "Junto" of 1727 to formal institutions like the Library Company of Philadelphia founded in 1731 (the first American subscription library), he realized a virtuosic array of civic projects in which the private pleasures of curious conversation and the public purposes of useful knowledge seemed to converge effortlessly. As the Quaker capital grew from a town into a city in this era, it became a vital emporium of metropolitan culture, importing from Georgian London everything from the architecture of Bloomsbury and the West End to the science of the Royal Society. If anywhere embodied the confluence of commerce and enlightenment in America, it was Philadelphia, with its straight, paved streets lined with brick and flagstone sidewalks. Being "the City nearest the Centre of the Continent-Colonies, communicating with all of them northward and southward by Post, and with all the Islands by sea," it was also the logical site, Franklin argued, for the gathering and exchange of knowledge at the American Philosophical Society.[13]

In 1743, Franklin attended a public demonstration given by an itinerant lecturer from Edinburgh named Dr. Archibald Spencer. Electricity had first "surprized and pleased" him, Franklin later recalled, during a visit to Boston, where he had seen Spencer perform. In his lecture on natural philosophy, Spencer included a brief discussion of the "electric fire" (as it was commonly known) as a subtle fluid, a weightless but material entity that exerted a force on material bodies. This electric fire was "diffused thro' all Space," and was "contained in, and [could] be produced from all bodies." To illustrate the point, he employed the spectacular techniques of Gray, Wheler, and Dufay, making "Sparks of fire . . . fly from [the] face and hands" of an electrified boy suspended in midair. Franklin was so taken with this performance that he helped Spencer to advertise his shows in the *Pennsylvania Gazette,* and later acquired his apparatus after the Scot's career ended in the 1750s.[14]

Serendipitous exposure to such performances doubtless afforded Franklin an immediate appreciation of the corporeal techniques used in experimental demonstration, but they offered little in the way of sustained exchange with the European sources of scientific activity. More important than these early contacts, therefore, was the two-way axis of intellectual patronage and communication that developed privately between Franklin and Peter Collinson, a Quaker merchant and fellow of the Royal Society who lived in London. Collinson was the most active metropolitan patron of British-American science in his day and had both a scientific and economic interest in American natural resources. A correspondent of Linnaeus, he encouraged the efforts of leading American botanists like Bartram and Colden, with whom he was in regular contact. Although not a botanist himself, Franklin's attempts to establish an American learned society in the 1740s connected him to the naturalists' network and to Collinson, and the Philadelphian effectively became one of Collinson's colonial promotions. In 1745, Collinson, who supplied books and journals for the Library Company of Philadelphia, sent a glass tube for reproducing recent European experiments as described in an article in the *Gentleman's*

Magazine of London, to which the Library Company subscribed. The article, entitled "An Historical Account of the Wonderful Discoveries . . . Concerning Electricity" and written by the Swiss physiologist Albrecht von Haller, emphasized electricity's sensational bodily effects by describing the performances of the Leipzig impresario Georg Matthias Bose, who had set spirits on fire with electric sparks drawn from human fingertips, and drawn sparks from female lips—a display he called the "Venus electrificata." "What proceeds from the electrified body," von Haller explained, "is really the production of fire."[15]

While journal articles described continental experiments to colonial audiences, Collinson's glass tube supplied the technology for experiment. Long tubes measuring twenty-seven to thirty inches provided the easiest way to generate static electricity, by holding and rubbing them with one's hands—or, as became the norm, with fur or buckskin—while standing on an insulating stand to cut off communication with the ground (the earth functioned as a giant conductor, so insulation was necessary to prevent the loss of charge). Given the virtual absence of colonial glass production at the time, Collinson's gift was significant. Franklin's fascination with electricity became so intense, however, that it stimulated something of a local glass industry. Although he complained of Philadelphia in 1747 that "there are no workmen to be had here," Franklin subsequently began to obtain tubes, jars, and panes from Caspar Wistar's New Jersey company, Wistarburg Glass. "Electricity is so much in vogue," he was to note, "that above one hundred [tubes] have been sold these four months past." Soon he proudly announced, "I am satisfy'd we have Workmen here, who can make the Apparatus as well to the full as that from London." By the late 1760s, Wistar's son Richard was advertising the sale of "electrifying globes and tubes" in the *Pennsylvania Gazette,* noting that "the abovementioned glass is of American manufactory . . . [and] consequently clear of the [Townshend] duties Americans so justly complain of." Thanks to Franklin's wide circle of interested acquaintances, Wistar exported glass for experimentation to Boston, New York, New

Haven, South Carolina, and Jamaica. Just a few years later, John Elliott established a "glass ware-house" at Market and Second Streets in Philadelphia, which also sold "electric globes and cylinders."[16]

Franklin busily set about his experiments at the Library Company, and in his own house, with the help of several friends: Philip Syng, a silver- and goldsmith who became treasurer of the city of Philadelphia in 1759; the lawyer Thomas Hopkinson; and Ebenezer Kinnersley, an unemployed Baptist minister. They began by examining the properties of electricity as a matter of private amusement and intellectual curiosity. Yet to these civic-minded gentlemen, private intellectual pleasure ultimately needed a public justification. In his *Autobiography*, Franklin retrospectively cast his personal conduct as a model for balancing private pleasures with public service, contrasting his own use of leisure for self-improvement and civic projects with the selfish indulgences of more idle diversions. "Reading was the only Amusement I allow'd my self," he wrote of his youth; "I spent no time in Taverns, Games, or Frolics of any kind." Public-mindedness was the path of virtue, whereas privacy resulted in self-absorption and vice. "I was never before so engaged in any study that so totally engrossed my attention and my time as this has lately done," he confessed to Collinson in early 1747; "I have, during some months past, had little leisure for any thing else." Turning electricity to some public advantage could redeem such an expense of private leisure, but practical applications initially proved elusive. With reluctance, Franklin reported to his English patron that he and his colleagues felt "chagrined a little that we have been hitherto able to produce nothing in this way of use to mankind." Electricity was enthralling entertainment. But how to make it useful?[17]

Nature's Electric Book

It was in the course of experimenting with glass tubes and bodily electrification that Franklin developed his new theory of positive and negative electricity, and the notion of an economy of electrical charge, which were to have such broad and enduring influence. In a letter to Collinson written during the summer of 1747, Franklin related a series of experiments using the tubes to demonstrate the communication of electricity between different bodies. Person A stands insulated on a wax "cake" and rubs a glass tube, he wrote. Person B, also standing on wax, brings his finger close to the tube (without touching), and "draws fire" from it. Finally, person C, standing on the floor, observes them and sees a spark issue from both electrified bodies, A and B, as he goes to touch them: "He will perceive a spark on approaching each of them with his knuckle." That communication of electricity between bodies had occurred was sensibly evident from the sparks. But how to account for this movement? Franklin and his colleagues believed (as had Spencer) that the electric fire was a "subtle fluid" that existed in ordinary matter. This notion had been derived from Newton's speculations in the *Opticks* early in the century that the "electric spirit" was one of the active powers of nature, with a status akin to that of the ether—a force that mediated between the material and immaterial realms of a divinely ordered universe. Franklin and his associates thought, however, that despite electricity's status as a subtle fluid, that is, as something neither strictly material nor immaterial, its behavior could best be understood in terms of the fluid mechanics of microscopic particles acting between material bodies, rather than "action at a distance"—the alleged power of matter to act where it was not, a notion respectable mechanical philosophers shunned as occult and magical. The leading electrical philosophies of the 1740s reflected this mechanical emphasis; the Abbé Nollet's theory of electrical efflux and afflux, for example, argued for the existence of different kinds of electricity (depending on the material by which they were generated) constantly streaming in and out of gross matter as microscopic particles. "Common matter is a kind of spunge to the electrical fluid," Franklin himself wrote, employing a characteristically tangible image.[18]

Unlike Nollet, however, Franklin began to conceive of electricity as a single fluid whose behavior in and between bodies was governed by differences in quantity. When matter contained "as much of the electrical as it will contain within its substance" and more still was added, he argued that the result was an electrical surplus that formed an "atmosphere" around the body in question. In the experiment with the tubes described earlier, Franklin reasoned that each of the three bodies began with an "equal share" of electricity as a "common element." But by rubbing the tube, person A "collects the electrical fire from himself into the glass": "the parts of the tube or sphere that are rubbed do, in the instant of the friction, attract the electrical fire, and therefore take it from the thing rubbing." Electricity remains in the glass until B, "passing his knuckle along near the tube, receives the fire which was collected by the glass from A." Because they both stand insulated on wax, A and B remain in the electrical state that has resulted from this communication: A with less electricity, B with more. C, meanwhile, standing on the floor, possesses "only the middle quantity of electrical fire," and so "receives a spark upon approaching B, who has an over quantity." At the same time, C gives some of his "middle quantity" to the electrically deficient A, redressing his "under quantity." If all persons had been in physical communication during the rubbing of the glass, all would have been electrified to the same degree: the electricity would have circulated equally among them. Alternatively, one could also "accumulate or subtract [electricity], upon, or from any body, as you connect that body with the rubber or with the receiver, the communication with the common stock being cut off." Franklin framed his new language of electric charge in terms of a self-balancing economy. "We say B, (and bodies like circumstanced) is electrified *positively; A, negatively.* Or rather, B is electrified *plus; A, minus.* And we daily in our experiments electrise bodies *plus* or *minus,* as we think proper."[19]

Franklin used this economic model of electricity to explain the workings of the Leyden jar in a series of experiments also conducted in 1747. The jars that Franklin used ranged from small glass bottles up to a foot in length to much larger containers with a capacity (hence, "capacitor") of seven or eight gallons of water (the Philadelphians also used granulated lead). Sealed at the top, these bottles terminated in a conductive metal wire, and were lined with a plate of tin foil, both inside and outside the glass. They were charged by connecting them to electrostatic generators. The Philadelphians possessed at least two such machines in this early period: one sent as a gift to the Library Company in 1747 by another London patron, Thomas Penn, the proprietor of Pennsylvania, and a second, cruder device made locally by the artisan Syng. Based on the model developed at the Royal Society by Hauksbee in 1709, the leading European machines by midcentury consisted of a wooden frame perhaps four or five feet tall, which housed a wheel four feet in diameter (Figure 4). A winch turned the wheel to rotate a glass globe, connected to the wheel by means of a pulley. Before the Leyden jar, experimenters charged and discharged themselves directly and charged and discharged others by having them touch the rotating globe; Leyden jars subsequently made it possible to store such charges. The American machines operated on the same principles, only by direct rotation of the globe via a winch, without the pulley. Franklin's description of Philadelphian technique emphasized the artisanal dexterity that was the basis of experiment: "Our spheres are fixed on iron axes which pass through them. At one end of the axis there is a small handle, with which you turn the sphere like a common grindstone. This we find very commodious, as the machine takes up but little room, is portable, and may be enclosed in a tight box, when not in use." Franklin reasoned that exciting the glass through friction (typically with buckskin) did not *create* electricity as others had supposed, but *pumped* it out of the earth into the globe, ready for communication to a "prime conductor," an insulated metal rod either standing on wax or suspended from above by nonconductive silk cords. Leyden jars could be charged when held by insulated experimenters in communication with the conductor, or by simply hooking the bottles onto the prime conductor with metal chains. Once bottled in this fashion, electricity could be discharged through bodies, animate and inanimate, at will.[20]

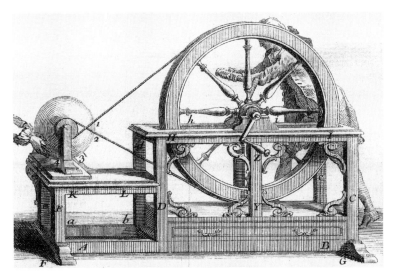

4. Generating Enlightenment

This classic electrostatic generator of the mid-eighteenth century required laborious cranking to procure visible sparks, a reminder that experiments during this era relied on both mechanical labor and artisanal dexterity. Turning a large wooden wheel in a frame rotated a glass globe (later replaced with plate glass) from which electricity could be transferred to the body through the hands. Later machines were smaller, used fur rather than hands to generate friction against glass, and had metallic "prime" conductors attached, on which Leyden jars were hung to collect electricity for controlled discharges.

The Leyden jar confounded electricians (as electrical experimenters were then called). Believing that glass was a permeable and conductive material, they could not understand how the Leyden jar could be grounded, yet remain charged, and how its glass could be handled without discharging this electricity when, if touched through a metal wire, it delivered a stunning electrical blow. Reasoning backward from observed effects to invisible causes, Franklin's key insight was to suppose that glass had to be a nonconductive material that could accumulate electricity but not discharge it of its own accord. He then applied his system of positive and negative charge, arguing that the shock experienced on handling both the glass and its metallic wire conductor resulted from the completion of an electrical circuit through which a sudden restoration of equilibrium was effected between the different charges in the jar. In a charged jar, the positive charge was located not in its water (as had been supposed) but on the inside of the glass. Since glass was a nonconductor, the charge could not pass to the negatively charged outside of the jar. Handling only the external glass body did not, therefore, establish a physical relationship between the jar's opposed charges—the prerequisite for electrical communication. It was only by handling the glass and the wire together that the human body itself completed the circuit between the unequal charges within and without the glass. The internal positive charge passed through the water and out of the metal conductor through the experimenter's body before reaching the negative outer coating of the jar. Franklin's conclusion was that all such shocks and explosions, with their attendant sparks, demonstrated an inherent tendency in electricity to seek a natural equilibrium of plus and minus charge between proximate bodies.

Experiments performed by the Philadelphians to demonstrate this tendency of electricity to find an equilibrium included letting a small cork ball, suspended on a thread, play between the jar's positively charged conductor, and a parallel wire connected to the negatively charged outer (or bottom) coating (Figure 5). The ball would dart be-

5. Illustrating Electricity

Compared to the flamboyance of most Enlightenment-era electrical imagery, the frontispiece to Franklin's book on electricity is restrained, favoring literal representations of elements from a variety of his experiments, including the electrical kite (figure X), the sentry box (IX), and the electrified gold-leaf book (V). It also shows demonstrations of the conservation of charge in the Leyden jar (I-IV). In figures I and II, a suspended cork ball plays between the jar's positive conductor and a parallel wire connected to its negative outer (or bottom) coating. According to Franklin, electricity sought an equilibrium of charge, so the ball would only stop moving when it had equalized the charges in both parts of the jar.

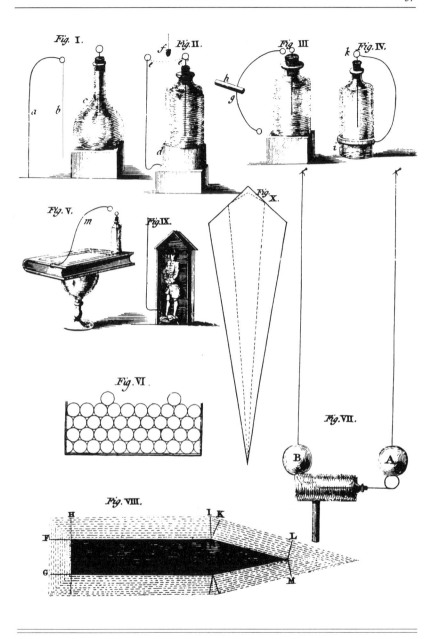

tween these two points, redistributing the charge, until it had restored the equilibrium between them; at that moment, the ball would be at rest. Other demonstrations used the bodies of volunteers to make the same point. "Place a man on a cake of wax," directed Franklin, "and present him the wire of an electrified phial to touch, you standing on the floor, and holding it in your hand . . . As often as he touches it, he will be electrified *plus;* and any one standing on the floor may draw a spark from him. The fire in this experiment passes out of the wire into him; and at the same time out of your hand into the bottom of the bottle." If you give him the jar, Franklin explained, he will be electrified *minus* and may draw sparks from those standing on the floor, restoring charge to the jar's outside or bottom. The startling power of the Leyden jar was thus channeled into controlled convulsions for experimental purposes, in carefully staged demonstrations of electricity's behavior as a lawful and predictable natural power. The proof, Franklin exuberantly reminded Collinson, was bodily experience of the most unambiguous sort: "If any one should doubt whether the electrical matter passes through the substance of bodies, or only over and along their surfaces, a shock from an electrified glass jar, taken through his own body, will probably convince him."[21]

That Franklin's written accounts of his experiments seem spontaneously commonsensical, even self-evident, is a tribute to the skill of his self-presentation, because it conceals the tacit conventions of early modern knowledge-making on which he drew to make his work credible to potentially skeptical readers. These conventions played a decisive role: they helped Franklin to overcome the division of labor between colonial observer and metropolitan interpreter, effectively allowing him to proceed as though no such division existed. In particular, by replicating the corporeal protocols of electrical experiment already established in Europe, Franklin exploited conventional early modern wisdom that unlike those who were socially dependent, such as women and men of inferior social rank, propertied gentlemen were masters of themselves and their passions and so could tell the difference between real phenomena and false ones. Being financially independent, they had no motive not to tell the truth. Franklin did not explicitly justify why he as an experimenter in America was qualified to observe and interpret interactions of electricity and body; he simply observed and interpreted and in so doing asserted his authority and trustworthiness as a gentleman, in effect eliding the issue of his American provenance.[22]

Franklin's Americanness was also made less visible by the idiom of natural philosophy as a discourse of universal knowledge. Whereas chorographic natural histories tended to emphasize in situ observations by actors in specific local geographies, natural philosophers after Newton tended to submerge issues of local provenance in discussing global physical forces. As Newton had written in the *Principia*, "The falling of stones in Europe and America" was evidence that "the causes assigned to natural effects of the same kind must be . . . the same." Natural philosophy afforded an idiom through which Franklin could make universal claims about electricity, and in which his colonial status was rhetorically marginal. From its title page, readers of his *Experiments and Observations on Electricity* (1751) would have known that these were undertaken "at Philadelphia, in America." But the text between the covers cast Franklin first and foremost as a gentleman, rather than an American (the title page of the second edition of the book in 1754 identified him simply as "Benjamin Franklin, *Esq.*").[23]

Franklin also took care to portray himself as a modest (colonial) witness of nature rather than a prideful (metropolitan) theorist. His early enthusiastic reception in France, it has been persuasively argued, depended on his ability to exemplify a "philosophical modesty" characterized by an openness to experience, a stance that militated against the kind of arrogant knowledge claims associated with "Cartesian rationalism." While this may be true, Franklinist philosophical modesty was a function of geopolitics, too. In his *Autobiography,* composed later, between 1771 and 1788, Franklin distinguished between Nollet's "*Theory* of Electricity" (a "*System*" the Frenchman was busily "defending") and his own "Observations and Experiments." Rather than reply to Nollet, the American famously demurred, insisting that his "writings contain'd only a Description of Experiments, which any one might repeat and verify, and if not to be verify'd could not be defended; or of Observations, offer'd as Conjectures, and not delivered dogmatically, therefore not laying me under any Obligation to defend them." Framed as a nonsystematic collection of mere observations, and presented as letters sent to a metropolitan patron, Franklinist electricity did not directly challenge the colonial order of knowledge. Ironically, then, before Franklin was embraced by his French admirers as a model of modesty for all natural philosophers, he was first obliged to play this role by the demands of epistemological hierarchies operating in the Atlantic world.[24]

Social factors also buttressed Franklin's credibility, and lent his work a metropolitan imprimatur. Informal connections between individuals and institutions were vital to Britain's Atlantic empire as a whole, and reflected the extension overseas of the relatively decentralized structure of British public life. Collinson provided Franklin with the crucial link: he not only supplied his Philadelphia client with apparatus, but also opened paths of communication back to the Royal Society and London publishers. The title page of the first edition made this clear to readers, presenting Franklin's experiments as "*communicated in several letters to Mr. P. Collinson of London, F.R.S.*" By the fourth edition in 1769, Franklin himself was identified as "L.L.D." and "F.R.S.," and the fifth edition in 1774 stated his membership in the Royal Academy of Sciences in Paris, the Royal Society at Gottingen, and the Batavian Society in Holland (as well as his presidency of the American Philosophical Society). Although Franklin later complained that his letters were initially "laught at by the Connoisseurs" at the Royal Society, there was more to scientific print circulation in Georgian London than the venerable pages of the society's *Philosophical Transactions*. Collinson and Dr. John Fothergill gained Franklin a reading public beyond the Society through Edward Cave, editor of the *Gentleman's Magazine,* who undertook to publish the *Experiments and Observations* as a volume in its own right. The results of this publication were an unprecedented success for an American author. With some notable exceptions like Nollet, Franklin's single-fluid concept of electricity as a self-regulating economy, described in the experimental accounts contained in his book, were to command broad assent among Europe's leading electricians. Astonishingly, the lexicon of Franklinism—positive and negative charge, plus and minus—eventually became the most widely adopted language of electricity in the Enlightenment.[25]

But where had this lexicon come from? In introducing Franklin's "rational experiments" to the British public, Cave stressed that the Philadelphian's signal virtue was to have led the reader through a scene of "pleasing astonishment . . . by a train of facts" to "probable causes." But in contrast to the image of commonsense experimentalism with which Franklin surrounded himself, and which historians have long embroidered into a paradigmatic antihypothetical "Newtonian" empiricism, these experiments relied on both hypotheses and teleological principles. His account of electricity as a self-regulating economy demonstrated both the rationality of a book-balancing tradesman—the "book-keeper of nature" one scholar has christened him—and the kind of teleological reasoning for which the mechanical philosophers had blasted Aristotelian enemies since the seventeenth century (and which self-described "Newtonians" carried over into their fights against dreaded "Cartesian rationalists").[26]

Franklin was not the straightforward tinkerer that his plain style suggested. In the first place, he did not possess the means to demonstrate the existence of this economy empirically by measuring quantities of electricity in bodies. The notion of a functional electrical economy was, therefore, a purpose that he ascribed to the electric fire, a working hypothesis he applied to provide a plausible account of the effects he was observing. In this respect, Franklinism was not good mechanical philosophy: electricity was not simply matter in motion but an active power characterized, in almost Aristotelian fashion, by innate tendencies. The impenetrability of glass was less a fact established by experiment than an a priori assumption to enable his theory—despite Franklin's taking Nollet to task, the idea of positive and negative charges was indeed a theory to account for the behavior of electricity in the Leyden jar. The notion of equilibrium restoration was a conjecture and a form of deductive reasoning, rather than a case of inductive reasoning founded on particular observations.[27]

Reasoning by analogy was a characteristic intellectual method of the Enlightenment, and it was fundamental to Franklinist electricity. Some, like those in the French school of political economy known as Physiocracy (literally "the rule of nature"), made explicit arguments about the need to remodel what they saw as the artificiality of mercantilist economics according to the more natural ebbs and flows of free trade that would find their own, better equilibrium with less human intervention. In Franklin's case, the direction of influence seems reversed: he developed his notions of economy and circulation writing about moral and social themes before applying them to electricity, although his thought clearly suggests a commitment to circulation and economy as natural and hence fitting models for society and self. In his *Dissertation on Liberty and Necessity, Pleasure and Pain* (1725), the young Franklin reduced pleasure and pain to a quantitative calculus of "portions" and "degrees" to be actively balanced for individual happiness (although he later discarded this as a libertine "erratum," his subsequent injunction to "use venery" in the *Autobiography*, among other things, shows that he remained convinced of the naturalness of passionate seminal circulation). Similarly, in *A Modest Enquiry into the Nature and Necessity of a Paper-Currency* (1729), he discussed the question of a currency in terms of the disadvantages of "wants" or shortages of money—that is, the more available and free-flowing the currency, the better able it is to keep prices down, promote the commercial circulation of commodities, and encourage labor. This interest in natural circulation proved lifelong, as his investigations into the Atlantic Gulf Stream attest. The passions, money, the Atlantic Ocean, and electricity: he conceptualized them all in terms of circulation and economy. In a sense, Franklin's Leyden jar *was* the Atlantic Ocean: both were natural economies of circulating fluids whose powers could be managed to human advantage. Unlike the Physiocrats, he eschewed programmatic statements justifying nature as an analogical model for social relationships. His style was typically more pragmatic than programmatic. The bookkeeper of nature identified the electric fire as a quantitative circulating entity that would find a natural state of equilibrium. But he never offered a definitive justification for why it was legitimate to conceptualize nature, society, and morality in terms of circulation, economy, and equilibrium. It is hard to tell if he was analogizing from nature to society or from society to nature, so circular is the logic of his thought. Like the entities it described, Franklinist analogy functioned in circulation, mediating between the natural and the social and navigating with spectacular success among different domains of electrical experiment.[28]

Unlike Newton's formulas for computing the force of gravity and the motion of bodies in the *Principia,* Franklin's system did not offer a way of mathematically accounting for electricity. It offered a description of electrical phenomena that could predict effects, but it did not make a direct metaphysical or theological statement about why electricity behaved like an economy. In the eighteenth century, inspired by the model of Newton's "laws of motion" (or "laws of nature"), both the natural and the human sciences increasingly borrowed the language of law to confer the status of inexorable and universal physical necessity on their accounts of cause and effect. Searching for laws became conceptually and rhetorically central to the definition of science in the Enlightenment, and soon enough, Franklin's program was promulgated as the "laws of electricity" by followers like John Winthrop IV, professor of natural philosophy at Harvard. But how did Franklin himself regard his achievements? Franklin presented his own work in a spirit of cheerful pragmatism and philosophical humility. In a caustic comment to Collinson, his choice of example beautifully illustrating the genteel worldliness of his rationalism, he dismissed questions like "Why was nature lawful?" as irrelevant. It is not "of much importance to us to know the manner in which nature executes her laws; it is enough if we know the laws themselves. It is of real use to know that china left in the air unsupported will fall and break; but *how* it comes to fall, and *why* it breaks, are matters of speculation. It is a pleasure indeed to know them, but we can preserve our china without it." According to Franklin, there are laws of nature and there may be explanations of those laws, but first things first: practical power over nature is what counts.[29]

In Chapter 4, we will see how American revolutionaries turned Franklinist electricity into an icon of enlightenment and of the power of reason to subject nature to law, as well as a nationalist myth of American mastery. If we return to Franklin's own writings on electricity, however, we discover that the master electrician and the bookkeeping rationalist was in fact convinced that electricity was still a wonderful power. This tension lay at the heart of the Franklinist enlightenment: the ambition of rational economic mastery versus an abiding sense of nature's power to override human control. Franklin displayed this humility in his experimental accounts by repeatedly exhibiting electricity's capacity to surprise mind and body. Even at the height of his success in accounting for electricity's different charges in the Leyden jar, he confessed: "So wonderfully are these two states of Electricity, the *plus* and *minus,* combined and balanced in this miraculous bottle! situated and related to each other in a manner that I can by no means comprehend! . . . [H]ere we have a bottle containing at the same time a *plenum* of electrical fire, and a *vacuum* of the same fire; and yet the equilibrium cannot be restored between them but by a communication *without!*" Did electricity reveal the cosmic order or confound it? Did it serve any clear purpose? Franklin admitted that "the beneficial uses of this electric fluid in the creation we are not yet well acquainted with, though doubtless such there are, and those very considerable." Were there a greater quantity of electricity in the universe, he hypothesized, the mutual repulsion between electrified bodies would clog the air with foreign matter, making it "unfit for respiration." That such a state of affairs had not come to pass urged the utmost reverence for divine power: "This affords another occasion of adoring that wisdom which has made all things by weight and measure!"[30]

The bookkeeper who made a balance sheet to contain nature's own commerce in electric charge, and who claimed only to observe nature's effects, also published metaphysical speculations that connected the electric fire to a vital "inflammable" principle underlying all of nature. Wasn't the ability to kindle fire and produce heat from bodies in so many different ways, including electricity, a sign "that the fire existed in the body, though in a quiescent state, before it was . . . excited, disengaged, and brought forth to action and to view? May it not constitute part . . . of the solid substance of bodies? If this should be the case, kindling fire in a body would be nothing more than developing this inflammable principle, and setting it at liberty to act in separating the parts of that body." This was no deist's account of the universe as a self-regulating machine nor a theological statement about electricity's role in the divine cosmic order. It was, rather, a radical vitalist cosmology in which nature was fundamentally a volatile entity with a latent tendency to destructive separation. This was Franklin's electrical cosmos: one that dynamically pitted the ambitions of experimental mastery and the rational design of the Creator against a naturalized volatility. Elsewhere, he dramatized the same point rather more comically. In one memorable misstep, where he had set about tenderizing a turkey, he accidentally stunned himself with the discharge of a battery of connected Leyden jars: "I then felt what I know not how well to describe; an universal Blow thro'out my whole Body from head to foot which seem'd within as well as without; after which the first thing I took notice of was a violent quick Shaking of my body." "That the progress of the electrical fire is so amazingly swift, seems evident from an experiment you yourself . . . made," his friend the Boston merchant James Bowdoin wrote him, "when two or three large glass jars were discharged through your body. You neither heard the crack, was sensible of the stroke, nor, which is more extraordinary, saw the light." Even in the master electrician's laboratory, the quickness and force of electricity defied the senses and vexed the body. With his signature modest mastery, the bookkeeper of nature marveled at electricity's wonderful powers.[31]

Franklin's active career as an electrical experimenter was destined to last only a short period of his life, essentially the few intense years of labor in the late 1740s and early 1750s. After that, public service for colony, empire, and ultimately nation denied him the time for systematic experiment, although he continued to correspond with natural philosophers and offered opinions and speculations on philosophical matters throughout his life. The short season of his experiments nonetheless proved pivotal in altering the course of electrical science, and in creating a market for electricity in North America. Experimental access to electricity depended on access to apparatus. Although electrical effects seemed like a disembodied commodity, the ability to generate them depended on the world of consumption and material goods. Joseph Priestley, a leading English electrician widely read in America, impressed on his readers that there was no science of electricity without experiments and the means to make them. "A mere reader," Priestley wrote, "has no chance of finding new truths." He urged "all persons who propose to understand the subject of Electricity, though they have no expectation of making discoveries, to provide themselves with an electrical machine, or at least desire some of their friends to show them the experiments . . . the acutest philosopher in the world could not converse about electricity without making many mistakes, and perhaps gross ones, after reading every book he could meet with upon the subject, if he had seen few or no experiments."[32]

Enticed by the prospect of learning about electricity firsthand, British Americans began to consume experimental science in the form of books and apparatus imported from Britain. Electrical instruments began their career in America as genteel goods, coming ashore as luxurious expenses made available by wealthy patrons like Penn, or increasingly, by merchants importing polite commodities. Franklin's circle, for example, used goldleaf books, "fine silken thread," and silk cords imported from London, as well as china plates, wine glasses, cork balls, lead wire, and silver bells. By the 1770s, the Boston merchant Cyrus Baldwin was importing "tea, coffee and chocolate, cotton-wool, best French indigo, electrical globes, crow quills, hautboy reeds . . . with many other articles too numerous to be enumerated," while in Philadelphia Andrew Porter was offering "a most elegant pair of 15 inch [electrical] globes, mounted on best mahogany claw frames." In 1775, an anonymous author in the *Pennsylvania Magazine*, writing under the apt pseudonym Atlanticus, avidly reported to his readers the latest European designs for electrical machines that would "excite stronger than any other I have yet seen." "There is no place where the *study* of electricity has received more improvement than in Philadelphia," Atlanticus proffered, but confessed that "in the *construction* of the machines the European philosophers have rather excelled," so it was as well for Americans to continue as active importers. For the literati, the latest apparatus was a hot commodity. A decade later, Yale president and sometime electrician Ezra Stiles recorded his envious curiosity at the mention by Thomas Jefferson, then governor of Virginia, of "a new simple apparatus for Electricity left by a British Officer in Philad[elphia]." "One person in Philad[elphia] had the secret," Stiles noted after Jefferson's visit, "& sells the cake or Amber-like plate at Eight Dollars." Such machines were the sine qua non of electrical enlightenment and as such were highly prized.[33]

Americans continued to import some of the finest machines from abroad well after the Revolution. As a colonial agent in Britain, Franklin had played the central role in shipping instruments made by London craftsmen to American friends. But local artisanal production of apparatus in America (as with Syng's machine) followed soon thereafter. Significantly, technological proliferation brought cultural diversification. Moreover, electricity was a commodity that circulated beyond elite circles, and for purposes other than formal experiment, such as public entertainment and commercial electrotherapy. John M'Cabe, for example, was a Dublin-born horologist in Baltimore who, like many of his fellow clockmakers, engaged in a variety of crafts productions. By the 1770s, he was selling electrical machines and other mathematical and philosophical apparatus (orreries or mechanical planetariums, air pumps, microscopes, and telescopes), as well as navigational instruments (land and sea compasses and mariners' clocks), in local newspapers. Another horologist, William Claggett, and the goldsmith Joseph Hiller, both Bostonians, made electrical machines and performed demonstrations with them during the 1740s and 1750s. Electricity moved through a variety of social spaces and class settings where philosophical knowledge and artisanal dexterity converged in surprising ways.[34]

Why did Americans import and build their own electrical machines? What did they do with electricity? For the moment, interactions between body and electricity lacked reciprocity: the body displayed the behavior of electricity, but how electricity affected the workings of the body remained unclear. The fascination of physical experience with electricity, however, would bring the electrified body heightened public notoriety and intensified philosophical scrutiny. Beyond this, it would open up new visions of self and society, nation and cosmos. As will become clear, the electrical enlightenment in America was about much more than simply describing the behavior of the electric fire.

CHAPTER TWO

Lightning Rods and the Direction of Nature

The difference between the atmosphere of Europe and America is the quantity of the electric fluid, with which the latter is much more highly charged.

CONSTANTIN VOLNEY, 1804

WHILE in exile in the United States at the end of the eighteenth century, the French historian and naturalist Constantin Volney observed a signal difference between the nature of America and the nature of Europe: the American atmosphere was, literally, more electric. This continental difference "may be made perceptible to the senses at any time, without any complicated apparatus," Volney explained, advancing as evidence eyewitness observations of the electrification of silk, the "superior vividness" of lightning, and the loudness of thunder. "When I first saw thunderstorms at Philadelphia," he noted, "the electric fluid appeared to me so copious, that all the air was on fire. Its arrowy and zigzag lines were broader and longer than any I had ever before seen." The superior force of American electricity impressed itself unmistakably on the senses. "So strong were the pulsations of this fluid, that they seemed to my ear and my face like the wind produced by the wings of some passing bird."[1]

Suggesting an innate electrical difference between the eastern and western hemispheres might have formed the basis for blaming (or praising) American nature, as was so often the case in early modern disputes of the New World. Certainly, this was a rare instance when electricity, typically unmentioned in such debates, appeared to testify to hemispheric difference. Instead of pressing any such claim, however, Volney linked the violence of American lightning to issues of technological, and thus cultural, abilities. He noted that seventeen deaths by lightning had been reported in newspapers between June and August 1797, as well as eighty "severe accidents." Such misadventures were "frequent in the country," he claimed, "especially beneath trees." The suggestion that Americans were not masters of their electrical environment was provocative, not least to Volney's own translator—Charles Brockden Brown. In a footnote to Volney's *View of the Climate and Soil of the United States* (1804), Brown patriotically defended the progress of that icon of American enlightenment—the lightning rod—that the Frenchman had ignored. "From the use of conductors, or from some other cause," Brown wrote back, "accidents from lightning are rare in the American cities. One death, from this cause, in twenty years, in New York or Philadelphia, would be a liberal calculation." American cities, at least, were not unenlightened provinces. In fact, eighteenth-century evidence would suggest quite the opposite. In 1772, Franklin, designer of the first protective lightning rod, could boast with some justification that, by contrast with the mother country, conductors were "so common" in America "that Numbers of them appear on private Houses in every Street of the principal Towns, besides those on Churches, Publick Buildings, Magazines of Powder, and Gentlemans Seats in the Country." Lightning protection was one pursuit of enlightenment in which the colonial periphery led metropolitan Europe.[2]

The invention and adoption of the lightning rod is a classic example of the supposed disenchantment of nature in the Enlightenment. Well into the twentieth century, historians of Franklinist electricity made common cause with eighteenth-century polemicists, embracing as their own the triumphal claim that the lightning rod shattered superstitious fears of the destructive power of lightning, becoming in the process one of the most celebrated icons of technological enlightenment. This iconic interpretation is misleading, however, because it overestimates conflicts between science and religion in the Enlightenment, and misrepresents the character of the religious discussions that accompanied the introduction of lightning rods in America. In his "willingness to rise above the superstition of his age," wrote I. Bernard Cohen, the author of the standard twentieth-century history of Franklin's work in electricity, "and particularly in the ease with which he ignored the possibility of a Prometheus-like fate and the wrath of the father's rod," Franklin proved himself "an emancipated spirit and a herald of our modern age." According to this heroic account, the awful voice of divine punishment that had traditionally been heard in lightning was silenced by a homespun secular ingenuity that fearlessly propelled mankind toward modernity. Lightning rods made enlightenment: protecting the human body from devastation freed the mind from fear and opened it to reason.[3]

But Franklin's simple metal rod was much more ambiguous, both theologically and technologically, than this mythical account suggests. Conductors were a dramatic application of the economic laws of electric charge, and through their protection of the human body they were a crucial instance in which the experimental science of electricity became significant to the public at large. But this protective capacity implicitly challenged the longstanding status of lightning as an instrument of divine punishment. Zealous sermonizers balked at a device that seemed to rob the Divine of his ability to direct nature to the moral end of punishing sin. Franklin's wielding of godlike powers was later to become a stock image of Enlightenment and Revolutionary propagandists. But American natural philosophers were themselves pious men who reaffirmed the divine direction of nature: their quarrel with theologians was not over whether God directed nature, but over the manner in which He did so, and the character of the natural order over which He presided. Thanks to their new technology, these natural philosophers could insist that this order was stable, intelligible, and predictable, but such rational claims were themselves vexed by the troublesome fact that in the eighteenth century, lightning rods frequently failed as protective devices. Did lightning in fact defy enlightenment? Either way, technology and theology were related, not separate, concerns.

Lightning as Electricity

Philosophers had speculated since antiquity on the causes of thunder and lightning, and by the mid-eighteenth century, conjectures on the relation between the two were not uncommon. Franklin himself observed that the sparkles on the water he noticed when crossing the Atlantic were electrical, and reckoned that electricity was produced by the friction of water molecules in the ocean, which then passed into the atmosphere via rising vapors, producing thunderclouds possessing varying quantities of charge (he also speculated that the aurora borealis, or northern lights, was produced by the motion of rising and descending electric vapors). The Philadelphian was the first, however, to go beyond mere conjecture and devise an experiment to demonstrate the electrical nature of lightning. The decisive move once again relied on the use of analogy to connect different theaters of physical experience, in this case the play of electric charges in the Leyden jar and the behavior of lightning with respect to the earth, translating mastery of laboratory forces into mastery of natural ones.[4] Franklin had argued that the discharges from these "wonderful bottles" were produced by the completion of a circuit between the jar's positively charged interior and its negative outer coating. This "explosion" resulted from the equalization of charges in the jar. Franklin then hypothesized that lightning strikes might also be electrical discharges, only on a much larger scale, that redistributed charge from positively electrified clouds to the negatively charged earth (or more typically, he thought, vice versa, from the positive earth to negative clouds). In so doing, he conceptualized the operations of nature in terms of the operations of his private laboratory, and set about contriving a persuasive demonstration.[5]

Such a demonstration was later to suggest the possibility of constructing protective lightning conductors, but in the beginning, metal conductors were used to draw down and collect lightning as electricity for the purpose of conducting experiments. Franklin's first attempt at a demonstration required an iron rod twenty to thirty feet tall, whose apex would "draw off" static electricity from the atmosphere and conduct it down the length of the rod to the inside of a sentry box ("Fig. IX" within Figure 5). This structure afforded the experimenter some shelter in case of a violent storm. Any communication between the bottom of the rod and the earth had to be cut off, so that charge would accumulate in the rod. To this end, nonconductive glass bottles were used to insulate the stand where the conductor terminated. Sparks could then be drawn off the rod to show that electricity had been collected from the atmosphere: "If the electrical stand be kept clean and dry, a man standing on it when such clouds are passing low, might be electrified and afford sparks, the rod drawing fire to him from a cloud." According to Franklin, this trial was entirely safe: "If any danger to the man should be apprehended (though I think there would be none) let him stand on the floor of his box, and now and then bring near to the rod the loop of a wire that has one end fastened to the leads, he holding it by a wax handle; so the sparks, if the rod is electrified, will strike from the rod to the wire, and not affect him." A simple extension of laboratory technique—the use of an insulated C-shaped conductor—allowed the experimenter to play with the awesome power of lightning itself.[6]

In taking pains to reduce the danger of such experiments, Franklin was motivated by anxiety over the difficulty of mastering electricity in such a powerful form. Despite the neatness of the analogy, manipulating lightning as electricity was dangerous and unpredictable; there was no way to be sure that it could be safely controlled on such a scale. Similar risks attended Franklin's famed kite experiment, though heroic descriptions barely even suggested this. According to Joseph Priestley, the English Dissenter and electrical philosopher to whom Franklin later confided his version of the experiment, Franklin believed that no building in Philadelphia was sufficiently tall for the sentry-box test, so he proposed an alternative. "It occurred to him," Priestley wrote, "that by means of a common kite he could have better access to the regions of thunder than by any spire whatever." In his own published instructions, Franklin recommended using a kite made of two crossed strips of cedar and covered in nonconductive silk (to withstand stormy weather and accumulate charge), terminating in a pointed wire. A key was fastened to the lower end of the twine string, by a silk ribbon. As with the sentry-box test, the experimenter holding the kite string should "stand within a door or window, or under some cover, so that the silk ribbon may not be wet." Nor should the twine or ribbon touch the window or door, lest some of the charge should accidentally be lost. "As soon as any of the thunder clouds come over the kite," Franklin observed, "the pointed wire will draw the electric fire from them, and the kite, with all the twine, will be electrified, and the loose filaments of the twine will stand out every way, and be attracted by an approaching finger. And when the rain has wet the kite and twine, so that it can conduct the electric fire freely, you will find it stream out plentifully from the key on the approach of your knuckle."[7]

This was a singular moment of enlightenment through self-evidence, in which the experimenter's body was the crucial instrument of knowledge. The identity of lightning and electricity was demonstrated through a sensational analogy—that is, by the physically similar experience of both forces. Moreover, unlike seventeenth-century experiments that were powerfully shaped by codes of Christian asceticism, here was a characteristically eighteenth-century view of enlightenment as worldly pleasure. The proof, Priestley exulted, was in the feeling. "Let the reader judge of the exquisite pleasure he must have felt at that moment . . . the discovery was complete. He perceived a very evident electric spark" (Figure 6). The walls of separation between the laboratory and nature were dissolved: by transferring the experimental protocols of the laboratory to the field, Franklin had effectively transformed nature itself into a giant laboratory, whose forces could be manipulated through experimental gestures. All the usual experiments with artificial electricity could now be replicated with electricity collected from the atmosphere: "Thereby the sameness of the electric matter with that of lightning [is] completely demonstrated."[8]

It was not long thereafter that Franklin hit upon the possibility of a protective conductor. Reasoning by analogy suggested that if lightning bolts were electrical discharges, then their movement could be manipulated using large metal rods, just like laboratory wires. It was in France, rather than America or Britain, that the sentry-box test was first executed with success. Franklin would later complain that the Royal Society mocked his early letters, seeming to close one important avenue to the European reception of his experiments. But although the Society was Georgian London's leading scientific institution, Britain's thriving literary marketplace meant that Franklin's writings could find important venues other than the pages of the *Philosophical Transactions*. Once more, Collinson (together with Dr. John Fothergill) was crucial, facilitating the collection of his electrical letters in a single volume in 1751 published by Edward Cave, editor of

6. Heroic Self-Evidence

Together with Fragonard's view of Franklin as Prometheus/Jupiter, this dramatic vision of Franklin the experimenter is the most often reproduced. Franklin does not appear as a god, however, but as a wizened American sage (his Americanness signaled by the putto who wears what seems to be a native American headdress). The artist exploits Franklin's use of his own body in the kite experiment to lift him into nature as a fearless heroic pursuer of electricity—the American man of science as man of action.

the *Gentleman's Magazine* (where electrical articles by Franklin also appeared). This visibility proved decisive: a translation of the volume that appeared the following year, ordered by the leading French naturalist Georges-Louis Leclerc, Comte de Buffon, was the handbook for the lightning trials successfully carried out in Marly-la-Ville near Paris (Buffon eagerly adopted the Franklinist system in opposition to that of the Abbé Nollet, who reportedly suspected "Franklin" was an invention of his enemies). In May 1752, the Abbé Mazéas reported to Stephen Hales at the Royal Society that, under the guidance of Buffon, d'Alibard, and a gentleman named Delor, a military man named Coiffier and a local priest called Raulet had indeed performed the sentry-box test, drawing sparks from the end of an insulated lightning rod reaching forty feet into the sky. Because of the slowness of transatlantic communication, Franklin did not know of the success of this experiment before he likely carried out the kite experiment in Philadelphia in June the same year. Following the logic of his own experiments, however, he then became the first person known to have mounted a grounded protective conductor, on his Philadelphia house some time in June or July.[9]

Though the utility of electricity itself was to remain highly ambiguous in the eighteenth century, protective lightning rods were astonishing proof of the usefulness of electrical research. They also seemed to imply a spectacular reversal of the intellectual division of labor that separated the early modern world into metropolitan natural philosophers and provincial natural historians. The "Philadelphia experiments," as they became known, were soon repeated by the leading European electricians: Nollet in France, John Canton in England, Pieter van Musschenbroek in the Netherlands, and Giambattista Beccaria in Italy. With the identity of electricity and lightning established, and the effectiveness of protective conductors accepted, the Royal Society, compensating for its earlier mixed reception of Franklin, awarded him its Copley Medal in 1753—an unprecedented honor for a colonial. But the awarding of the medal was not perceived by contemporaries as a revolutionary rupture in the geopolitical dichotomy between European explanation and American observation. In fact, it rhetorically reinforced this hierarchy. President Macclesfield's award speech made a paternalistic show of British cosmopolitanism, embracing Franklin as a fellow Briton whose genius redounded to the glory of the greater British nation: though he was "not a Fellow of this Society nor an Inhabitant of this Island," he was yet "a Subject of the Crown of Great Britain," and his invention ought to be "universally diffused" to the "honour of this Society and of the British Nation." Invoking Virgil, Macclesfield continued: "Whether he be Trojan or Rutulian, I shall regard him without discrimination," for all savants were "Constituent Parts and Fellow-Members of one and the same illustrious Republick." Franklin, meanwhile, was careful not to overstep the bounds of provincial modesty, distinguishing his "observations and experiments" of nature's wonders from the prideful rationalism of Nollet's "Theory of Electricity," a "System" the Frenchman was busily defending. Although claims about the political significance of the lightning rod were made a generation after its invention, in the ideological ferment of the American Revolution, at the time of his achievements, Franklin cast himself as a modest colonial witness of nature, not a revolutionary natural philosopher.[10]

Yet behind the naïve pleasures and seemingly effortless mastery of Franklin's experiments, powers of destruction and disorientation remained. Several American electricians after Franklin, among them Professor John Winthrop IV of Harvard and President Ezra Stiles of Yale, mounted conductors on their houses, passing wires through their domestic chambers so as to collect electricity for experimental testing. As Mason Chamberlain's 1762 London portrait of Franklin shows, this experimental rod was deliberately broken in the middle, where a pair of bells was mounted (Figure 7). The bells rang when the conductor became electrified, causing brass balls to communicate a charge from the upper rod to the lower. Sometimes the electricity was "very faint," Franklin noted, and the sparks jumped between the bells only intermittently. At other times, "the fire pass[ed], sometimes in very large, quick cracks from bell to bell . . . in a continued dense, white stream, seemingly as large as my finger, whereby the whole staircase was enlightened as with sunshine." Although portraits like Chamberlain's portrayed scenes of nonchalant mastery, proximity to conductors placed the experimenter in potentially mortal jeopardy, as the death of Georg Wilhelm Richmann made abundantly clear. An electrician in St. Petersburg, Richmann was killed when lightning struck a conductor in his laboratory in 1753. William Watson, the Royal Society's leading electrician, explained in the *Philosophical Transactions* that because Richmann's rod was not grounded, the lightning bolt passed into his body as "the nearest non-electric [conductive] substance in contact with the floor." Although would-be electrical martyrs like Priestley claimed to envy this heroic demise, Richmann's death gave philosophers pause. Ebenezer Kinnersley, Franklin's electrical colleague and the leading electrical showman in British America, took pains to deny that such misadventures were evidence that electricians were trespassing beyond divine bounds. "Some are apt to doubt the Lawfulness of endeavouring to guard against Lightning," Kinnersley explained while demonstrating model lightning rods, but lightning rods were not "inconsistent with any of the Principles either of Natural or Revealed Religion."[11]

Reassuring performances like Kinnersley's sought to play down the mortal dangers that lightning continued to pose to body and mind. As Ezra Stiles put it when lightning struck his house in 1789, his home was "surrounded by Death, or that profusion of the electrical Fluid." Let us here recall too that other kite experiment—Loammi Baldwin's. Baldwin's account of his 1771 restaging of Franklin's famous experi-

7. Franklin as a Bookish Natural Philosopher

Several portraits of Franklin made by British artists in the 1760s depict him as a European gentleman-philosopher rather than an intrepid American experimenter. Here Chamberlain places a bewigged Franklin, quill in hand, safely in his London study while a thunderstorm rages outside his window. There is an element of risk, however. The bells on the left are attached to a grounded lightning rod and are designed to ring when it is electrified. Franklin sometimes observed large, potentially dangerous, sparks passing between such bells.

ment vividly shows how uncertain human mastery of lightning-as-electricity remained, well after the triumphal invention of the lightning rod. Baldwin's account is exemplary in exhibiting the range of disorienting cognitive responses electricity could provoke, and in particular the problem of secular enthusiasm—the confusion of cause and effect—that it persistently raised. While raising his kite into the atmosphere, Baldwin froze at the appearance of a "rare medium of fire between [his] eyes and the kite," as his entire body appeared to become enveloped by a huge electrical flame. Two simultaneous struggles ensued, one physical, the other epistemological. Let us revisit Baldwin's description in detail:

> I cast my eyes towards the ground;—the same appearance was there.—I turned myself around;—the same appearance [of fire] still between me and every object I cast my eyes upon.—I felt myself somewhat alarmed at the appearance. I stood, however, and reasoned with myself upon the cause, for some time, but gained very little satisfaction,—the same fiery atmosphere surrounded me, only more bright and apparent. I was about to discontinue my experiments for that time; but reason accused imagination with error; and supposing it might possibly be only fancy, not knowing the cause of such an appearance, and feeling no very bad effects from it, I continued to raise the kite.

As "the fiery atmosphere was increasing and extending itself, with some faint, gentle flashings," he experienced "a general weakness in [the] joints and limbs, and a kind of listless feeling." But the source of this debility was unclear: was it the effect of electricity on his body or "only the effect of surprise" on his mind? Baldwin couldn't tell. His family, who had been witnessing all of this, were "vastly more surprised than I had been myself," seeing him engulfed "in the midst of a large bright flame of fire, attended with flashings," expecting to see him "fall a sacrifice to the flame." His neighbors confirmed this observation. "I shall make no remarks at this time upon the cause," he concluded agnostically, "but leave it for the present to the consideration of the learned."[12]

Baldwin vividly dramatized the unpredictable epistemological drama of electricity: the loss of physical control and the confusion of cause and effect. Reason and imagination were in conflict, with reason unequal to disciplining its darker cognitive rivals: imagination, or what he called "fancy" and "surprise." Reason could not determine whether such debility was the physical effect of electricity or the result of fear produced entirely from within the mind. With no separation between the experimenter's body and the phenomenon to be observed, Baldwin could only observe his own disorientation. In the long shadow of Franklin's heroic account of experimental self-evidence, his narrative showed that senses remained fallible when engaging with electricity. The Janus face of Franklinist self-evidence was the problem of enthusiasm.

"His Lightnings Shall Enlighten the World"

In cases where a metal conductor could not be used to redirect lightning away from bodies, Franklin counseled the reverse strategy: removing bodies from the electroconductive path of lightning. He cautioned against sitting near chimneys, for example, or mirrors or gilt-edged pictures during a thunderstorm. "The safest place," he wrote, conjuring a consummate image of comfortable bourgeois insulation, "is in the middle of the room . . . sitting in one chair and laying the feet up in another. It is still safer to bring two or three mattrasses or beds into the middle of the room, and folding them up double, place the chair upon them; for they not being so good conductors as the walls, the lightning will not chuse an interrupted course through the air of the room and the bedding." He continued: "Where it can be had, a hamock or swinging bed, suspended by silk cords . . . affords the safest situation a person can have." This, Franklin confidently promised, left one "quite free from danger of any stroke by lightning."[13]

Once again, a rhetoric of common sense masked the drama of Franklin's intervention. This corporeal self-possession was really an act of repossession. In removing the body from the path of lightning, conductors seemed to remove human beings from a divine system of moral regulation through punishment (and, indeed, the fear of punishment). According to Calvinist views of natural phenomena that survived well into the eighteenth century, lightning strikes were a form of special providence, interventions by God to maintain the order of His universe. In describing his kite experiment, Baldwin offered an account of the disintegration of rational human agency that might result from interaction with electricity, but he did not point toward a divine agency overruling reason (hence, his narrative's eerily agentless quality). Divine agency was, however, the central and explicit issue for Protestant commentators on lightning who insisted on a close relationship among bodily insecurity, awe at divine power, and moral striving. An unmediated relationship between lightning and the body worked to regulate human affairs: vulnerability to natural destruction deterred moral backsliding. Logically, protecting the body from catastrophe undermined the incentive for moral discipline.

Theological commentary on the sacred purpose of lightning was especially voluminous in colonial New England. By the turn of the eighteenth century, ministers like Increase and Cotton Mather sought to align their religious faith with natural philosophy in the era of Newton, essaying a delicate balance between Puritan preaching in New England and scientific correspondence with the Royal Society back in London. The result was a schizophrenic cosmology—one that alternated between awe at divine ability to surpass the ordinary course of nature through providential wonders, and an emphasis on the orderliness and intelligibility of nature as a result of divine design. This emphasis on rational design, known as "physico-" or "natural theology," was a pious tradition of Protestant natural philosophy in the late seventeenth and eighteenth centuries. Physico-theology found cause for devotion in the study of nature but, unlike the radical Puritanism of the English Revolution, it was also a doctrine of social moderation.

Where Restoration critics made wonder at the presence of the divine in nature a scapegoat for the mystical enthusiasm and bloody strife of the civil wars, physico-theologians like Gilbert Burnet, John Ray, and William Derham insisted on preserving an important role for wonder at the perfect rationality of the created universe. Reason and piety were linked in a similar fashion for Protestants across the Atlantic, particularly for philosophically literate New Englanders like Cotton Mather, whose *Christian Philosopher* (1721) was a virtual manifesto for physico-theology.[14]

In the case of lightning, however, ministerial pronouncements routinely challenged physico-theology's emphasis on the rational stability of nature. In his *Essay for the Recording of Illustrious Providences* (1684), Increase Mather's accounts of lightning strikes on human bodies and structures were matter-of-fact descriptions of lightning's destruction of matter. Like the innumerable stories of persons struck and killed by lightning that appeared in American newspapers during the colonial period (and like Loammi Baldwin's narrative), he depicted scenes of devastation, but chose not to tackle the issue of causation. Yet he did include "Philosophical Meditations" that offered a "Theological Improvement" of his lightning stories: lightning produced "stunning," "amazing," and "wonderfull" effects that were evidence of divine sovereignty over human life. A generation later, Cotton Mather insisted, like his father, that God's ways were ultimately "wonderful" (inscrutable) to man: *"The Thunder of his Power who can understand?"* By contrast with philosophers who thought they could silence the skies, the devout were deafened by the meanings they heard in thunder and lightning. Thunder was "a Note prepared for the Songs of the Faithful," the sound of God speaking to his flock: "There is this *Voice* most sensibly to be heard in the *Thunder, Power belongeth unto God.*" "*Let me,*" Cotton Mather implored, *"be in good Terms with one so able to destroy me in a moment!"* Vulnerable bodies made for moral selves in the face of an absolute divine agency. The Presbyterian

Gilbert Tennent was actually struck by lightning during the revivals known as the Great Awakening in 1745. "The Flash of the Lightning," Tennent recounted, "was so violent as to tear my shoes to Pieces, twist one of my Buckles, and melt a little of two Corners of the other." He offered no naturalistic interpretation, declaring bluntly that in striking him, "God hereby glorifie[d] his *Sovereignty*." The "Rod of God" called saints and sinners alike to account.[15]

The rod of Franklin dueled with the rod of God. As devices that physically redirected nature, lightning conductors seemed to contest divine direction of the universe by frustrating His power to punish through direct devastation of the body. In 1755, an earthquake struck New England. Mather Byles, Cotton Mather's nephew, clarified the implications for his Boston congregation. Although there were verifiable "Laws of Nature," God "over-rules them all, suspends or alters them as he pleases." Earthquakes "may be accounted for on Philosophical Principles" but were still "the Effect of Almighty Power." When God visited "the Convulsions of the Earth, [and] the Ruins of dissolving of Nature" on a people, "they should know the Rod, and who hath appointed it." The earthquake of 1755 also brought lightning and electricity into public debates over God's sovereignty over nature. Like Byles, the Reverend Thomas Prince of Boston's South Church took pains to subordinate considerations of *"Natural, Instrumental or Secondary Causes"* to "the *First* and *Principal Cause* in these great Affairs." An explicit theological link was again made between bodily vulnerability, divine agency, and the passions of cognition. "I need not describe the Surprize, the Anguish and Horror that seizes and distresses our Minds," Prince sermonized, "when we are feeling our Houses rocking over our Heads and . . . the Foundations of the Earth moving under our Feet." These great convulsions were not undirected catastrophes, but God's moving the matter of the earth in an "intelligent Manner" to moral ends. *"Tremble thou Earth at the Presence of God,"* preached Prince, quoting the Book of Psalms. "His Lightnings shall enlighten the World, And the Earth shall see and tremble."[16]

Prince was no philosophical illiterate. He had in fact studied at Harvard, and like the Mathers, had pursued a serious interest in natural philosophy. This led him to revise his earthquake sermon of 1727 to include an appendix "concerning the Operation of God in *Earthquakes* by Means of the *Electrical Substance*," in which he reflected on the relationship between divine and natural agency. The year, let us recall, was 1755, just a few years after the invention of the lightning rod and the publication of Franklin's *Experiments and Observations on Electricity*. Acknowledging that Franklin's discoveries had "greatly surpriz'd" the learned world, Prince accepted that "this *Electrical Substance* seems to be one of the mightiest Agents we know of among material Substances in this lower World." The earth, he justly observed, was "the grand Source from whence [electricity] rises and to which it returns." But even as he included electricity with "*sulphureous, nitrous, mineral, watery* and *airy* Substances" as a secondary cause of earthquakes, he insisted on returning to considerations of God as the prime mover. In the operations of electricity, all "may plainly see therein the *Traces*, the Directions and the *Actions* of An Intelligent Being, who is every where and always present with that wonderful Substance of his Creation, Sustenation and Co-Agency." Prince could accept Franklin's economic model of lightning-as-electricity, agreeing that electrically charged thunderclouds appeared to "act like *perfectly knowing* and *spontaneous Agents*." But as mere matter they must be "moved, guided and proportioned by a Being who knows them." Electricity was "intelligent," but purely because of the divine impulse that animated it. And there was a sting in the tail. Perhaps, Prince suggested, lightning rods were actually to blame for the 1755 earthquake, by conducting atmospheric electricity into the earth? Noting that more conductors had been erected in Boston than anywhere else in New England, he wondered "whether *any Part* of the *Earth* being fuller of *this terrible Substance*, may not be more exposed to *more shocking Earthquakes*"? "O!" he sighed, in fateful conclusion, "There is no getting out of the mighty Hand of God! If we think to avoid it in the *Air*, we cannot in the *Earth*: Yea, it may grow more fatal." Only "*Almighty Friendship* through *Christ* the *Mediator*" could offer salvation—not artificial machines like conductors.[17]

This charge was met not by Franklin but John Winthrop, Isaac Greenwood's successor as professor of natural philosophy at Harvard and a committed Franklinist. Propagandistic praise of Winthrop's rebuttal has always received more attention than the content of the reply itself, and this has had a significant impact on the historiography of lightning rod controversies. To some, Winthrop's refutation was an unqualified triumph of science over superstition. Young John Adams led the jeering. Penning his own response to Prince in the margins of his copy of Winthrop's *Lecture on Earthquakes* (1755), Adams disdainfully noted that Prince's exclamation about the "mighty hand of God" was "very popular, for the Audience in general like the rest of the Province, consider Thunder and Lightning as well as Earthquakes, only as Judgments, Warnings &c and have no Conception of any Uses they can have in Nature." For Adams, belief in nature's unintelligibility denoted vulgarity of mind. He could not tell whether popular terror at nature "derived from real revelation" or was "artfully propagated," although once again the enthusiastic "Imagination" seemed a prime candidate, because it inflamed fear without true cause. Whatever the reason, lightning rods had unfortunately met with "Opposition from the Superstitions, Affectations of Piety, and Jealousy of New Inventions." Winthrop concurred: "How astonishing is the force of prejudice, even in an age of so much knowledge and free inquiry!" In 1760, at a meeting of the Junto, Franklin's Philadelphia group of "leather-apron men," members discussed the question while Franklin was in London: "May we Place Rods on our Houses to guard them from Lightening without being Guilty of Presumption?" Their answer was unequivocal. "Tho the rending Peal of Thunder may fill the Minds of the Ignorant with Terror who from an Ignorance of its Cause and their natural Superstition may imagine that it is the immediate Voice of the Almighty and the Streaming Lightening are Bolts launched from his Right Hand and commissioned to execute his Vengeance yet in Reason's Eye Lightening or Thunder is no more an Instrument of Divine Vengeance than any other of the Elements." Here was an American branch of the "Party of Enlightenment" engaged in creating a polemical dichotomy between rational enlightenment (lightning rods and their advocates) and fearful superstition (their religious critics). "So far is it from being Presumption to use [the lightning rod]," the Junto concluded, that "it appears foolhardiness to neglect it."[18]

This dichotomy between reason and superstition did not mean that natural philosophers opposed themselves to all religious interpretation of nature, however. This was not a battle between secular science and religious orthodoxy. On the contrary, advocates of conductors insisted on divine agency, too, albeit of a different character than that described by Prince. All positions on the morality of lightning rods were theologically informed and, in the end, the concerns of philosophically literate theologians were not so very different from those of theologically sensitive natural philosophers. Certainly, Winthrop contested the technical basis of Prince's aspersions. A key issue was the movement of electricity through the earth. Winthrop correctly pointed out that it was impossible for electricity to collect in any one terrestrial location, as Prince had suggested. The electricity that lightning rods drew into the earth would not accumulate; it would diffuse. Despite the well-observed coincidence of lightning and terrestrial convulsions such as volcanic eruptions, inferring causal relations between electricity and earthly commotions remained speculative. Terrestrial accumulations of electricity would be truly wonderful, Winthrop reasoned, a violation of the lawful chain of cause and effect in nature, effectively undermining the very project of natural philosophy. "To have recourse to *miraculous* interpositions of the 'Divine Direction,'" as he alleged Prince had had, "is to put an entire end at once to all reasonings about electricity, or earthquakes, or any other natural phenomena." Nature did receive divine direction, Winthrop emphasized, but "according to the established laws of electricity." Without regular chains of cause and effect, there could be no physical science.[19]

In rejecting Prince's "miracles" in favor of natural laws, Winthrop nonetheless situated his philosophy of electricity and earthquakes in a theological framework. "The consideration of a Deity is not peculiar to *Divinity* but belongs also to *natural Philosophy*," he wrote, whose "main business" was to "trace the chain of natural causes from one link to another, till we come to the First Cause . . . continually actuating, this whole chain and every link of it." Winthrop, like Prince, held that discussions of second causes alone were unpardonably *"atheistical."* Nor did he disagree with Prince's view that electrified thunderclouds reflected "the continued agency of God." He did dispute, however, that this agency manifested itself in a miraculous role for electricity in earthquakes. Earthquakes were provided for in the fabric of nature by the Almighty, but as part of its ordinary course. To impute the causes of these disasters to human intervention through the use of conductors was an absurd and even impious conclusion, one that challenged the notion of divine omnipotence and dissuaded common people from protecting themselves, needlessly "fill[ing] with unnecessary terrors the minds of many persons, who were not well enough acquainted with the laws of electricity." "I cannot believe," he confessed, "that there is so much as one person, who is so weak, so ignorant, so foolish, or, to say all in one word, so atheistical, as ever to have entertained a single thought, that it is possible, by the help of a few yards of wire, to get out of the mighty hand of God." If anyone was guilty of presumption, it was Prince. He agreed with the reverend that God was sovereign in nature, but in line with the main currents of physico-theology, Winthrop's deity operated within his own laws; He did not have the power to override them.[20]

While historians have often agreed with Adams and depicted Winthrop's defense of Franklin as an epochal victory for progress against superstition, Winthrop's victory was not universal. The idea of an absolute divine sovereignty that could overturn natural laws continued to be cherished throughout the eighteenth century by zealous interpreters whose most urgent concern was to insist on the direct moral regulation of man by God. The Connecticut pastor Ebenezer Gay, for example, resolutely described the death by lightning of three young men in 1767 as "Lessons of divine Providence." "Prepare, and always be ready for death," he advised his parish. "Have you any more Security of Life than those who lie there had, a Moment before the *forked Lightning* struck them breathless and motionless?" Yet toward the end of the century, a growing number of works demonstrated the influence of technology on theology. The lightning rod did not obviate divine agency; it simply revealed its true face as a benevolent force that sustained a stable and intelligible cosmos amenable to human control. "What idea am I to form of thunder and lightning?" the pupil asked his teacher in *The Catechism of Nature* (originally 1774–1779), a didactic dialogue by the Dutch Mennonite pastor Johannes Martinet reprinted in Philadelphia in 1790. The teacher just happened to be an exemplary Franklinist. Lightning was useful for purifying the atmosphere, and fear of thunder resulted from "a vulgar mistaken idea." Thank you, answered the student, "I shall not be afraid of the lightning, as I used to be." In his *Discourses on the Marvellous Works in Nature* (1791), the German Lutheran Charles Reiche, a correspondent of Franklin's and a Pennsylvania settler who published under the imprimatur of the science faculty at the University of Pennsylvania, held that despite the natural ferocity of lightning, there lay behind it "a direction, and . . . a directing power" of ultimate benevolence. "When you behold in the clouds the traces of a benevolent intelligence and supreme direction," Reiche exhorted, once more in the idiom of a father addressing his children, "can you doubt that there must really be an all intelligent and all-directing Being? . . . Under his government ye may smile at the menaces of lightning and of thunder, and see, even in the most tremendous flashes, an eye, that watches over you."[21]

The Prince-Winthrop controversy was not unique. In the same year, a writer going by the name Benevolus in the *South Carolina Gazette* wrote of Dr. John Lining's struggle to mount lightning rods against local religious opposition in Charleston. "Those *Anti-Electricians*," he lectured critics, "who pretend to be actuated in their censure by a religious principle would do well to consider that the fruit of religion is charity." "Let us, my friend, raise our *sharp Points*," Benevolus concluded, undaunted. Like Winthrop, he encouraged readers to "bless God, for [this] new and wonderful discovery." Ironically, yet tellingly, the clergy themselves were among the first to raise "electric spires" to protect church steeples—the tallest, most vulnerable structures on the British-American landscape. Charles Woodmason, an Anglican itinerant who toured the Carolinas, praised the "guardian point" precisely for its protection of houses of worship. "No falling steeple trembles from on high, / No shivered organs now in fragments fly," Woodmason chimed, noting that "the steeple and organ of St. Philip's church in Charles Town, [had formerly] been twice damaged by lightning." For many, electricity and divinity dwelled in the same house. In August 1785, Ezra Stiles was lecturing on natural philosophy in the Yale College chapel, where the electrical apparatus was kept. While speaking, a thunderstorm passed overhead. Stiles improvised a discussion of the electrical nature of lightning and "Dr. Franklin's pointed Metallic Rods for the Defence of Buildings and Ships." Even the Reverend Prince himself, Winthrop delighted in pointing out, had written that he *"never was against erecting"* lightning rods. In fact, Prince welcomed them, provided their adoption was accompanied by "a due submission to the sovereign Will and Power and Government of God in nature, and in humble Hopes of greater Safety." Winthrop would have concurred without hesitation.[22]

Thunder Houses

When Ebenezer Kinnersley set out to lecture on electricity in public in the 1750s (the story of the next chapter), his coup de théâtre was the "thunder house"—a small model house or church, usually made of wood, used for demonstrating the lightning rod (Figure 8). The use of physical models derived again from the analogical method on

8. The Exploding Thunder House

Reasoning by analogy was a characteristically enlightened intellectual method, and one fundamental to Franklinist electricity. A prime example of analogy is visualized in this engraving: as lightning destroys the church at left, so electricity from a Leyden jar can be made to destroy a model "thunder house" at right. By implication, those who can master electricity in the laboratory with metal wires can master lightning with conductors. Here, however, the emphasis is on exquisite destruction: the image captures the pleasure of blowing up the model rather than protecting it.

which the Franklinist system depended: protecting model houses from electric sparks could convince audiences that conductors would secure real buildings from the full force of lightning. The demonstrator would attach a small chain to the end of a wire running down the front or side of the model. When a spark was applied from a generator, the charge ran safely down the wire and the chain. With the chain removed, however, a second discharge passed into the thunder house, igniting a small pile of gunpowder, blowing its walls apart. The analogy between laboratory and natural lightning cut both ways. While electricians could appear as impressive masters of electricity in the laboratory, their mastery of lightning beyond the laboratory often seemed less sure, placing the analogy between the theaters of experiment and nature in question. Could such simple small-scale experiments really bring control over nature itself? In reality, conductors regularly failed to protect the buildings on which they were mounted. As one writer put it with disarming nonchalance in 1796, "Some late accidents in Europe seem to prove that conductors do not always secure buildings from lightning."[23]

Who actually used lightning rods? The democratic myth surrounding the protective conductor has always insisted on its universal adoption. The evidence suggests, however, that the lightning rod was a public technology of characteristically genteel eighteenth-century character—of "universal" benefit for humanity, yet employed primarily by gentlemen. Franklin did not seek to patent his invention, and widely disseminated instructions for constructing and mounting conductors in print, most notably through *Poor Richard's Almanack* in 1753. These instructions were repeated in numerous newspapers, editions of the *Experiments and Observations on Electricity*, the *Transactions of the American Philosophical Society* (begun in 1771), the *Memoirs of the American Academy of Arts and Sciences* (launched in 1785), and a variety of American literary magazines. Perhaps even more persuasive were the spectacular demonstrations by itinerants like Kinnersley: according to one South Carolina writer, locals took to fixing "sharp pointed Iron Rods on their Churches, Houses, [and] Granaries" after seeing his shows. So conductors were in public circulation, replicable without legal restriction, and apparently easy to install. "Philanthropos," who published a set of instructions in the *Connecticut Courant*, noted that protecting buildings from lightning was "much easier and cheaper than securing them from rain." People of modest means, however, might not have been able or willing to pay for such installations. The American Philosophical Society's minutes show that mounting enough lightning rods to protect a building in 1792 cost 6 pounds, 15 shillings—no trifling sum.[24]

It is certainly possible that poorer Americans mounted their own crude conductors. But in both the Americas and Europe, lightning rods mattered most to the wealthier members of society and to the state—those whose property was tall enough to need protection against the elements. The Royal Navy and East India merchants, for example, were particularly interested in lightning rods for their ships, attaching to the tops of their masts long metal chains that trailed into the sea. Captain James Cook and Joseph Banks both praised the chains that saved their ships from lightning strikes during thunderstorms in the Pacific Ocean and the East Indies. Conductors protected British military installations such as gunpowder magazines and forts in West Africa and the Caribbean. The Caribbean, racked with thunderstorms, was a natural market for conductors. On French Saint Domingue, itinerant electricians from Paris made a tidy profit installing conductors in the 1780s, protecting official buildings, powder magazines, and plantation houses. The Jamaica slave owner and obsessive diarist Thomas Thistlewood mounted a lightning rod on his home in the 1760s (he also owned a copy of Franklin's treatise on electricity). Lightning rods mattered in British America because they could protect the new structures that went up—literally up into the sky—as a result of increased colonial wealth in the eighteenth century: new churches with lofty spires; public buildings like the Academy, Statehouse, and Bank of the United States in Philadelphia; and the houses of increasingly propertied gentlemen. From New England merchants like John Brown of Providence to Southern planters like Landon Carter and Thomas Jefferson of Virginia, gentlemen mounted conductors to protect the mansions they had built on the profits of slavery. No coincidence then that Kinnersley's demonstrations before genteel audiences used houses to demonstrate the importance of conductors. Lightning rods were a telling symbol of New World enlightenment.[25]

When lightning struck these new American mansions, they became life-sized thunder houses, as electricians set out to understand the behavior of lightning in matter. This was a characteristically enlightened response to lightning and marked a significant departure from older folk beliefs. In the seventeenth century, lightning strikes on houses in New England were often seen as manifestations of spiritual invasion or demonic possession, phenomena unamenable to naturalistic explanation. Houses struck by lightning were porous bodies penetrated by supernatural agents like witches and devils. Logically, therefore, lightning protection took symbolic rather than practical form: New Englanders "dressed" their houses with talismans (poppets, for instance) at doors and windows to ward off spirits. As in the sensational narratives in colonial newspapers, recycled by ministers like the Mathers, lightning strikes were properly a source of terror and awe, offering pious moral lessons but defying rationalization. By contrast, the new houses of the eighteenth century reflected and advanced a much bolder claim to physical permanence and (potentially) spiritual security in British America. Built of brick rather than wood, the finest American residences rose to two or three stories, following recent British styles, and housed the most affluent members of colonial society. More imposing and seemingly less permeable than their predecessors, these houses projected the social authority of a rising colonial gentry. Fine houses (often adorned by well-kept gardens) played their part along with conversation, dress, and entertainment in a culture of refinement founded on improvement of self and environment. As protectors of these privileged spaces, conductors themselves became signal technologies of improvement.[26]

When lightning struck, the results were unpredictable. In a dramatically counterintuitive conjecture, Franklin had speculated that in lightning strikes positive charge from the earth more commonly traveled upward to strike negatively charged thunderclouds rather than vice versa, so that "for the most part, in thunder-strokes, *it is the earth that strikes into the clouds, and not the clouds that strike into the earth.*"

This question of the literal direction of lightning preoccupied American electricians. To establish the truth of Franklin's hypothesis, the American Academy of Arts and Sciences in Boston asked "to have every circumstance communicated, which carries evidence to show the direction of the charge." In reply, John Lathrop, a Boston minister deeply interested in electricity, discussed the case of a grave digger struck in 1798. According to Lathrop, an electrical equilibrium had been established by positive charge traveling from the earth to the sky, citing the "bursting open of the [man's] canvas trousers, the tearing of the hat, and the breach in the man's head" as evidence of the charge's direction. The electricity had passed "from the earth, directly under the man, [and] tore asunder the dry linen trowsers, which were open at the bottom, and tight at the top."[27]

Different American communities made sense of lightning according to their different priorities. Late-century accounts such as Lathrop's evidenced the emergence of a secular discussion of the physical direction of lightning, alongside ongoing theological discussions of its spiritual direction. By excluding theological considerations, American electricians gave a materialistic account of lightning, casting it as an intelligent and self-directing agent whose movements were neither arbitrary nor divinely predetermined, but a function of its relationship to the variable electroconductivity of matter. Just as Franklin had turned his house and the sky into laboratory spaces, American philosophers now converted colonial mansions into laboratories—at least on paper—by analyzing their electroconductivity. The result was literally a map of the path of lightning through matter. With his colleague John Jones, David Rittenhouse, president of the American Philosophical Society in the waning years of the century, described a lightning strike's effects on adjacent kitchens behind two houses in Drinker's Alley, Philadelphia. The account included a draftsman's sketch of the direction lightning had traveled through both structures, reconstructed through the trail of physical destruction caused

(Figure 9). Lightning entered both chimneystacks at once, collapsing the north wall of each (unseen). Conducted by an iron hook in the left-hand structure (a woman at the fireside saw "a large ball of fire"), it emerged at the hearth (A), where it made a cavity "as large as a man's head." In the right-hand structure, the lightning emerged at the roof (B), before it "entered the chamber [C], tearing off some of the ceiling and plaistering of the wall" (nonconductive matter was damaged most because it resisted the movement of electricity). Conducted along the gilt frame (E), the lightning passed down the wall to the lower hinge of a closet door, which was "thrown to a distance by the explosion," before dispersing via rivets and a nail in the floor to the kitchen below. Pewter plates lay melted where they touched and a coffee pot was blasted from the shelf over the fireplace. The backboard of a hand bellows, suspended from a nail, was "split through, apparently with great violence" (F). Both kitchens were "filled with smoke, soot, and ashes." Surprisingly, no one was injured.[28]

Rittenhouse and Jones's narrative made it clear that metal was the conductor that determined the lightning's path through the right-hand building. If the lightning's path seemed clear enough, however, its direction (upward or downward) remained in question. Observation of effects alone was not enough to establish direction. According to Rittenhouse and Jones, the woman who had witnessed the fireball swore it had moved upward, based on the appearance of the cavity near the hearth (A). But they overruled her, insisting there was no "appearance which could determine whether its progress was upwards or downwards." In another case reported by Lathrop, a young girl seated in a parlor was killed when lightning struck the house. Here, Lathrop admitted, "I cannot determine from any circumstances related whether the discharge was from the cloud, or from the earth." Lightning was simply too quick for the senses. In a further case, Lathrop recounted an incident where two men were struck in a basement. Samuel Carey, the house owner, claimed "that the discharge was from the

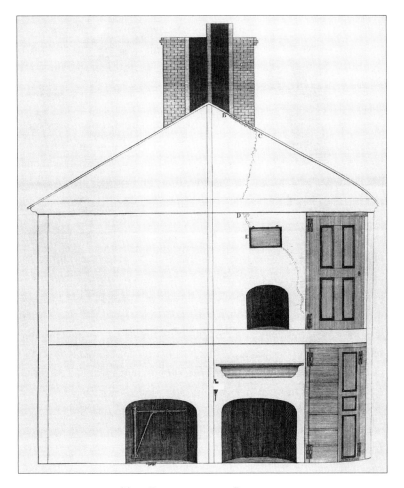

9. The Direction of Lightning

In the eighteenth century, lightning rods mattered most to those with significant property to protect: ships, government buildings, churches, and fine houses. While religious ministers continually insisted that lightning was a form of divine punishment, a secular conversation started among gentleman-philosophers regarding the direction and nature of electroconductivity. This diagram charts the path of a lightning-bolt that struck the chimney (B-C) of a Philadelphia house.

cloud, [as] he says was evident to people, who saw it." Lathrop begged to differ. "I would query," he commented, "whether the motion of lightning be not too quick for any eye to discover its direction, whether from the cloud, or from the earth." "Optical deception" of some form was more likely.[29]

If it seemed impossible to establish lightning's direction, lightning-strike narratives could yet show that these were neither random acts of nature nor specially directed acts of God, but rationally predictable phenomena. Why, for example, had only a single person been killed in the incident in the parlor described by Lathrop? The answer was entirely electroconductive. In the cellar directly beneath the girl who was killed, Lathrop intriguingly explained, there was an iron spit, one end of which had been resting on the cellar wall. Passing through the wall and through the spit (melting it in the process), the lightning bolt "took the nearest good conducting matter, which happened to be the unfortunate child, sitting on the floor directly over the spit." Death was instantaneous. The accident showed "the danger of placing ourselves in the course, between different portions of conducting matter, in the time of a thunder shower." As Franklin had suggested, the mystery of why lightning struck certain individuals and others lay not in divine moral discrimination, but in the variable electroconductivity of matter. In several incidents described in American journals, lightning zigzagged from one chamber to another in the same house. The reason? The metal wires of domestic-service bell systems, which created a metallic communication between the different rooms. In one case, lightning smashed a large mirror in one room and followed the bell wire to the fireplace. Then, "as the cord which hung down within the tassel, made a bad conductor, the charge flew off from the wire opposite to the corner of a large picture of president Washington." Running down the picture's gilt edge, it passed through a "small opening between the breastwork and the marble mantle piece, entered the chimney and went off." "The manner in which the bell-wires are distributed in a house, is [thus] of great moment," concluded Arthur Lee, after describing the effects of a similar stroke on the house of William Shippen, the prominent Philadelphia physician. "They ought always to be disposed with a view to the possibility of their becoming conductors." That is, one should think of a house as being like a thunder house.[30]

Returning to Philadelphia in the mid-1780s, Franklin proudly reported to Marsilio Landriani, an Italian advocate of *parafulmini* (lightning rods), that he found "the Number of Conductors . . . greatly increased, their Utility having been made manifest by many Instances of their good Effect in preserving Buildings." Noting that his own house had been so protected in his absence, he assured Landriani that if properly installed and maintained, conductors were infallible. "Mr. Rittenhouse, our Astronomer, informs me," Franklin wrote, "that having inspected with his excellent Telescope many Conductors that are within his View, he finds that [although] the Points of a number of them have also been melted," yet "we have no Instance of any considerable Damage done to any House that was furnished with a compleat Conductor." Lightning rods did fail, however, and with some regularity. Lathrop's discussion of bell wires is a case in point, an example of "a brick house, and with a good conductor" nonetheless damaged by lightning. "It may be a question," he conceded, "why the whole charge did not go down the conductor to the earth?" The key was the proximity of the lightning rod to the bell wire mechanism: because they were only eighteen inches apart, electricity flew easily between the two. A second factor was insufficient grounding: "the rod did not enter sufficiently deep into the ground"; nor was the earth sufficiently wet to conduct the electricity away. Rittenhouse and Francis Hopkinson, another active member of the Philosophical Society, reported the "remarkable" case of the three-story, two-chimney house of the tobacco merchant and land speculator Thomas Leiper, near Chester, Pennsylvania, whose roof was damaged by lightning despite being furnished with two conductors. Why did these rods fail? They were well made, insisted the authors, "being screwed together" into one continuous length "and not connected by hooks," a design sometimes blamed for conductor failures. Rittenhouse and Hopkinson opted for rocky soil as the main culprit. "Had the earth been sufficiently moist," they concluded, "it is likely we should not have seen any effects of the lightning." Again, the lightning rod itself was not to blame.[31]

Vexing questions of cause and effect presented themselves: Did lightning rods protect houses from lightning or did they actually cause lightning strikes? Were lightning rods in fact dangerous? According to Rittenhouse and Hopkinson, the Leiper incident suggested that conductors "may sometimes invite a discharge of the electric matter, which would otherwise have passed elsewhere." But they confidently rejected the idea, advanced by Nollet in Paris and Franklin's rival Benjamin Wilson in London, that pointed conductors were a menace. Preventing future mishaps was essentially a question of perfecting design. American journals repeated what Franklin had already insisted: sinking the rod several feet away from the foundation of the building, into moist earth or water, and regularly replacing rusted or melted points (like the ones Rittenhouse spied through his telescope) would do the trick. Improving conductors became something of a minor obsession. Robert Patterson, professor of natural philosophy at the University of Pennsylvania, was awarded the American Philosophical Society's Magellanic Premium in 1792 for the modifications he made. Noting that iron points melted after conducting lightning, Patterson (calling himself "Philo-Franklin") advised coating conductors' points with black lead, and rust-proofing the bottoms with tin or copper.[32] In Britain during the 1770s, political and scientific careers turned on bitter disagreements over lightning rod design: Benjamin Wilson sought to demonstrate the danger of Franklin rods in order to replace them on royal buildings with his own knob-ended conductors. But in loyally Philo-Franklinist America, there was no such controversy. Because the community of American electricians functioned as a community of gentlemen, debating lightning-rod design in America was neither a commercial nor political issue, but a polite experimental conversation.[33]

Lightning rods continued to fail, however. In June 1788, the house of one General Pickens in Augusta, Georgia, was "struck with lightning [and] took fire, even though it was furnished with a conductor," with the result that it was "totally consumed." Strikingly, those who reported such disasters almost always exonerated the lightning rod, sometimes with absurd results. In one such case, Hugh Williamson, a natural philosopher at the University of Pennsylvania, blamed cattle for "licking away the earth at the end of the house, so that the rod did not touch the earth by six inches." Among the enlightened, belief in the lightning rod thus became ironically talismanic: what was remarkable was not that it worked, but that it should ever fail.[34]

These failures did not provoke broad controversy in America, but they did affect understandings of the relationship between human and divine agency. Here we return to the intersection of technology and theology so important in the Prince-Winthrop debate of 1755. Machines like conductors carried important spiritual implications. As we have seen, for many natural philosophers the success of the lightning rod helped to characterize both the universe and its Creator as benevolent, while for some theologians this technological innovation was an act of presumption that changed nothing—God's will was still sovereign, his capacity to punish still absolute and terrifying. It has been argued that the establishment of all technologies involves a process of "black-boxing," through which they move from an initial phase of contestability and instability to the point where their efficacy becomes an unquestioned fact. The pages of early American scientific journals strove to black-box the lightning rod. But other responses were possible in the same period, and there are some extant clues that this box could, and did, reopen with unsettling technical and cosmological consequences.[35]

One such clue is a passage from the lengthy diary kept by the Virginia planter Landon Carter. A wealthy slaveholder in Richmond County, Virginia, Carter figured prominently in local politics as a member of the House of Burgesses. Educated in London and deeply interested in scientific agriculture, he was also an unusually devout member of the Anglican Church. In his voluminous journal, he described the failure of conductors to protect his imposing mansion, Sabine Hall, during a thunderstorm in April 1773. As in the accounts examined earlier, Carter attempted to reconstruct the lightning's path through his home. It seemed to have entered at the chimney and run down the eastern side of the house before dividing and entering a number of different rooms, smashing windows and shivering frames as it went. Dividing again, it then ran onto the piazza and several passageways, knocking down three of Carter's slaves, whose stricken bodies provided evidence of the bolt's progress. "Winny I saw in a Suppliant Posture, but stupid, Betty recovering but quite useless in her lower limbs, and Poor Joe dead to every Appearance" (all three in fact recovered).[36] Issues of social and natural mastery converged in Carter's version of events. Although his dependents (both white and black) looked to their master for help and protection, Carter painted himself as the lost patriarch who, for all his social might, saw himself as impotent in the face of overwhelming natural forces. He invited his imaginary reader to picture the "dismal" scene, in the wake of the strike, where his grandchildren, of "sorrowful countenance," cried "with Concern." His "poor slaves crowd[ed] round and follow[ed] their master, as if protection came only from him, and yet quite void of senses enough to assist me." Writing privately in his diary, he confessed his belief that ultimately such protection could come only from God. The failure of his lightning rods proved not only this but, he concluded, the vanity of all "pretended" philosophy as well:

Electricity only teaches us when this etherial fire is collected, or in what bodies it abounds most; and the invention of points, only imagines an attractive atmosphere which perhaps this instance will shew to be mere imagination or that power of attraction is to be overruled according to the quantity of fire collected; for I have points on the western corner chimney, and yet they never discharged any of this lightning. Besides the variety of directions that this flash took, seems to puzzle the philosophy of all conjectures; but as to relief when bodies are afflicted by it, I don't read of one solid or just conclusion. Besides, how came the wench Betty's left hip to be burned and down her right leg and under her cloaths and stockings, when not a garment or stocking was so much as singed? Again, by what impulse does any explosion divide, and run in different directions, and then meet again as this evidently did? Does not this show that this pretended discovery of the etherial fire, is but mere conjecture which perhaps solves an instance or two but cannot go any farther? Therefore to God alone let every application be made, and if his mercy sees fit, we shall be protected.[37]

More than the resolution of technical debates turned on the success or failure of conductors. Where philosophers read the lightning rod as evidence of the controllability and benevolence of divine nature, the failure of Carter's points convinced him that the very notion of an electrical science was vain "imagination." God's nature was not reducible to laws established by reason: lightning's multiple directions "puzzled" philosophy, "overruling" ostensible laws like attraction. The failure of conductors suggested that electricians peddled conjectures with limited ability to predict effects in the real world. In Carter's view, Franklin's optimistic analogy between manipulating laboratory electricity and guiding the path of lightning was a fallacy. The philosophy of electricity was "hocus Pocus" because it could explain why lightning melted conductors in some cases, but not all, and why conductors worked sometimes but not always. Perhaps, he speculated, echoing the debates in Europe, conductors might even cause lightning strikes? This was "one of those occurrencies," he judged, "that ought to correct the lives and behaviour of men." Carter was not a Calvinist, and did not consider lightning rods presumptuous. "Let us [not] fancy [God's] Judgments," he counseled; "we are in the right to guard against" lightning. Yet the best "security as to preternatural causes" was ultimately not a lightning rod but "goodness in heart, life, and Conversation." Lightning thus remained part of a moral rather than merely natural order, and in striking his house it destroyed more than his metal rods. The failure of his conductors ironically served a higher purpose: reaffirming the awesome mystery of divine sovereignty. Franklin rods, which aimed to direct the path of lightning, were no match for the Almighty, who directed nature itself.[38]

CHAPTER THREE

Wonderful Recreations

Electricity has one considerable advantage over most other branches of science, as it both furnishes matter of speculation for philosophers, and of entertainment for all persons promiscuously. Neither the air pump, nor the orrery; neither experiments in hydrostatics, optics, or magnetism; nor those in all other branches of Natural Philosophy ever brought together so many, or so great concourses of people, as those of electricity have done singly.

JOSEPH PRIESTLEY, 1775

IN the eighteenth century, demonstrations of enlightenment often took place in the dark. In British America, as in Europe, philosophical showmen demonstrated natural phenomena to paying audiences through spectacular visual effects designed to surprise and charm the senses. The aesthetic economy of what the French called *la physique amusante* demanded that electricity be rendered as diverting and beautiful as possible. Electrical showmen thus often darkened their demonstration rooms to accentuate the sudden illumination provided by electric flashes. During the late 1740s, Benjamin Franklin routinely opened his home, a makeshift demonstration theater as well as laboratory, to delight friends with such performances. Of one, which seemed to make the point that experiments made better knowledge than books alone, he wrote, "We electrify upon wax in the dark, a book that has a double line of gold round upon the covers, and then apply a knuckle to the gilding; the fire appears everywhere upon the gold like a flash of lightning" ("Fig. V" in Figure 5). It was Franklin's co-experimenter, however, Ebenezer Kinnersley, who became a virtuoso of electrical performance, the leading such showman in all British America. Between 1749 and 1774, Kinnersley exhibited a panoply of electrical effects before a devoted public. Regular displays included an "artificial

Spider... animated by the Electric Fire, so as to act like a live One," a representation made to resemble "Fishes swimming in the Air," and, with an Orientalist nod to the magic arts that electricity often seemed to mimic, "a Leaf of the most weighty of Metals [gold] suspended in the Air, as is said of *Mahomet's* Tomb."[1]

Unlike other scientific demonstrations, such as those involving mechanical orreries (which depicted the position and movement of the planets) or magic lanterns, electrical performances were not merely displays of visual effects, but fully engaged the senses of participants through direct physical experience. It was common, for example, for ladies and gentlemen in the audience to be invited to hold hands and take simultaneous electric shocks in a human circuit. In so doing, their bodies became sites where truths about "the amazing Force of the Electric Fire" might be learned. "Electric mines" were made to surprise the unwary; coins clenched between volunteers' teeth were sent flying by electrical discharges; air was "set on Fire by a Spark from a Person's Finger" and made to burn "like a Volcano." Kinnersley suggestively advertised one demonstration simply as "a diverting Experiment, which cannot be well understood, but by those who try it." Active powers made for participatory, not passive, spectacles—a show that never seemed to end. In one memorable performance, demonstrators drew "Spirits kindled by Fire darting from a Lady's Eyes," while "the Salute Repulsed" involved "Fire darting from a Lady's Lips, so that she may defy any Person to salute her." This show, Latinized as the "Venus electrificata" among Europeans, was known as the "electric kiss" in the colonies.[2]

Interactive performances like these, more than textual accounts of experiments in books and journals, catapulted electricity into the public sphere in the 1740s. Electricity was the preeminent public science of the eighteenth century because it offered an experience of enlightenment. Spectacular bodily electrical effects became de rigueur in the courts, salons, and coffeehouses of Europe; according to one commentator, electricity took the place of quadrille in some circles. "Wonders" were ubiquitous in advertisements for such performances. But exactly what kind of wonders were these? According to one historical interpretation, electrical displays exhibited "wonders tamed": electricity as a predictable and explainable phenomenon subject to rational laws. This was enlightened wonder—the recognition of order in nature. At the same time, demonstrators continued to exploit the prodigious sense of wonder inspired by electricity's capacity literally to surprise the body. Especially after the 1760s, such wonder-mongering seemed politically dangerous to authorities in Britain and France. In an era of mounting challenges to the ancien régime, those who claimed legitimacy as interpreters of nature's wonders might use their command of active powers to legitimate public criticism of political and religious establishments. Some leading experimenters, notably Franklin himself and his close English associate Joseph Priestley, were men of middling rank known for being Dissenters, liberals, and ultimately, opponents of British state policy. "The English hierarchy (if there be anything unsound in its constitution) has . . . reason to tremble even at an air-pump or an electrical machine," Priestley wrote in the 1770s. On this reading, experimental demonstrations involved competing projects for political control: that of radical experimenters seeking to use wonder to rouse the passions of the populace, versus that of establishment commentators and institutions hoping to subdue these passions by mobilizing public opinion against the "charlatanry" of heterodox science, religion, and politics.[3]

Electricity also became a fashionable spectacle in the coffeehouses, libraries, courthouses, taverns, and genteel homes of colonial British America. The "political control" thesis, with its emphasis on demonstrators using wonder to manipulate rather than enlighten audiences, is sustainable in the few colonial situations where, as we shall see, mastery of electricity was projected as a form of racial superiority. But it does not readily apply to the majority of demonstrations that took place in British America, where institutions for regulating entrepreneurs were lacking. American demonstrators sought to advance individual agendas, in Kinnersley's case by disseminating the fundamentals of the Franklinist electrical system. But these performances involved much more than any simple notion of the "democratization of science" or the "transmission of enlightened rationalism." Audiences' consumption of electricity was bound up with the pursuit of several important cultural commodities. Kinnersley situated Franklinist electricity in a carefully crafted program for participation in public science as a path to refinement. Here was a distinctively provincial mode of engaging with enlightened science—enlightenment as genteel self-improvement—that opened up three interwoven paths for audiences: the cultivation of reason through natural philosophy, the inculcation of piety through contemplation of the divine, and the pursuit of polite sociability. These were not straightforward routes to enlightenment but tortuous ones along which reason vied with sensory disorientation, pious contemplation with worldly spectacle, and polite sociability with impolite bodily display. In exploring this uncertain boundary between the work of experiment and the pleasure of play, demonstrations did not unambiguously assert the agency of reason and experiment over the active powers of nature, but raised questions about the limits of such agency. Electricity unfolded "a most amazing scene of wonders"—but were these wonders truly tamed?

Provincials at Play

As a branch of natural philosophy, electricity appealed to British Americans of middling and higher social status as a form of "rational recreation." Although this genre of philosophical entertainment had been prominent since at least the Renaissance, it was more widely embraced than ever during the eighteenth century as part of the commercialization of leisure. Since the sixteenth century, texts on natural magic, above all Giambattista della Porta's *Magia Naturalis* (1558), had invited aristocratic readers to "search out the Causes of things which produce wonderful Effects," encapsulating rational problems in diverting entertainments. This tradition was carried forward by a number of European writers in the early modern period, notably Jesuit philosophers like Athanasius Kircher who fashioned practices like mathematics, optics, and "pyrotechny" into instructive amusements.[4] By the Enlightenment, electricity enjoyed a central place in this tradition and was part of a thriving commercial market for knowledge as pleasure through public demonstrations and the consumption of philosophical apparatus for experimenting at home. This commercial domestication of experiment is evident from the marketing of apparatus to bourgeois consumers, and from the handbooks written to accompany such devices, such as Benjamin Martin's *Essay on Electricity* (1746) and John Neale's *Directions for Gentlemen, Who Have Electrical Machines, How to Proceed in Making Their Experiments* (1747). Electricity also became prominent in rational-recreational compendia. Acknowledging its debt to della Porta, as well as more recent incarnations of the genre, William Hooper's *Rational Recreations* (1782–1783) drew on the many electrical handbooks available since the 1740s, with much of its third volume devoted to electricity. The reader, Hooper promised, "will readily discover at the same time he admires the phenomena, the source from whence they proceed, and learn, that far from being marvellous or incomprehensible, they are the regular and necessary effects of the laws of nature." Such texts implicitly invoked the spirit of Castiglione's *Book of the Courtier* (1528), in which one of the courtiers declares, "To find the truth out about a certain thing, I would like to find it out through a game." In the Enlightenment, experimental play was one important key to unlocking nature's rational secrets.[5]

First in Britain, and then in the Netherlands, France, and beyond, the emergence of experimental demonstrations from courtly culture into more public settings was driven largely by attempts to promote applications of Newton's mathematical laws of bodily motion to projects in practical mechanics. As we saw in Chapter 1, in this respect the 1720s and 1730s were a crucial period, when textbooks like that by Desaguliers used a combination of demonstration and illustration to urge the material utility of Newtonianism before an increasingly broad public. Electricity's usefulness was by no means clear, however. As a form of experiment that relied on artificial apparatus, electricity was clearly part of the same milieu as early industrial machinery, interesting to many of the same sorts of people (Desaguliers himself published on electricity). But because the utility of electricity remained uncertain, it straddled the worlds of useful labor and playful entertainment. Demonstrations of electricity remained closer to the spirit of the court rather than the factory, a curiosity more useful for personal cultivation than a resource for practical application. As a cultural rather than material commodity, electricity thus made perfect sense in British America—a colonial context where industrial manufacturing was restricted by parliamentary legislation obliging Americans to consume British goods rather than produce their own.[6]

The possibility of earning a living by demonstration encouraged British and American itinerants in the 1740s to turn natural philosophy to commercial advantage. Perhaps the first to do so in America was Isaac Greenwood, who had been dismissed from Harvard for drunkenness in 1738 and who, with Franklin's help, came to Philadelphia in 1740. Since the 1730s, Franklin had been reprinting extracts on natural history and natural philosophy from journals like the Royal Society's *Philosophical Transactions* in the pages of the *Pennsylvania Gazette*. In 1740, the *Gazette* ran advertisements for Greenwood's "Course of Philosophical Lectures and Experiments," to be offered for "Gentlemen" at the Library Company, with Franklin organizing subscriptions for the event. Greenwood used an air pump; most other shows at the time utilized optical machines like the microscope and camera obscura. One *Gazette* advertisement in 1743 told of magic lantern shows at Joseph Barber's Temple Bar on Second Street, Philadelphia, offering for only sixpence "wonderful Curiosities, as made the Ignorant believe were done by Magic, Conjuration or Witchcraft." The Scot Archibald Spencer, whom Franklin saw in Boston during 1743–1744, was probably the first to cross the Atlantic and offer electrical demonstrations along the Eastern seaboard. He was by no means the only one. One of the most successful itinerants was another Scot, the blind Henry Moyes, who toured the American republic in the mid-1780s, charging two shillings for lectures on natural history and natural philosophy (including electricity). Spencer had remained in America, but Moyes apparently "made so much profit . . . that on his return to Scotland he was enabled to buy . . . a place near Edinburgh." The most fantastic itinerary was surely that of the "Transylvanian Tartar" Samuel Domjen. Having traveled through Germany, France, Holland, and England, Domjen received instruction in electricity from Franklin while in Philadelphia, and subsequently toured Maryland, Virginia, the Carolinas, and Jamaica, intending to make his way home via Havana, Mexico, Manila, China, India, Persia, and Turkey. "A strange project!" observed Franklin. "He wrote to me from *Charles Town* that he had lived eight hundred miles upon electricity, it had been meat, drink, and cloathing to him." Announcements in the *South Carolina Gazette* show Domjen asking twenty shillings in late 1748—a sizeable sum—of those who wished to be "electrified." Wonderful experience paid handsomely in the market for diversion.[7]

The master demonstrator in British America, however, was Ebenezer Kinnersley. Ever attuned to economic possibility, and cognizant that his "ingenious Neighbour" was "out of Business" in the late 1740s, Franklin "encouraged [Kinnersley] to undertake showing . . . Experiments for Money." Together, they drew up two lectures that Kinnersley was to deliver, with only minimal variation, during the quarter century after 1749 (Figure 10). Born in Gloucester, England, Kinnersley had come to Pennsylvania with his family in 1714 when he was only three years old. Having been ordained as a Baptist minister like his father, he became known in Philadelphia when he publicly decried what he called the "Enthusiastick Ravings" of revivalist preachers like George Whitefield, for which he was denied his own parish. Subsequently, Kinnersley became an integral part of Franklin's experimental circle, and made valuable contributions to electrical research in his own right (inventing, for example, a thermometer for testing the temperature of electrical discharges). A gifted speaker, he was made professor of English and oratory at the College of Philadelphia in 1753 (again with Franklin's assistance), the venue where he most often performed his demonstrations. He also toured as an itinerant. In the early 1750s, for example, he traveled the entire length of British America including the Caribbean islands, performing in genteel private houses, libraries, colleges, and courthouses in Boston (Faneuil Hall), Newport, New York, Philadelphia, Annapolis, Charleston, Bridgetown (Barbados), and St. John's (Antigua). Charging five shillings for admission per performance (seven shillings, sixpence for couples), in 1753 he may have made as much as £200 on his West Indian tour alone. The public imagined for such demonstrations was a well-to-do rather than universal one: the five-shilling admission, though less than Domjen's, was still more than a day's wages for most American laborers in the mid-eighteenth century. The London Coffee House in Philadelphia, where tickets for his performances were sold, was the "most genteel tavern in

10. The Spectacle of Enlightenment

Public demonstrations of electricity explored cultural dichotomies that fascinated the eighteenth century: reason and wonder, politeness and vulgarity, piety and commerce. This broadside notice for a demonstration by Ebenezer Kinnersley, like others of the era, entices audiences with descriptions of a variety of "wonderful," spectacular displays.

Newport, March 16. 1752.

Notice is hereby given to the Curious,

That at the COURT-HOUSE, in the Council-Chamber, is now to be exhibited, and continued from Day to Day, for a Week or two;

A COURSE of EXPERIMENTS, on the newly-discovered

Electrical FIRE:

Containing, not only the most curious of those that have been made and published in *Europe,* but a considerable Number of new Ones lately made in *Philadelphia* ; to be accompanied with methodical LECTURES on the Nature and Properties of that wonderful Element.

By *Ebenezer Kinnersley.*

LECTURE I.

I. OF Electricity in General, giving some Account of the Discovery of it.
II. That the Electric Fire is a real Element, and different from those heretofore known and named, and *collected* out of other Matter (not created) by the Friction of Glass, *&c.*
III. That it is an extreamly subtile Fluid.
IV. That it doth not take up any perceptible Time in passing thro' large Portions of Space.
V. That it is intimately mixed with the Substance of all the other Fluids and Solids of our Globe.
VI. That our Bodies at all Times contain enough of it to set a House on Fire.
VII. That tho' it will fire inflammable Matters, itself has no sensible Heat.
VIII. That it differs from common Matter, in this: its Parts do not mutually attract, but mutually repel each other.
IX. That it is strongly attracted by all other Matter.
X. An artificial Spider, animated by the Electric Fire, so as to act like a live One.
XI. A Shower of Sand, which rises again as fast as it falls.
XII. That common Matter in the Form of Points attracts this Fire more strongly than in any other Form.
XIII. A Leaf of the most weighty of Metals suspended in the Air, as is said of *Mahomet*'s Tomb.
XIV. An Appearance like Fishes swimming in the Air.
XV. That this Fire will live in Water, a River not being sufficient to quench the smallest Spark of it.
XVI. A Representation of the Sensitive Plant.
XVII. A Representation of the seven Planets, shewing a probable Cause of their keeping their due Distances from each other, and from the Sun in the Center.
XVIII. The Salute repulsed by the Ladies Fire ; or Fire darting from a Ladies Lips, so that she may defy any Person to salute her.
XIX. Eight musical Bells rung by an electrified Phial of Water.
XX. A Battery of eleven Guns discharged by Fire issuing out of a Person's Finger.

LECTURE II.

I. A Description and Explanation of Mr. *Muschenbroek*'s wonderful Bottle.
II. The amazing Force of the Electric Fire in passing thro' a Number of Bodies at the same Instant.
III. An Electric Mine sprung.
IV. Electrified Money, which scarce any Body will take when offer'd to them.
V. A Piece of Money drawn out of a Person's Mouth in spite of his Teeth ; yet without touching it, or offering him the least Violence.
VI. Spirits kindled by Fire darting from a Lady's Eyes (without a Metaphor).
VII. Various Representations of Lightning, the Cause and Effects of which will be explained by a more probable Hypothesis than has hitherto appeared, and some useful Instructions given, how to avoid the Danger of it : How to secure Houses, Ships, *&c.* from being hurt by its destructive Violence.
VIII. The Force of the Electric Spark, making a fair Hole thro' a Quire of Paper.
IX. Metal melted by it (tho' without any Heat) in less than a thousandth Part of a Minute.
X. Animals killed by it instantaniously.
XI. Air issuing out of a Bladder set on Fire by a Spark from a Person's Finger, and burning like a Volcano.
XII. A few Drops of electrified cold Water let fall on a Person's Hand, supplying him with Fire sufficient to kindle a burning Flame with one of the Fingers of his other Hand.
XIII. A Sulphurous Vapour kindled into Flame by Fire issuing out of a cold Apple.
XIV. A curious Machine acting by means of the Electric Fire, and playing Variety of Tunes on eight musical Bells.
XV. A Battery of eleven Guns discharged by a Spark, after it has passed through ten Foot of Water.

As the Knowledge of Nature tends to enlarge the human Mind, and give us more noble, more grand, and exalted Ideas of the AUTHOR *of Nature, and if well pursu'd, seldom fails producing something useful to Man, 'tis hoped these Lectures may be tho't worthy of Regard & Encouragement.*

☞ *Tickets to be had at the House of the Widow* Allen, *in* Thames Street, *next Door to Mr.* John Tweedy's. Price Thirty Shillings *each Lecture. The Lectures to begin each Day precisely at* Three *o'Clock in the Afternoon*

America," according to John Adams. Public advertisements offering the use of experimental apparatus were often restricted to the "Use of the Members only" or to "gentlemen, who have formerly been educated in this seminary" (the College of Philadelphia).[8]

Kinnersley's career was made possible by the quickening of communications across the Atlantic, and between Britain's mainland and Caribbean colonies, in the first half of the eighteenth century. Faster Atlantic crossings and the mushrooming of colonial newspapers meant easier travel and widespread newsprint advertising—the preconditions for an emergent culture of itinerancy. The most spectacular proof was the evangelical revivals of the 1720s to 1740s, known collectively as the Great Awakening, during which preachers like Whitefield reached an unprecedented audience of thousands on both sides of the Atlantic. Ironically, given his outspoken dismissal of the revivals, Kinnersley (and his imitators) would co-opt some of the Awakeners' key techniques: the use of personal correspondence networks, newspaper advertising, and the crafting of a studied public showmanship. The art of display was essential in electricity. In this respect, there was an unmistakable continuity between Kinnersley's demonstrations and those of the leading Europeans like Nollet in Paris and William Watson in London. All made electrifying the human body their centerpiece. The circulation of styles of performance was as crucial to the circulation of electricity around the Atlantic as was the movement of books or apparatus.[9]

What distinguished Kinnersley's shows from the metropolitan performances of Nollet and Watson, however, was that they tied electricity to a program for improving the provincial self. "Improvement" was an early modern discourse of progress that derived from the practice of agriculture both as a material and moral endeavor. Since the Fall and the loss of Eden, zealous Protestants in particular had sought spiritual redemption by attempting to reconvert the wilderness into a garden. Improvement of the self (return to the state of grace) naturally accompanied the improvement of land (the return to Eden). Improvement became a refrain in colonial situations, to justify English impositions of settled agriculture and civil Christian government over "ignorant" and "barbarous" local populations from North America to South Asia. In British America, improvement also connoted the polishing of rough colonials into provincial cosmopolitans conversant with metropolitan cultural life. Familiarity with the sciences was therefore increasingly seen in the eighteenth century as a path to enlightenment.[10]

The cultivation of reason through understanding of cause and effect in nature was one of Kinnersley's central appeals. Like Greenwood and Spencer, Kinnersley was not just a spectacular or commercial entertainer; he used his demonstrations to support specific interpretations of natural phenomena. While he advertised his performances as an opportunity to witness "entertaining and astonishing Wonders of Nature," he stressed that electricity's wonders were rational. These were not exhibitions of inexplicable marvels, but demonstrations that British Americans, like their European counterparts, were intellectual masters of wonder who could rationally explain the operations of nature. Although styled "philosophical demonstrations," these displays were not open-ended experiments with unpredictable results, but artificial recreations of natural effects. While true wonders baffled and astonished the mind, electrical demonstrators artfully manufactured wonderful sensory experience as a commodity with which to dazzle but also instruct audiences. It was no contradiction then to call these "methodical LECTURES" on a "wonderful Element," as Kinnersley did.[11]

Following Franklin, Kinnersley adopted Newtonian conceptions of a universal subtle fluid to describe the ontological status of the electric fire for his audiences. He began by delineating briefly electricity's general physical characteristics and experimental history. "A real Element, and different from those heretofore known and named," the electric fire was "an extreamly subtile fluid" that pervaded the universe, "intimately mixed with the Substance of all the other Fluids and Solids of our Globe." He discussed attraction and repulsion but also included the most recent experiments, repeating the remarkable claim that "our Bodies at all Times contain enough [electricity] to set a House on Fire." Franklin rather than Newton dominated the intellectual content of his displays, and demonstrations of Franklin's economy of positive and negative charge were featured front and center. One key display, which replicated Franklin's own experiments, involved standing an audience member on an insulating stool in order to pump electricity out of his body through a metal chain. When approached by a second volunteer, who remained in communication with the earth, a visible spark jumped between them, indicating an equalization of charge. Kinnersley explained: "If any one . . . shou'd approach the Person on the stool with their Knuckle, they will give him [electric] Fire; for the Fire being pump'd out of the Person on the Stool . . . he has then less than his common or due quantity, and consequently strongly attracts it from other things." Bodily experience made the Franklinist system persuasive in public demonstrations—just as it had made private experiments so compelling. Likewise with "Mr. *Muschenbrock's* wonderful bottle": the Leyden jar. Imitating some fabled (perhaps even apocryphal) European performances, such as Nollet's simultaneous electrification of two hundred Carthusian monks in a Paris monastery, Kinnersley invited audience members to form human circuits to experience "the amazing Force of the Electric Fire, in passing thro' a Number of Bodies at the same Instant," often increasing the quantity of electricity by charging an entire battery of Leyden jars joined by a wire. He also discharged electricity through vacuums created by air pumps and ten-foot water-filled troughs, to show that all matter was pervaded by this subtle fluid.[12]

Despite these literally electrifying spectacles, such performances took pains to stress the ability of reason to dispel wonder at the seeming miracles of electricity. Of the two lectures Kinnersley routinely delivered, only one survives as a full script, the other being merely an outline used for advertisements. But the manuscript lectures written by Joseph Hiller, a Boston goldsmith and contemporary of Kinnersley's who constructed his own apparatus and gave demonstrations in the 1750s, provide a full narration of both lectures. Since Hiller's first is an almost verbatim copy of Kinnersley's first, it is reasonable to suppose that the second follows Kinnersley closely as well (the relationship between the two men is unclear, however). Like Kinnersley, Hiller explained how to use a Leyden jar to demonstrate the interplay of positive and negative charges in Franklinist terms. In so doing, he repeatedly emphasized that electrical phenomena were not the "miracles" they might once have seemed. "Another Miracle! the Ignorant would say," he declared, on displaying the suspended model lid of Mahomet's tomb. Through the doctrine of attraction and repulsion, however, "the Mystery [was] . . . understood." Bodily emanations seemed prodigious indeed. The firing of spirits with electricity from human fingers might, in earlier ages, have "passed for a miracle," Hiller cautioned, "and an Imposter might [have] used it among the Ignorant to Establish a false Doctrin[e] or Overthrow a true one." "I don't pretend to [the] working of miracles," he insisted. Opposing imposture did not preclude instruction through entertainment, however. "Be pleased Gentlemen and Ladies to take each other by the hands," Hiller declared, preparing his audience for a circuit shock from the Leyden jar. "I shall give the Bottle But a very gentle Charge." In "amazing" the audience, such displays were nonetheless designed to illuminate the nature of electricity.[13]

Performances like these opened a path to the cultivation of reason, and in the case of the lightning rod, made dramatic claims to the ability of experiment to command natural powers beyond the demonstration chamber. Kinnersley's shows promised "not only the most curious of those [displays] that have been made and published in *Europe*, but a considerable Number of new Ones, lately made in *Philadelphia*." Kinnersley and Hiller both repeated Franklin's demonstration of the identity of lightning and electricity, and his analogical assertion that lightning could be manipulated as a form of electricity using conductors. Hiller drew out the lesson for his audience: Franklin had shown natural lightning to be as lawful as laboratory electricity—"the Electric matter and the matter of Lightning are the same and subject to the same Law's." Flashing electric discharges were modeled as "Flash[es] of Lightning from [a] painted Thunder-Storm." "Observe the Brightness of the Flash!" cried Hiller. "What a Crack was there!" Thunder houses were used to show "how lightning, when it strikes a house or ship, may be conducted to the earth, or water, without doing the least Damage." This was a particularly impressive display, one calculated to showcase provincial American ingenuity. When the miniature rod was attached to the thunder house, and an electrical charge passed through its roof, no harm resulted. After removing the lightning rod, however, and applying a second charge, the house was blown apart when the electric spark ignited the gunpowder. What did audiences make of such displays? Audience reactions to a twentieth-century restaging of the thunder house are suggestive: a gasp, then laughter, then applause.[14]

Such masterly displays delighted American audiences. "The experiments Mr. *K*. has exhibited here have been greatly pleasing to all sorts of people that have seen them," Franklin learned from James Bowdoin II, the wealthy Boston merchant and sometime electrician who saw Kinnersley perform at Boston's Faneuil Hall in 1751. "His experiments are very curious, and I think prove most effectually your doctrine of electricity." In South Carolina in 1753, it was reported that "a curious Gentleman there, in the Presence of many People, has lately met with extraordinary Success in the Experiment of the Electric Kite." Because of the "very tremendous Thunder Storms which often do considerable and irreparable Damage," the writer explained, "Carolinians are . . . fully convinced of the Reasonableness of the Method proposed by the Electriciors, for preventing, in a great Measure, the dreadful Effects of that destructive Element." As a direct consequence of witnessing electrical displays, the letter went on, Carolinians were "now fixing sharp pointed Iron Rods on their Churches, Houses, Granaries, &c. for that laudable Purpose."[15]

Beyond these fragmentary comments on lightning rods, only one firsthand American account of audience response to electrical demonstrations appears to have survived: a letter from a "gentleman" of Newport, Rhode Island. But because Kinnersley published the letter as part of his newspaper advertisements, it is safer to read it for what it reveals about what Kinnersley wanted his public to know about his performances, rather than as a spontaneous private response (Kinnersley may even have taken a leaf out of Franklin and written the letter himself). The Newport gentleman praised Kinnersley for disclosing "a most amazing Scene of Wonders" that afforded the utmost pleasure as rational improvement. The "astonishing Properties and Effects of this wonderful Element" were "beautiful and surprising"; displaying them had "feasted the philosophic Genius with the highest Gust of rational Pleasure and Delight." His explanations of the cause of lightning, "which had puzzled and perplex'd the Philosophy of every Age," dispelled all previous errors. "The Truth of this Gentleman's Hypothesis, appear'd in so glaring a Light, and with such undeniable Evidence, that all my former pre-conceiv'd Notions of Thunder and Lightning, tho' borrow'd from the most sagacious Philosophers, together with my Prejudices, immediately vanish'd." The Rhode Islander ended with a flourish of provincial pride: the "Mystery" of electricity, for so long "wrapp'd up in Clouds and thick Darkness," had finally been illuminated "by ingenious *Americans*." Here was the twist in the tale of British-American improvement: these provincials were not slavish apers of metropolitan fashion but participants in European cultural life who, whenever possible, wished to outplay the Europeans at their own games.[16]

Exalting the Author of Nature

In the increasingly affluent and leisured settler society of British America, criticism of the harmful effects of public entertainments found recurrent expression in Puritan New England, Quaker Pennsylvania, and Anglican Virginia, as well as in the secular project of building the American republic. Early in the eighteenth century, Calvinist ministers in Massachusetts like Benjamin Colman counseled parishioners to reject wanton amusements that undermined the "gravity of reason" in favor of "sober mirth" and "sanctifying recreations . . . governed by reason and virtue." In Pennsylvania, William Penn's Great Law of 1682 declared gaming a sin "against nature; and against government, as well as . . . God." After 1700, those who engaged in "rude and riotous sports as Prizes, Stage-plays, Masques, Revels, Bull-Baiting, [and] Cockfightings" could be fined or imprisoned. The evangelicalism of the Great Awakening, meanwhile, often succeeded in curtailing entertainments where legislation did not. "I was never in a place so populous where the gout for public gay diversions prevailed so little," the genteel clubber and antirevivalist Alexander Hamilton complained of Philadelphia in 1744, in the wake of Whitefield's barnstorming tour. During the 1760s and 1770s in Virginia, fervent Baptist attacks on the mores of the liberal Anglican gentry prompted one Loudoun County planter to fear the "intire Banishment of *Gaming, dancing* and Sabbath-Day Diversions." Finally, in 1774, while encouraging the nonimportation of luxurious British manufactures, the Continental Congress incorporated Christian austerity as a republican virtue to fortify the nascent American nation, enacting legislation to "discountenance and discourage every species of extravagance and dissipation, especially all horse-racing, and all kinds of gaming, cock-fighting . . . and other expensive diversions and entertainments." An American tradition of religious resolve to govern the gratification of the passions through the public regulation of leisure came to serve the very founding of the republic.[17]

In a culture where religious and political authorities sought to police leisure, the moral aspect of electrical entertainments was crucial to securing public legitimacy. Kinnersley, who from his training as a Baptist to his run-in with the Awakeners was fully aware of the importance of religion in American public life, made the cultivation of reason and piety inseparable. His advertisements quoted the Newport gentleman of 1752 confirming that his demonstrations possessed "the highest Christian Character": Kinnersley "endeavours to make this new Branch of *Natural Philosophy* subservient to the true Intent of all Knowledge, both natural and reveal'd, *viz.* to lead us to the first Cause, by refining, enlarging and exalting our Ideas of the great Author and God of Nature." He drew conspicuously on the rationalist lexicon of physico-theology to frame his displays. "The Knowledge of Nature," he wrote, "tends to enlarge the human Mind, and give us more Noble, more Grand and Exalted Ideas of the AUTHOR of Nature." Displaying electricity as one of God's active powers, and explaining the laws that governed its behavior, demonstrated the rationality, and hence divinity, of nature. The "Entertaining and Astonishing Wonders of Nature" were not vulgar exhibitions but demonstrations that drew the spectator's mind up to the power and goodness of the Creator. Natural knowledge was useful spiritual knowledge. Samuel Williams, who succeeded Winthrop at Harvard and lectured on electricity in the 1780s, continued this train of thought, arguing that phenomena like earthquakes (linked by many to subterraneous and atmospheric electricity) were "mighty works of nature," the contemplation of which would inspire "philosophic mind[s to] rise in admiration and reverence, to the First Great Cause Of All." All "true philosophy agrees with the holy scriptures in ascribing all such events to his agency," Williams declared, and "it was no doubt with a view *ultimately* to *moral* purposes, that the laws of nature were first established." Williams was a classic physico-theologian. "There is the same harmony, rule and order," he insisted, "the same general and stated laws, in the causes and operations of earthquakes, as there are in all other events of nature." Even natural catastrophes were part of a harmonious divine plan.[18]

Electricity was a rational wonder: it confirmed the lawfulness of God's creation. As an active power that mediated between the material and immaterial realms, it naturally invited discussion of divine will. Yet this rational wonder did not eradicate all pious Christian sense of mystery at electricity. American commentators drew back at the idea of reducing divine sovereignty to a force of utmost predictability. Despite achievements like the lightning rod, human beings still depended on the pleasure of the Almighty, who might sustain or destroy His creation at will. This was not a religious critique aimed at enlightened science; it was the view of the enlighteners themselves. After a mid-century electrical demonstration in England, John Wesley, the founder of Methodism and an ardent medico-electrician, was astonished: "How came issues out of my finger, real flame, such as sets fire to spirits of wine? . . . How these and many more as strange phenomena arise from the turning round a glass globe? It is all mystery." As electricity coursed through the bodies of zealous Protestants on both sides of the Atlantic, it provoked religious passions that transcended the moderation of conventional physico-theology, inspiring awe at God's sheer power to move matter. Electrical demonstrators made a point of indicating the limits of what their philosophy explained. Franklin's economy of charge accurately described how electricity behaved, Hiller told his audience, but like Newton's account of gravity, it stopped short of explaining why electricity behaved the way it did. As good Protestant demonstrators, Kinnersley and Hiller both poked fun at "miracles" like the suspension of Mahomet's tomb, distancing their performances from the credulousness of Catholicism and Islam. But they did not disdain using the language of the miraculous to evoke their own sense of divine mystery at electricity's behavior. Positive and negative charges were "*wonderfully* . . . Ballanced in this *miraculous* Bottle," Hiller confessed, borrowing Franklin's words, "situated and Related to Each other in a manner that we can by no means comprehend." At the heart of the Franklinist system, a wonderful metaphysical mystery remained concerning the ultimate cause of electricity's behavior.[19]

Certain electrical phenomena seemed to defy all explanation. In 1773, Kinnersley started exhibiting a model of the electric eel, "on touching of which, while in the water, an electric shock may be as sensibly felt, as from a live one." The eels Kinnersley mimicked were not rational wonders but truly astonishing ones: for years, none could understand their ability to do naturally what experimenters so labored to do artificially—generate and store electricity. When real eels were exhibited in Charleston and Philadelphia, naturalists were floored by their "*amazing* power of giving so sudden and so violent a shock to any person who touches it." It was a powerful reminder, to those already convinced, of the fallibility of human knowledge and the folly of curiosity. For in British America, even polite commentators traditionally insisted on divine limits to what reason could achieve. "We see only the Outside and Film of Things," wrote Joseph Pope in the *American Magazine and Monthly Chronicle,* "not the Thousandth Part of what is so." "Almighty God hath hid all the rest from our Eyes, to baffle our foolish Curiosity, to raise our Admiration of his Power, and to excite our Homage and Adoration to him the great Author of all Things." Wonders like the lodestone, the rattlesnake's power of "fascination," and sleepwalking were "impenetrable to our vain and fruitless Inquiries." We are "groping in the Dark," Pope insisted; "*Curious*

Impertinents in the Case of Futurity." This was not a problem, however, but a spur to humility and virtue. "We cannot be Knowing," Pope counseled, "but we can be Virtuous."[20]

Belief in limits to knowledge was related to belief in an immanent rather than withdrawn deity. Spectacular natural philosophy borrowed a religious idiom of *demonstration* and *display* for its experimental performances, a telling semantic link because philosophical showmen tended to share the preacher's goal of exhibiting the almighty power of God. Nature might function according to regular laws, but because of their dependence on the deity, those laws were neither fully revealed nor fully stable. No one stressed the connection between electricity and divine will more than Joseph Priestley, author of *The History and Present State of Electricity* (1767), which circulated widely in British America.[21] During the heights of anti-Jacobin paranoia during the 1790s, a mob was to destroy Priestley's Birmingham laboratory, and he ultimately immigrated to Pennsylvania. But Priestley was no atheist; despite his philosophical materialism, his Unitarian views were devoutly held. He was also a leading exponent of the religious utility of experiment. Electrical machines, he wrote, "exhibit the operations of nature, that is of the God of nature himself, which are infinately various." Electrical displays dramatized divine sovereignty. "A life spent in the contemplation of the productions of divine power, wisdom, and goodness, would be a life of devotion," he counseled. In one show, he recreated the power of God by simulating the effects of an earthquake, using an electric discharge to destroy a model town. The performance was evidently popular in America. One popular encyclopedia described it years later as follows: "Small sticks, cards, models of houses, or the like, should be placed on the surface of the body over which the explosion is to be sent, so as to stand very light. They will never fail to be shaken, and will often be overturned by the explosion." Was this playing God or dis-playing God? Priestley would have insisted on the latter. Such spectacles encouraged pious circumspection, even as human ability to manipulate nature increased. The "cultivation of piety [is] useful to us as *men*," Priestley encouraged the faithful. "It is even useful to us as *philosophers:* as true philosophy tends to promote piety, so a generous and manly piety is reciprocally, subservient to the purposes of philosophy."[22]

It was with deliberate theological sensitivity, therefore, that Kinnersley, Hiller, and others demonstrated the lightning rod, that quintessential technology of enlightened reason regarded by some as an attempt to defy divine punishment. Even before Reverend Prince of Boston argued that lightning rods caused earthquakes by conducting electricity into the earth, Kinnersley had sought to explain to audiences that "endeavouring to guard against Lightning" was "not chargeable with Presumption, nor inconsistent with any of the Principles either of natural or revealed Religion." Demonstrators did not place electricity beyond religious consideration; rather, they insisted that lightning rods were compatible with divine sovereignty. Surely, Hiller implored, God intended "every Rationall every pious mind" to grasp the moral good of saving human lives? It was God's "kind providence, who is *good to all and whose tender mercies are over all his works,*" that communicated this "Usefull knowledge to mankind." Kinnersley's advertisements took the same approach, including an approving quotation from the Book of Proverbs: "A prudent Man foreseeth the Evil, and hideth himself, but the Simple pass on, and are punished." In British America, the best defense against accusations of spiritual presumption was a scriptural one.[23]

Kinnersley had openly opposed the emotionalism of the Great Awakening as enthusiasm run amok. "What Spirit Such Enthusiastick Ravings proceed from I shall not attempt to determine," he had written of the popular reaction to Whitefield in 1740, "but this I am very sure of, that they proceed not from the Spirit of God, for our God is a

God of Order, and not of such Confusion." Hoping that God would "cause rational Religion to appear in its native Charms and Comeliness," he rejected the appeal to the passions by those "pretending to have large Communications from God . . . to have been encompass'd as it were, with Flames of Lightning." Yet Kinnersley the Baptist was no arch-rationalist. In particular, he was not above exploiting the power of strong emotion. "I am not against the Preaching of Terror, in order to convince prophane impenitent Sinners of their awful and tremendous Danger, provided it be prudently managed," he had written in pious criticism of Whitefield.[24]

So enlightenment, of a properly zealous kind, was not incompatible with terror. Kinnersley and Hiller both featured representations "of the seven Planets, shewing a probable Cause of their keeping their due Distances from each other, and from the Sun in the Center." They hypothesized that the sun held the planets in its gravitational orbit because it acted as a positively electrified body attracting negative bodies. Fortunately, the Creator had endowed the planets with enough "projectile force" to "Counter-Ballance the attraction of the electrified Sun," preventing them from falling into it. This was demonstrated by electrifying a pair of spheres to different degrees, one of them coated in gilt to represent the sun, the other representing one of the planets. Thanks to the planet's divinely regulated speed and distance from the sun, it "Rowl[ed] on thro' unrefining ether in a Regular Orbit and perform[ed] perpetual Revolutions round its Center." Yet God could terminate these operations if he chose. The precise source of celestial regularity remained obscure. If the electrical hypothesis of these planetary orbits turned out not "to be true, there may notwithstanding be something Instructive" in it, Kinnersley maintained, for it was a reminder "that Cause is in the Hands of the Great Maker and Governor of the Universe and more absolutely in his power than this Electric Repulsion seem's to be in Mine." The god of active electric powers was, moreover, a morally vengeful one. "Shou'd he then for the Punishment of our Sins decree in his Anger to withdraw" the cause of celestial regularity, "all the Planets and their Satellites, all the massy Globes of this System wou'd rush suddenly together, dash each other to pieces, and form one mighty ruin." The god evoked by this electrical eschatology resembled a Calvinist's more than a deist's. God demanded virtue from human beings, and enjoyed the power physically to punish their sins through extraordinary actions transcending or terminating the regular laws of nature. Zealous audiences approved: the Congregationalist minister Andrew Eliot of Boston despised the immoral display of the theater but, tellingly, was delighted at Kinnersley's performances. Earthquakes and other "convulsions of nature," Samuel Williams would later argue, created "strong impressions of the power and majesty of God," forcing human beings to "see and feel their dependence upon their Creator," leading them to "a steady course of virtue" and "an habitual trust in his providence and protection." Contemplating the sheer destructiveness of God, whether through electricity or earthquakes, was salutary, for it awoke "attention to morals." Or as the Pennsylvania Lutheran Henry Muhlenberg declared after attending one of Kinnersley's demonstrations, "Great are the Lord's works of creation and preservation, but even greater is the work of redemption!"[25]

Electric Kisses

"The great Business . . . and Hurry of the World, is nothing else but Diversion, and a Way of wasting the Time," lamented the *American Magazine* in 1744, in an article entitled "Man's Life a Continual Round of Hurry and Amusement." Electricity, however, provided a moral alternative to mere "idle" leisure beginning in the 1740s. In addition to cultivating reason and piety, electrical demonstrations were also made to promote interactions between gentlemen and ladies for the exchange of rational and pious knowledge through heterosocial conversation. Unlike vicious homosocial behaviors such as card-playing, drinking, and gambling, philosophical demonstrations combined entertainment with the pursuit of reason, piety—and social virtue.[26]

Driving the cultural and material reproduction of polite sociability among the American gentry—refined demeanour, verbal displays of wit and learning, elegant dress, fine houses—was the desire for identification with gentility and civility. Polite culture was underwritten by social ideology. For conjectural historians like David Hume, politeness figured as a specific stage in the development of human societies, an advanced phase of elegance and learning driven by the comity of commerce rather than the uncertain conquests of war. The participation of women was not incidental but integral to this form of social life because heterosocial interactions were understood to be the foundation of civility, refining male conduct and uniting the economic and political agents of society, whose competing interests might otherwise come into conflict. James Forrester's *Polite Philosopher* (1734) was one of several British commentaries on the topic reprinted in British America. "The very essence of politeness," Forrester insisted, was the interaction of ladies and gentlemen. "The acquaintance of the ladies only . . . can bestow that easiness of address, whereby the *fine gentleman* is distinguished." "Being accustomed to submit to the ladies" gave a "new turn" to male behavior, opening a "path to Reason" and checking the impulse to "contend, destroy, and triumph over one another." The rational conversation of gentlemen, meanwhile, improved the minds and morals of ladies, fitting them for companionate marriages and the responsibilities of motherhood. "Your duty, my dear," a fictional mother instructed her daughter in Charles Allen's *Polite Lady* (1760), "is to sit mute, and to profit and improve by [the] wise reflections" of gentlemen, and so progress to "knowledge, virtue and piety."[27]

The commodification of science through the sale of apparatus, handbooks, and demonstrations opened up an important intersection between science and politeness for the bourgeoisie. Polite science was a tradition that emerged from a culture of elite heterosocial conversation in the second half of the seventeenth century, and was propagated through texts like Bernard de Fontenelle's *Conversations sur la Pluralité des Mondes* (1687) and Francesco Algarotti's *Il Newtonianismo per le Dame* (1737). These works helped to rehabilitate the image of natural philosophy, formerly seen as the antisocial preserve of solitary gentlemen, recasting it in the form of didactic but diverting dialogues between the sexes. This tradition spanned the Atlantic across the second half of the eighteenth century, from the importation of works such as Benjamin Martin's *Young Gentleman and Lady's Philosophy* (1755), to American reprints of texts like Priscilla Wakefield's *Mental Improvement; or, The Beauties and Wonders of Nature and Art* (1794), a popular work on natural history at century's end. Some of the first general science texts published in North America were indeed polite science texts, *"for the use of students of both sexes."*[28]

The wonders of electricity provided a singular conversation piece around which ladies and gentlemen could gather for mutual improvement. In 1769, The *New-England Almanack; or, Lady's and Gentleman's Diary* noted that lightning rods were of sufficiently controversial moment to have *"become a topic of conversation"* in polite circles. Both at home and in public, ladies of genteel social rank enjoyed access to electricity as a polite and improving diversion. In her diary, the Quaker Elizabeth Drinker mentioned declining to see Kinnersley at the College of Philadelphia in January 1760, but was subsequently "entertain'd with divers objects in a Micrescope; and with several expediments in Electricity" at the home of Mrs. Bringhurst. Museums of natural curiosities provided a more public venue for heterosocial interactions.[29] In Hannah Webster Foster's novel *The Coquette* (1797), Lucy Sumner describes Mr. Bowen's museum as "a source of rational and refined amusement," where "the eye is gratified, the imagination charmed, and the understanding improved." In May 1765, Ann Ashby Manigault, wife of the Charleston planter and merchant Gabriel Manigault, recorded in her diary that she had been out to see "experiments in Electricity"—probably William Johnson's at the Charleston Library Society. Charles Wilson Peale's Museum in the Pennsylvania statehouse and Gardiner Baker's in New York later in the century also prominently displayed functioning electrical machines.[30]

Gentleman demonstrators entertained and instructed their children as well; polite science often meant domestic science (Figure 11). The fictional Tom Telescope's exegesis of the *Newtonian System of the Universe* (1761) exemplified the hope that children would learn rational moral lessons from philosophical diversions. Later reprinted in the United States in a version edited by Robert Patterson (1808), professor of natural philosophy at the College of Philadelphia and sometime electrical demonstrator, *Tom Telescope* offered a Franklinist account of electrical machines, Leyden jars, and lightning rods in a morally zealous setting where experiments were presented as an "innocent amusement" that would "not only divert the mind, but improve the understanding." Natural philosophy thus guided moral decisions about leisure, with electricity providing an alternative—one "suitable to the taste of all the company present"—to the "covetousness and cheating" of pastimes like card playing and gaming.[31]

If the presence of ladies at polite demonstrations was fundamental, encouraging women to attend was not necessarily an act of sexual egalitarianism. In his *Brief Retrospect of the Eighteenth Century* (1803),

---

11. SCIENCE IN THE PARLOR

Electricity became a public spectacle in the eighteenth century, but the commodification of science through the sale of smaller, increasingly affordable machines meant that experiments could also be conducted in the home. This image captures the family romance that helped drive the domestication of science. Although electricity was commodified as leisure, its genteel promoters claimed that unlike less virtuous pursuits such as card playing or visiting the tavern, it could instill important moral lessons of reason, piety, and polite sociability.

A New ELECTRICAL MACHINE for the TABLE.

the Presbyterian minister Samuel Miller emphasized that the progress of female education had been one of the signal advances of the Enlightenment, crucial to the cause of refinement and civilization. Such progress, however, did not necessarily contradict the belief that women were "obviously destined to *different employments and pursuits* from men." To many gentlemen, familiarizing ladies with natural philosophy was a necessary antidote to their predisposition to panicky unreason. According to the Newport gentleman quoted by Kinnersley, electrical shows were sometimes too startling for women. "Pray make my Compliments to Madam your Spouse, and assure her from me, she will be highly-delighted with this Gentleman's Experiments, especially with those on Lightning, in the *Second Lecture,* which seem to be the only exceptionable Experiments with the Ladies." Advertising electrical demonstrations in 1793, Isaac Greenwood III, artisan and surgeon-dentist (and a descendant of Harvard's first natural philosophy professor) likewise gallantly bowed to female sensibility, offering his compliments "particularly to the Ladies, and Youth, of both Sexes," and reassuring "them that no *Shocks* (as they are called) will be given on any Account." Ladies were rhetorically positioned as needing gentlemanly reason to soothe female fears. Rebutting charges that lightning rods caused earthquakes, John Winthrop sought expressly to reassure not just ministerial critics but "a great number of others, especially of the more *timorous Sex* . . . who have been thrown into unreasonable terrors, by means of a too slender acquaintance with the laws of electricity." By demonstrating predictable chains of cause and effect, natural philosophy performed a special service to the fairer sex, turning panic at the inexplicable into wonder at the lawful. "Philosophy extends our view of the universe," wrote the author of the *Compendious View of Natural Philosophy . . . for the Use of Students of Both Sexes* (1796): "[Our understanding] of its various phenomena, and of the principles and powers by which they are produced . . . . banish[es] prejudice and superstition."[32]

This was a decidedly ambivalent program of female enlightenment. While urging the initiation of women into the causal explanations of natural philosophy, gentlemanly performances reinforced the notion that female nature was driven by passion, not reason. Audiences were encouraged to see the natures of women and electricity as equally passionate through mutually constitutive display. In the "Venus electrificata," or electric kiss, invented by the German electrician Georg Bose in the 1740s, the demonstrator invited a lady from the audience onto an insulating stool, connected her to the prime conductor, and charged her with electricity. Retaining her charge thanks to her insulation, she painfully repulsed the "salute" of approaching suitors, as electricity discharged between the two bodies. Hiller's narration emphasized that the force of female passion was on display here as much as that of the electric fire. "You are now madam filled with fire, a fire of the purest kind, it gives you no pain while you keep it in your own Breast, But you suffer when you communicate it to other's." "No man," he warned, "can approach your hand or Cheek without Smarting for his presumption." Kinnersley advertised spirits "kindled by Fire darting from a Lady's Eyes (without a Metaphor)." "Much has been sung of the fire in ladies eyes," Hiller commented, "that it strikes through the heart and kindles the spirits into an amorous flame." This was "not mere hyperbole," for "some ladies may really have fire enough in their eyes to kindle spirits of any kind." "If there is such fire in the mild and smiling," he concluded impishly, "what must there be in the angry one?" The joke with electricity was often to literalize the metaphorical, and this was abundantly clear when it came to platitudes about the power of female passion. As passionate witnesses, women couldn't produce evidence about electricity on their own; it took others (gentlemen) to turn their bodies into evidence about electricity as well as the passions themselves.[33]

Playful though this Venus was, by the end of the century she was cutting a more radical figure. Parading the female body as a wondrous fountain of electric fire played on long-standing associations between natural and erotic powers. Articulated classically in Lucretius's *De Rerum Natura,* these were revived in the late Enlightenment by the self-styled Lucretian philosopher-poet Erasmus Darwin, a leading light in the Lunar Society of Birmingham (and sometime medical electrician). Since the 1730s, some British commentators had balked at what they perceived as the lewdness of Linnaean natural history, with its emphasis on plant classification according to sexual characteristics. By contrast, Darwin's poetry made a celebration of plant sexuality, seeming to condone thereby the free play of human sexual urges; his association with the leading English feminist Mary Wollstonecraft strengthened his link to agendas of free love, sexual egalitarianism, and democratic politics in the era of the French Revolution. Darwin made the Venus part of this late-century radical Enlightenment in his poem about "the loves of the plants," *The Botanic Garden* (1789), also reprinted in America. As a sign of the times, the following passage, excerpted in the *New York Magazine and Literary Repository* in 1793, concerns an electrified woman kissing a young man, whose salute is not now repulsed:

> *Nymphs! your* fine hands ethereal floods amass
> From the warm cushion, and the whirling glass;
> Beard the bright cylinder with golden wire,
> And circumfuse the gravitating fire.
> . . . *Or,* if on wax some fearless Beauty stand,
> And touch the sparkling rod with graceful hand;
> Through her fine limbs the mimic lightnings dart,
> And flames innocuous eddy round her heart:
> O'er her fair brow the kindling lustres glare,
> Blue rays diverging from her bristling hair;
> While some fond youth the kiss ethereal sips,
> And soft fires issue from their meeting lips.
> So round the virgin Saint in silver streams
> The holy Halo shoots its arrowy beams.

> ... Starts the quick Ether through the fibre-trains
> Of dancing arteries, and of tingling veins,
> Goad's each fine nerve, with each new sensation thrill'd,
> Bends the reluctant limbs with power unwill'd.[34]

The Venus fascinated the male gaze because she incarnated a set of fluid relations between electricity, gender, and sexuality. Some invocations were purely metaphorical: von Haller's account of Bose's demonstrations included a uniquely gallant version of the two-fluid theory of electricity, describing "the *male fire,* which is attended with crackling, and the *female fire,* which is a luminous emanation without violence or percussion" (a variation on alchemical sexual polarities). "Could one believe," asked von Haller, "that a lady's finger, that her whale-bone petticoat, should send forth flashes of true lightening, and that such charming lips could set on fire a house?" In John Cleland's novel *Fanny Hill* (1749), not to mention a stream of erotic poems about electric eels, electricity became the vital force of sexual libertinism. But as ever, electrical ideas went beyond mere metaphor. There were practical jokes; for example, to see whether eunuchs would conduct, the experimenter Joseph Sigaud de la Fond electrified *castrati* in Paris in 1781.[35] There were also serious attempts to theorize sexuality in terms of electricity. The French natural philosopher Charles Rabiqueau saw sexual reproduction as an electrical process: "The feminine sex is the depository of the tiny human spheres which are in the ovaries. These little spheres are an electrical substance, inert and lifeless; like an unlit candle or an egg ready to receive the spark of life, or ... the spirit of fire." The London physician John Shebbeare held the unorthodox view that male erections resulted not from a distention of blood vessels but from the action of a "vital fire" akin to electricity; conversely, male "dejection" arose from "the want of a sufficient quantity of this fire." Medical electricity often took the form of sexual therapy. The commercial therapist James Graham invited couples to increase their fertility in his electrified "celestial bed" in London in the

1770s, while others recommended shocks for specifically female complaints. James Walker, a Virginian who studied physiology at the College of Philadelphia in the 1790s, wrote his dissertation on the "*causes of sterility in both sexes; with methods of its cure.*" Attributing female sterility to menstrual blockages, Walker counseled that "the uterus may be stimulated by shocks passed through the pelvis." One American doctor quoted Graham as saying that "women may, with propriety, adore electricity."[36]

In threatening cherished wisdom about the seemingly natural order of male reason to female passion, female natural philosophers like Laura Bassi of Bologna were often cast as prodigies, while more politicized figures like Wollstonecraft became proverbial monsters. Fears of gender-confused "manly women" tormented the conservative mind in the late Enlightenment, and Darwin had helped remake the Venus figure in this radical image. But the wonderful spectacle of the electrification of Venus remained ambiguous. At once the demonstrator's puppet and the fountainhead of active powers, her discharges revealed female passion, yet also female ability to defy male will. To many at electrical demonstrations, the Venus was probably little more than an amusement. Alexander Anderson Jr., a Columbia College medical student who attended lectures on electricity at the university in the 1790s, often visited the museum of his friend Gardiner Baker in Lower Manhattan. "I had the pleasure of waiting upon Miss N[ancy] who was much delighted with the experiments in Electricity," he recorded in his diary. "My brother's attempts to kiss Miss Jane while insulated, excited no small mirth, when they were separated by the fire flying from their noses." To the more philosophical, however, the Venus represented the most fundamental of heterosocial affections. Thomas Green Fessenden, a Vermont belletrist, marveled at that "most wonderful of phenomena"—the moment when "the lady's eyes seem to be fountains of *animal electricity.*" In such moments, thought Fessenden, electricity revealed hidden connections between the powers of nature and society. "Light, heat, or calorick, electricity, Galvanism, Perkinism, animal spirits, the social feelings, especially when *love* is concerned, and the stimulus of society," he ventured, "are all intimately connected or different modifications of the same matter."[37]

An Impolite Science

While some demonstrators sought to discipline electrical experience by making it bear lessons of rationality, piety, and politeness, not everyone made electricity a path to improvement. Because these performances took place in an entrepreneurial culture largely free of institutional controls, the meanings of electrical display inevitably varied. The career of the Venus is a case in point: as we have seen, playful and seemingly innocuous displays of electrical/female nature could become associated with radical rethinkings of those natures. Electricity was a wonderful commodity, apt to be remodeled depending on the social identity, even the political moment, of the body it penetrated. Above all, electricity raised questions about human control over nature, beginning with bodily self-control. Even in polite demonstrations, the meanings of electricity were potentially contradictory, because the electrical body in performance enacted a competition among different discourses: rational understanding versus disorienting surprise; pious reflection versus worldly experience; polite self-control versus bodily contortion. Historians of science have argued that the Enlightenment rejected both wonder and play as legitimate modes of engaging with nature. While this may hold for elite natural philosophers, it does not adequately describe the behavior of middling and elite publics, neither learned nor vulgar strictly speaking, who engaged with electricity. Without doubt, wonder and play were central to electricity's appeal as an enlightened public science.[38] Paradoxically, what made electricity thrive as a form of polite science was its impolite character—its direct bodily experience mocked early modern conventions of corporeal restraint. The experience of enlightenment as wonderful play, and perceptions of the limits of reason, were thus in direct relation. As Darwin had written, electricity bent "reluctant limbs with power unwill'd." It was precisely as power "unwill'd," convulsing polite bodies against their will, that electricity enthralled eighteenth-century audiences.[39]

Franklin devised some of the grandest pranks. In 1748, he proposed to end the season's experiments "somewhat humorously, in a party of pleasure, on the banks of [the] *Skuylkil*" for "the "principal People" of Pennsylvania. "A turkey is to be killed for our dinner by the *electrical shock*," he informed Collinson, "and roasted by the *electrical jack*, before a fire kindled by the *electrified bottle:* when the healths of all the famous electricians in *England, Holland, France,* and *Germany* are to be drank in *electrified bumpers* . . . under the discharge of guns from the *electrical battery*." Kinnersley created representations of the solar system and the aurora borealis (the northern lights), rang a series of musical bells "by an electrified Phial of water," and turned the "Wheel of a curious Machine . . . by Lightning," on which he played a "Variety of Tunes." He poked fun at less virtuous pastimes, too, exhibiting "a new Property discovered in electrified Points by Means of which will be exhibited a Sort of electrical Horse Race," and ended his performance by discharging a battery of guns by electricity, "after it has darted thro' ten Foot of Water." These set pieces had no explicit intellectual or moral purpose—they were simply entertaining displays.[40]

In the hands of less philosophical (and less genteel) demonstrators, the sensual appeal of electrical wonders seemed like the pursuit of spectacle for its own sake, independent of rational, moral, or polite purpose. Spectacular electricity was marketable by anyone who could operate a machine and find a venue. Significantly, the first to offer public shows with an electrical machine in the 1740s (at a charge of ten shillings) was not Kinnersley but William Claggett, a Boston clockmaker who had seen von Haller's account of Bose's displays and built his own generator. Electricians obliged to compete in the world of public-house entertainment, like Claggett and the New York "posture-master" Richard Brickell, advertised the sheer corporeal thrill of "electerisation" to sell their effects: "to be shown, the most surprizing effects of phenomina, on electricity of attracting, repelling . . . particularly the new way of electrifying several persons at the same time, so that fire shall dart from all parts of their bodies." These electrical spectacles, a part of the public house's world of gaming and drinking, emphasized what the mother in Charles Allen's *Polite Lady* condemned as "odd feats of bodily agility" to "gratify the gaping curiosity of those who like to gaze and stare at strange sights." Virtuosic bodily performance of this kind could "never afford any rational amusement to people of sense," she warned, but they were nonetheless mainstays of colonial taverns, as well as coffeehouses. A variety of machines and "monsters" vied for the distracted gazes of tavern-goers: automata, music machines, optical devices, planetariums, wax figures, and models of cities (Malaga and Jerusalem were favorites). Lower-class inns like those on New York's Bowery featured bull- and bear-baiting, cock-fighting, even live alligators and lions. Electricity was part of this world, too. At John Williams's house in Boston, where Claggett performed, the entertainments included a musical machine, a "posture master boy . . . transforming himself into various shapes," and a "fiery lynx, or Tyger Lyon, worthy the sight of the curious." Even the genteel regularly sought the frisson of exotic curiosity. In Newtown, Maryland, Alexander Hamilton, a connoisseur of houses of entertainment along the Eastern seaboard, "was entertaind by the tricks of a female baboon." A half century later, the New York medical student Alexander Anderson Jr. wrote of attending chemistry lectures at Columbia in the morning before going in the afternoon to see "a *male* and *female Bison* and a child with two heads, the former alive and the latter preserved in spirits" on John Street.[41]

In the eighteenth century, polite humor typically connoted the control of wit and wordplay, but electricity's surprising contortions kept it closer in spirit to the grotesque ribaldry of Rabelaisian carnival. Electricity not only cultivated reason and piety; by violating the bounds of politeness, it also provoked laughter. This was important: making electricity diverting meant creating moments of disorientation through sensory scrambling.[42] Shocking the body, and undermining corporeal self-control, became the central strategy in electrical entertainment. The sober Priestley cautioned his readers on the need for "temperance" in the pursuit of science, whose pleasures were "but one degree above those of sense," but even he conceded that "a great deal of diversion is often occasioned by giving a person a shock when he does not expect it." Like contemporary card-playing manuals that taught readers to spot cheats, handbooks that unfolded the secrets of electricity made it possible to surprise others. "It was pleasant," Priestley wrote, "to see people start at the same moment, and compare their different reactions." Bose described one practical joke where he seated guests at his dinner table, rigging up their forks—the very tool of rational civility—to an electrical machine, to surprise his company when they began to eat. Watson at the Royal Society performed a similar stunt, laying an "electrical mine" underneath a rug, a trick recorded by Priestley and repeated by Kinnersley.[43]

Electrical enlightenment was emphatically a project in embodied knowledge, not disembodied reason: instead of politely removing bodies from physical experience and transmitting knowledge through textual or verbal exchange, electricity was an impolite science where enlightenment was experienced as convulsion.[44] This element of physical coercion was playfully handled, to be sure, but it also suggested more violent possibilities. Demonstrators routinely cautioned imitators to moderate their often painful shocks, yet from destroying thunder houses to shocking audience members, it was precisely the element of violence that made electrical performances suspenseful. "There are no bounds (but what expence and labour give) to the force man may raise and use in the electrical way," Franklin exulted in a vision of limitless power. "For bottle may be added to bottle *in infinitum,* and all united and discharged together as one, the force and effect proportioned to their number and size." Such force could not always be contained, however. Franklin regularly knocked himself unconscious with errant discharges, while the experimenter Georg Richmann was killed outright in St. Petersburg in 1753, when lightning struck his ungrounded instruments. "Too great a charge might, indeed, kill a man," Franklin admitted. Kinnersley strove to reassure audiences that his were still "agreeable" entertainments despite such misadventures. While promising to pass the "amazing Force of the Electric Fire . . . through a Number of Bodies at the same Instant," he made clear that "No One need feel the Effect of any of these Experiments, but those who choose it, and as the Quantity of Lightning made use of will be but small, the most timorous Person need not apprehend being either hurt or terrified."[45]

In fact, violence hedged the margins of genteel electrical performance, threatening to overstep polite taste. Per Kalm, the Linnaean naturalist who traveled through North America at mid-century, recorded in 1749 that Franklin accidentally sent a painful shock through a lady's nose at a demonstration, requiring the assistance of a doctor. While Franklin jovially bantered about electrically tenderizing turkeys, Priestley routinely killed mice, kittens, dogs, and rats with blasts from batteries of Leyden jars. Kinnersley, too, advertised "animals killed by [electricity] instantaneously" in his lectures. Whether ladies in particular looked away from such spectacles or not, chivalrous commentators made them seem to do so, in scenes reminiscent of Joseph Wright of Derby's painting *An Experiment on a Bird in the Air Pump* (1768), where a polite audience (including ladies and children) is dismayed at the apparent death of a bird in an artificially created vacuum. The Philadelphia physician Charles Caldwell, for one, swore off such destruction as a violation of sentimentality, recommending instead "easy, cheap and familiar" methods of experiment that the reader might adopt "without having his feelings hurt by the agonizing pangs of tortured and dying animals."[46]

Experimental violence was not limited to animals, however; in certain circumstances, it was carefully extended to human beings. As genteel and commercial performances, electrical demonstrations routinely excluded nonwhite Americans. There were occasional exceptions to this. In Providence, Rhode Island, in 1793, Isaac Greenwood III announced that he would set aside one night of his "electrical exhibition" in order "to give the Black People an Opportunity of being somewhat enlightened in that pleasing and noble branch of philosophy." But such exceptions appear to have been extremely rare. More commonly, when nonwhites were present at demonstrations, electricity was used to make distinctions between those who enjoyed mastery of nature's wonders and those who did not. According to Priestley, indeed, one of the great values of natural philosophy was that it demarcated the civil from the savage; its ability to produce "great inventions" was one of "the capital advantages of men above brutes, and of civilization above barbarity." No one saw electrical machines as engines of empire and the progress of white Christian civilization in the "untutored continent" of America more clearly than William Smith, Anglican provost of the College of Philadelphia and sometime experimental philosopher. "Providence . . . called Great Britain (a nation enjoying liberty, religion and science, in their purest and most improved state) to the possession of that part of America," Smith declared in a speech to raise funds for the College of Philadelphia in 1772, "which seems thus ordained to empire." "LIBERTY," he enthused, "will not deign to dwell, but where her fair companion KNOWLEDGE flourishes by her side." When it came to convincing private citizens to support the college, Smith cited its professorships and schools, and its proud possession of a "compleat APPARATUS for Experimental Philosophy; some parts of which, particularly the Electrical and Astronomical apparatus, are not be equaled, perhaps, in any other part of the world." Experimental performances could be especially useful for demonstrating the difference between the civil and the barbarous on colonial frontiers.[47]

In 1772, the *Pennsylvania Gazette* announced that Kayashuta, "the great Seneca Chief," had returned to Philadelphia to witness once more the electrical experiments that had "so much engaged his Attention and Admiration" on his previous visit. In the *Gazette*'s view, Kayashuta marveled not at the power of nature or God, but at the white man's art: "His first Enquiry at his Return was, whether he could have another Opportunity of seeing Thunder and Lightning produced by human Art?" (One French account from 1779 had the reaction of an "Indian" to electrical shows rather differently: "'These White-men are very clever rascals,' says he, without showing the least surprise or the slightest reflection.") While Europeans were anxious about the potential to use wonder to stir up the mob at home, in the colonies, they were eager wonder-mongerers, seeking to impress nonwhites with their superior civilization and convince them of their legitimacy as rulers. Edward Bancroft's novel *The History of Charles Wentworth* (1770) provides an early fictionalized account of the use of scientific instruments as instruments of colonial power, by a man named Gordon, a self-exiled European living among the Arawak in the interior of Dutch Guiana. The narrator describes the key scene thus: "I was led into a dark room, and made to observe an electrical apparatus, which was kept ceremoniously covered over; this, with some of the most surprizing phenomena of chemistry and natural magic, [Gordon] gave me to understand, that he sometimes used among the Indians, to inspire them with a belief in his being possessed of an unusual degree of knowledge, and a supernatural power." Electricity was valuable on colonial frontiers as white magic. In 1780, the *Virginia Gazette* printed a letter from a "Gentleman in Tillicheery" (Tellicherry on the Malabar Coast in the East Indies) in which the author bragged about "astonishing" the natives with a solar microscope and magic lantern. "They are ready to think me in a league with the Old Nick himself, when I introduce the devil, punch, and all the grotesque figures that accompany the magic lanthorn, dancing about on the wall just as I please." Electricity supplied the coup de grâce: "None have immortalized my name so much as that extraordinary faculty I have of drawing forth fire from a tube of brass, making bells ring that never are touched, and causing little paper men and women to get up and dance a cotillion."[48]

To impress audiences still further with the power of their philosophy, British Americans meted out violent electrical discharges on the bodies of Africans, both model and real. On his West Indian tour of 1752–1753, Kinnersley performed a new demonstration, one original to British America (it was later repeated on the mainland). This remarkable performance was described in a broadside advertisement printed at St. John's, Antigua: "A flash of lightning [is] made to strike a small house, and dart towards a little lady sitting on a chair." The lady is "preserved from being hurt; whilst the image of a Negroe standing by, and seeming to be further out of danger, will be remarkably affected by it." The model "negroe" was the colonial exemplar of unenlightenment. Kinnersley repeated his moral, now racially inflected: "A prudent Man foreseeth the Evil, and hideth himself, but the Simple pass on, and are punished." More disturbing still was the use of strong electric shocks on rebel slaves, rounded up in the aftermath of Tacky's Rebellion on Jamaica in 1760. Here military and cultural contests converged. Tacky's Rebellion was the most serious slave revolt in the eighteenth-century British Caribbean, and featured a public contest between the power of natural philosophy and African shamanism in the punishments and executions of the Obeah men who had been led by Tacky, a Gold Coast native. A British witness reported the scene to the House of Commons. An Obeah man "bid defiance to the Executioner, telling him that it was not in the Power of the White People to kill him." When he finally perished, his black audience was "astonished." The report also described what the author regarded as the successful use of electricity to display the superiority of European science over African superstition. "On the other Obeahmen, various Experiments were made with Electrical Machines and Magic Lanthorns, which produced very little Effect; except on one who, after receiving many severe Shocks, acknowledged his Master's Obeah exceeded his own." We cannot know for sure what the Africans thought, but the British clearly wanted to believe that electricity impressed them as both physical and spiritual power. (The British believed that displaying the severed heads of rebel slaves deterred uprisings, too, but this was probably only wishful thinking.) Whatever the reality, the British saw electricity as a way to project their superiority in the colonial social order. In this case, they valued it not for its power to provoke contemplative wonder, but for its ability to inspire spectacular terror.[49]

"Electricity has one considerable advantage over most other branches of science," Priestley noted in his *History and Present State of Electricity*, "as it both furnishes matter of speculation for philosophers, and of entertainment for all persons promiscuously."[50] As his remark implies, the playfulness and wondrousness of electricity exploited by demonstrators like Kinnersley and Hiller to serve their goals of reason, devotion, and politeness could be directed to different ends, as performers saw fit. By contrast with polite performances, in taverns sensual gratification, rather than the parlor's lessons of improvement, was the order of the day. Even enlightening demonstrations unfolded electricity's impolite alter ego, since their lessons relied on sudden convulsive shocks. These were demonstrations of ambivalence: was man the master of the electric fire, or electricity the master of man? Ironically, perhaps the most unequivocal affirmation of rational mastery came on British America's frontier, in demonstrations that were racial contests.

While Kinnersley and Hiller used mild shocks on genteel bodies to initiate polite audiences in the laws of electricity, Caribbean authorities unleashed violent shocks on Africans to mark the difference between master and slave. Both were demonstrations of enlightenment in British America.

CHAPTER FOUR

Electrical Politics and Political Electricity

Dr. Franklin . . . intends shortly to produce an electrical machine, of such wonderful force, that instead of giving a slight stroke to the elbows of fifty or an hundred thousand men, who are joined hand in hand, it will give a violent shock even to nature her self, so as to disunite kingdoms, join islands to continents, and render men of the same nation strangers and enemies to each other.

ANONYMOUS, 1777

IN 1748, Ebenezer Kinnersley designed a new electrical entertainment, one unique to British America. As we have seen, Kinnersley was adept at exploiting the theatrical potential of electrical display. His "magical picture," here described by Franklin, was a master stroke:

> Having a large metzotinto with a frame and glass, suppose of the KING (God preserve him) take out the print, and cut a pannel out of it near two inches distant from the frame all round . . . . With thin paste, or gum water, fix the border that is cut off on the inside of the glass, pressing it smooth and close; then fill up the vacancy by gilding the glass well with leaf gold, or brass. Gild likewise the inner edge of the back of the frame all round, except the top part, and form a communication between that gilding and the gilding behind the glass: then put in the board, and that side is finished. Turn up the glass, and gild the fore side exactly over the back gilding, and when it is dry, cover it, by pasting on the pannel of the picture that hath been cut out, observing to bring the correspondent parts of the border and picture together, by which the picture will appear of a piece, as at first, only part is behind the glass, and part before. Hold the picture horizontally by the top, and place a little moveable gilt crown on the king's head. If now the picture be moderately electrified, and another person take hold of the frame with one hand, so that his fingers touch its inside gilding, and with the other hand endeavour to take off the crown, he will receive a terrible blow, and fail

in the attempt. If the picture were highly charged, the consequence might perhaps be as fatal as that of high treason . . . The operator, who holds the picture by the upper end, where the inside of the frame is not gilt, to prevent its falling, feels nothing of the shock, and may touch the face of the picture without danger, which he pretends is a test of his loyalty.—If a ring of persons take the shock among them, the experiment is called, *The Conspirators*.[1]

In this entertainment—one dependent on bodily manipulations—a joke is made, and a relationship established, between the experimental management of active natural powers and the subversion of political powers. Having carefully dismantled and refabricated the king's portrait, Kinnersley inserts in its frame a gilt lining to act as a secret conductor, so that when invited to lift off the crown, a volunteer receives a shock as he or she unwittingly completes the circuit between crown and lining. Approaching the king's person without due reverence for royal authority results in a shocking punishment. The knowing Kinnersley, however, removes the crown without suffering a shock, by avoiding the gilt lining when he grasps the frame. He can even "touch the face of the picture without danger, which he pretends is a test of his loyalty." The moral? Forbidden gestures of desacralization may be carried out with impunity when concealed by shows of loyalty. The successful deconstruction and penetration of the royal facade are cloaked in an ostensible pledge of fealty, recalling the Jacobite custom of toasting the king by raising a glass "over the water"—a covert pledge of allegiance to Bonnie Prince Charlie, the Stuart pretender exiled in France. In the case of electricity, mastery of natural powers signaled the possibility of new political powers.

The relationship between science and politics in the Enlightenment is classically defined by the debate over the French Revolution and the allegation, associated particularly with Edmund Burke (and which was to reverberate so powerfully in Max Horkheimer and Theodor Adorno's twentieth-century critique of "instrumental reason"), that the atheistic rationalism of enlightened science fanned the flames of radical Jacobinism and produced the "Terror." According to this view, reason mutated into a secular enthusiasm, unchecked by religion, tradition, or sentiment, that resulted in the murder of thousands. Intersections between the languages of science and politics became telling: Burke used chemical metaphors to conjure images of revolutionary madness in his *Reflections on the Revolution in France* (1790). The relation between science and politics in the American Revolution has been greatly overshadowed by this narrative. Where scholars have addressed the American case, they have identified the Constitutional period as a moment when technological languages of politics became prominent in public discourse, and seen this as an expression of American Enlightenment as practical genius. Taking their cue from the enlightened claim that politics could be reduced to a science, they have read these metaphors as evidence that specific provisions, such as checks and balances, were put into place according to a "Newtonian" model of the universe as a perfect, self-running machine. But in so doing, they misjudged both the pragmatic process of early American state formation and the character of Newtonian natural philosophy, which described the universe not as a perfectly balanced machine but as a cluster of dynamic tensions, necessarily sustained by God.[2]

Languages that combine science and politics can be misleading if we assume that specific sciences determine the content of politics (or vice versa), but they do reveal polemical conceptions of the relation between human affairs and the natural world. The rhetorical career of electricity in the American Revolution, examined here for the first time, provides a case in point, and a vitalist counterpoint to accounts of mechanical constitutionalism. With the onset of revolution and war in the mid-1770s, electrical demonstrations disappeared from American public culture, but the electrified body did not. Chapter 3 showed the subversive potential of the electrical body in performance, as it bucked polite self-restraint and flirted with the impolite pleasures of coercive physical experience. By themselves, these demonstrations produced no discourse of social or political transgression. Patriot writers in the Revolution, however, developed the transgressive potential of the electrical body into a full-blown symbol of republican resistance. Staging the "magical picture" in 1748 may have been a prophetic act but, coming at a time when no quarrel existed between Britain and America, it was one of pure play. Two decades later, the Revolution generated sustained rhetorical engagements between electricity and politics that were no joke.

Electricity lent itself readily to political use because it provided a language of power through which to articulate the meaning of revolution. As such, it was a rhetorical commodity too valuable to go uncontested. While *electrical politics* represented the world of the political in terms of the electrical, discourses of *political electricity* insisted conversely that electricity was in fact a function of politics. Instead of dissolving American resistance into the will of nature or God, it emphasized the role of individual human actors. Political electricity also divided between positive and negative politicizations of the key figure of Franklin who, as a revolutionary statesman, was suddenly taken to embody a new relationship between science and politics. These politicizations turned, once again, on questions about human agency that were central to the Enlightenment. Was man the master of electricity or electricity the master of man? The first three chapters have shown that despite Franklin's storied achievements with the Leyden jar and the lightning rod, Americans did not apotheosize reason, but continued to acknowledge electricity as an awe-inspiring natural and divine force that defied complete control. Theirs was a wonder-full enlightenment. The Revolution now posed the same question in political terms: was American resistance commanded by God and nature, or the machination of human agents? Languages of electricity now became critical as a means of debating the agency behind enlightened politics.[3]

Electrical Politics

In the late eighteenth and early nineteenth centuries, American evangelical Christians began comparing the bodily effects of spiritual rebirth to those of electric shocks. Electricity made a ready metaphor for the throes of "awakening" because of the physical similarity between electrification and spiritual rebirth: the compulsion of the body by unseen forces. A language of ecstatic convulsion gradually established itself in conversion narratives, which had been an important part of American public discourse since at least the Great Awakening, if not before. One ex-slave, for example, described his conversion thus: "Then, like a flash, the power of God struck me. It seemed like something struck in the top of my head and then went on out through the toes of my feet. I jumped, or rather, fell back against the back of the seat." The fits and starts caused by electric shocks were rhetorically persuasive in such accounts because of associations between the "electric fire" and the purifying baptismal fires of Protestant discourse. Making electric shocks a metaphor for the bodily convulsions by which God compelled true believers drew on and amplified the notion that electricity, as an active power, was an essential means by which God imparted vital and spiritual motion to the universe. Invocations of electricity to describe God's movement of the body faithful appeared in North America at least as early as the 1780s, but they appear to have proliferated in the early nineteenth century, starting with the revivals of the so-called Second Great Awakening. Caleb Rich, a prophet in rural New England, wrote of one such occasion that he "felt as it were a shock of electricity, my lips quivered, my flesh trembled, and felt a tremour throughout my whole frame for several days." When John Bishop met the founder of the Shaker movement, Ann Lee, he recounted that "the effect was like the sudden operation of the electric shock," and that "he was instantly released from all his weariness." The Connecticut publisher Samuel Goodrich recalled the "sudden and irregular storms of fervor" at Methodist meetings, where "the very air . . . seemed impregnated with electric fluid."[4]

The electrical metaphors used to describe spiritual awakenings were not plucked from thin air; they resonated because of the material culture of electrical demonstration in the eighteenth century, described in Chapter 3. The dominant electrical metaphor in the American Enlightenment, however, was political rather than religious (although, as we shall see, the two were related).[5] The war for independence disrupted electrical demonstrations during the 1770s and early 1780s, but it also occasioned the use of the electrified body as a metaphor for the power of republicanism, and as a way of imagining a new national community.[6] In this discourse of electrical politics, bodily electrification functioned as a conceit to describe the effects of momentous political events on individual and community. The defense of liberty through the pursuit of republican virtue flowed from powerful feeling. Invoking bodily electrification naturalized, indeed celebrated, disobedience to Britain as obedience to natural and divine fiat, revealed through a radical libertarian sensibility that burned in the American breast. From the early years of resistance and war to late-century commemorations, Patriot writers staged and restaged the ostensibly spontaneous coming together of the nation. As the active powers of nature were conjured as the active powers of politics, the electricity of republican virtue brought a new American body politic to life.

Electrical politics was a discourse of revolutionary sensibility, of republican awakening as universal communion.[7] In 1766, in one of the first such usages, the *Boston Evening-Post* likened William Pitt the Elder's public approval of resistance to the Stamp Act of 1765 ("*I rejoice that America hath resisted*") to "the Voice of a God. Like an electric Shock it instantaneously pervaded the whole American Continent."[8] Sudden political awakenings now had the force of giant electric shocks. Electrical-political rhetoric was a form of action at a distance: it moved sympathetic audiences across geographical space. "The ties of blood like electrical conductors will convey and communicate the effects" of patriotic sentiments, echoed William Henry Drayton, the South Carolina judge turned revolutionary, in notes for a speech before the Continental Congress in 1778. Throughout the revolutionary era, the language of universal electrical conductivity functioned as a language of intense sympathetic feeling and fraternal participation. "The news [of an independent government in] . . . South Carolina has aroused and animated all the continent," John Adams wrote to James Warren on the eve of revolution in April 1776. "It has spread like a visible joy, and if North Carolina and Virginia should follow the example, it will spread through the rest of the colonies like electric fire." Human circuits of bodily electrification at experimental demonstrations became transformed into a figurative universal republican community, the electric fire of liberty now traveling from body to body. In 1783, while discussing Lord Sheffield's disparaging prediction of America's continued commercial dependence on Britain, John Francis Mercer insisted that Americans would feel the slight of such claims like "an electric Shock," and respond accordingly. "There is a great deal of good sense among the People of America and when once they feel— they will rouse—and those latent powers, which they once with so much glory and advantage exerted, will soon set things to rights." American freedom depended on American feeling. Looking back in his *Autobiography* (1821), Thomas Jefferson remembered the intensity of Virginian resistance in the Revolution as "a shock of electricity, arousing every man and placing him erect and solidly on his centre."[9]

Republican virtue behaved like the electrical virtue: it was a communicable property of proximate bodies. Writing in the *Pennsylvania Gazette*, a commentator named "Senex" urged legislators to "mix with the mass of the people, and get again electrified with a portion of that stern and republican virtue" necessary for serving them. Republican virtue resided in the bodies of the common people, just as electricity resided in common matter. The following year in the *Gazette*, "the

Honest Politician" used electricity in a negative sense not only to discuss the decline in American virtue, but also to portray the unity he claimed existed between Americans and their new French allies. "There is *half* that virtue in the towns and trading cities of these States there was at the beginning of the contest," the Politician wrote. Fluctuations in the value of the currency had encouraged Americans to speculate and seek personal fortunes more than the common good. "The idea catches like the electric fluid," he lamented. "Every man must scheme, cheat, be bribed, or speculate, in order to enrich himself." Yet if Americans at times appeared internally divided, they were yet united with their sudden Gallic brethren, with whom they had formally allied in 1778. It was in vain that Britons hoped that Americans and Frenchmen would never form a coherent coalition, just as like-charged electric bodies naturally repelled each other: "The dissimilarity of manners and the difference of language would then more powerfully disgust and separate, as some electrical bodies at a greater distance attract, at a lesser distance repel each other: For, to use the language of the philosophers, where the sphere of attraction ends, the sphere of repulsion begins." But the Politician insisted on the strength of Franco-American sentimental union. "Our tempers have commixed," and "the same interest has made our sentiments the same . . . a love for the alliance has pervaded every class of men, and . . . unanimity takes place amongst ourselves, and attachment to our friends rises up like a wall to repel the incursions of the Tartans."[10]

Martial enthusiasm was another powerful bond that electrified Americans in the Revolution. Radical republican sensibility was electrical feeling of such ardor it drove Patriots to the heights of manly virtue: military self-sacrifice in defense of the commonwealth. Virtue meant virility, and electrical politics above all meant sympathy among the men whose blood and action would constitute the new nation. The *râge militaire* of 1775–1776 produced a poetry of masculine sacrifice to animate American manhood. "Go act the hero, every danger face, / Love hates a coward's impotent embrace," ran one of many such verses. Martial animation was soon electrified. Significantly, in linking electricity to the defense of liberty, Patriots articulated a historic positive conception of enthusiasm. The imperative to fight was not a delusion or the work of a frenzied imagination—it was a form of revelation, as powerfully self-evident as an electric shock. "Glowing with the love of civil liberty," an anonymous writer in the *Pennsylvania Gazette* recalled the *râge militaire* toward the end of the war, "General Mifflin was undoubtedly the first man in this State who appeared in arms, and by his example kindled that flame of military enthusiasm, which, like electricity, immediately seized the inhabitants of this city." This was a stunning American reversal of British Protestant moderation: enthusiasm was no longer a danger, it was a sacred electric force. "By the force of his eloquence, at the town meetings, [Mifflin] roused the lukewarm, inflamed the brave, and animated the virtuous, more than any other speaker, to those acts and resolves, which laid the foundation of the Revolution and Freedom of Pennsylvania." William Duer, concerned with American vulnerability at Fort Ticonderoga, made the same connection at the Continental Congress, lamenting congressional delays in organizing defensive battalions. "Would to Heaven that the Spirit and activity which has of late animated the Councils of the State of New York would diffuse itself throughout the other States! A portion of their Electrical fire is certainly wanting."[11]

Electricity suggested sympathetic connections not only across space, but across time as well. In the early years of the republic, electrical politics functioned as a form of literary festival to glorify the Revolution's memory. In 1791, more than two hundred "respectable citizens" gathered when the Sons of Tammany celebrated the anniversary of the Columbian order. "In the most brilliant and pathetic language," Brother Josiah Ogden Hoffman "traced the progress of the liberty we enjoy," as well as the origins of both the Columbian order and the Society of the Cincinnati. One of the thirteen toasts that echoed round the tables was to "the memory of our immortal countryman, Benjamin Franklin, whose Philosophy drew electric fire from Heaven, and whose Patriotism cherished the first sparks of that freedom which now blesses his native land." Others remembered the Revolution as the singular moment of sympathetic national consolidation. In the words of the British natural philosopher Erasmus Darwin, friend of Franklin and America: "The patriot-flame with quick contagion ran, / Hill lighted hill, and man electrised man." In "Greenfield Hill" (1794), Timothy Dwight figured "freedom's living flame" as "electric, [and] unextinguishable." The American minister to Spain, David Humphreys, similarly recalled how in the Revolution "the flame of freedom ran / From breast to breast, and man electriz'd man."[12]

Humphreys used the language of electrical communication specifically to suggest the limitless reach of the binding intensity of true patriotic feeling. On the twenty-third anniversary of the Declaration of Independence, in "A Poem on the Love of Country" (1799), he asked, "At home or abroad, what breast is not then as it were electrified by sympathetic recollections" of the Revolution? "It is pleasant to reflect," he wrote from Madrid, "that on the same day, in all parts of the world where a few Americans are assembled, they are in the habit of rejoicing together . . . cherishing those social sentiments which were so feelingly participated in their common toils, sufferings and dangers." Only a "cold-blooded wretch" could fail to feel "the sacred flame of patriotism kindling with redoubled ardour, from the mingled remembrance and emotion which this festival forces on his mind." Self-conscious sentimental commemoration of the Revolution performed through the rhetoric of the staged spontaneity of festival would sustain the fragile Union. That this festival had been so "generally" and "cordially" celebrated was the result of a "unanimity" produced "by the concurrent feelings of a free people." No other occasion could be "better calculated than this spontaneous solemnity" to inspire Americans with "love of country and force of union."[13]

Revolutionary commemorations claimed a universal relevance around the Atlantic world. At the Crown and Anchor Tavern in London in 1791, before the Terror gripped France, an assembly of one thousand dined and toasted the storming of the Bastille to the words and music of Robert Merry's ode asserting "the hallow'd Rights which Nature gave": "Fill high the animating glass, / And let the electric ruby pass / From hand to hand, from soul to soul; / Who shall the energy controul, / Exalted, pure, refin'd, / The Health of Human kind." Even after the radical turn in France, republicans persisted in invoking electricity to describe liberty's international advance. To the poet St. John Honeywood, liberty-as-electricity connected the entire Atlantic world through radical sensibility. How "much we joy'd when Reason's potent call / Awoke to life the energy of Gaul; / When young Fayette, a lov'd and injur'd name, / From Freedom's altar caught the hallow'd flame; / From breast to breast th'electric ardour ran / And in full glare display'd the rights of man." On news of Washington's death in 1799, the American minister in London Rufus King reported how several Frenchmen expressed the singular debt owed the American by the French, "among whom the electric example of the American Revolution sowed the first seeds of the desire of Independence." The excess of encomiums grew tiresome to John Adams, however. "The history of our Revolution will be one continued lie from one end to the other," he complained. "The essence of the whole will be *that Dr. Franklin's electrical rod smote the earth and out sprung General Washington. That Franklin electrized him with his rod—and henceforward these two conducted all the policy, negotiation, legislation and war.*"[14]

The discourse of electrical politics, invoked throughout the revolutionary and early republican era, marked both a culmination and a rejection of technological languages of early modern politics. In his classic study of the relationship between early modern technologies and political metaphors, Otto Mayr identified and examined links between clockwork and authoritarianism in continental Europe, and feedback mechanisms and liberalism in eighteenth-century Britain. According to this account, continental princes and their servants seized on the clock as the ideal embodiment of a social order predicated on the action of a single decisive component: as the escapement governed the clock, so the prince ruled the authoritarian polity. Technological metaphors of the liberal eighteenth-century British state, by contrast, derived mainly from machines whose interacting parts sustained a dynamic and self-regulating tension. While continental European clocks were symbols of princely power, British feedback mechanisms like the steam engine "governor" were made to evoke the internal balances of the constitutional polity. Viewed teleologically, the discourse of electrical politics completed the whiggish metaphorical sequence implied by Mayr's account: continental clockwork authoritarianism gave way to the self-balancing mechanisms of British liberalism, only to be superceded by the vitalistic electrical language of revolutionary libertarianism in America. There is, however, a marked difference between electrical politics and its metaphorical precursors—the absence of any invocation of material technology. Electrical politics was a discourse of science and politics that omitted mention of the machines by which electrification was actually experienced.[15]

This omission was not absentminded but studied, and ideologically significant. Electrical machines embodied the paradoxical method of experimental philosophers: using artificial technologies to gain direct access to the active powers of nature. Machines put human beings into a physical relationship with active powers, but in so doing, they inevitably mediated the relationship between the human and the natural. It should by now be clear that direct interaction with electricity was in fact a highly mediated pursuit: technologically, economically, socially, and intellectually. It required access to experimental apparatus, and paying for admission to electrical performances; and ideas about the nature of electricity, and its role in the cosmos, were situated in very specific philosophical and theological traditions. Invoking electricity as a metaphor for self-evidence and directly revealed knowledge thus belied the heavily mediated nature of electrical experience. Such invocations served a useful polemical purpose, however: they represented resistance and republicanism as expressions of the will of nature and the divine. We shall see later that rhetorically parading the figure of the machine was a favored strategy of critics of American resistance, who sought to dismiss resistance to British authority as a work of artifice and self-interest. Precisely to the contrary, Patriot writers conjured bodily electrification as spontaneous rather than artificial, subsuming the social (hence, contestable) agency of individual rebels/Patriots into the redemptive and unquestionable agency of nature and the divine. The discourse of electrical politics could not incorporate the mediating technologies involved in making electrical knowledge because it sought to naturalize and sacralize revolution as the artless work of higher cosmic forces.[16]

Electrical politics was also a discourse on the cusp between Enlightenment and Romanticism, adapting the corporeality of electrical experience to figure Americans as an impromptu yet deeply felt community, the forerunner of the democratic "body electric" of which Walt Whitman would sing in the nineteenth century. It was a language of enlightened politics both original and surprising. From the pens of Patriots, electricity became for the first time anywhere a language of radicalism, republicanism, and revolution. Here was a dangerous yet exhilarating rhetoric of vitalist libertarian politics that preceded, and contested, the more conservative discourse of politics as a work of mechanical art produced by the constitutional period. Unlike the polemics surrounding the French Revolution, American electrical politics was not a language of revolution by reason, but by revelation: a celebration of wonderful physical experience as the engine of republican enlightenment. In America, where the language of personal evangelical experience enjoyed a public legitimacy it lacked in Europe, science could appear as part of an antirationalist culture. Where European authorities (in this case British critics in particular) fearfully saw the hallmarks of Levellerist delusion in American declarations of independence, Patriots elevated self-evident feeling to a central role in their revolution. Political enthusiasm thus became a celebratory, even necessary, act of national self-imagination. "It will be found universally true," conceded even the diehard rationalist Adams, "that no great enterprise for the honor or happiness of mankind was ever achieved without a large mixture of that noble infirmity."[17]

The source of this nation-building enthusiasm ultimately remains elusive, however. Idiomatically, electrical politics displayed all the trappings of evangelical discourse but consistently lacked explicit invocation of the agency of God. Yet it would be a mistake to regard it as a secular idiom. In the first place, electricity was embedded in a web of religious associations in the eighteenth century. Of particular relevance here are electricity's status as an active power by which God animated the cosmos, and the metaphorical relation between the effects of electric shocks and the pangs of spiritual rebirth. Second, electrical politics derived much of its rhetorical vigor from what was clearly an evangelical sensibility. Recent interpretations of the Revolution have persuasively explored the way in which the founding documents of the republic described secular events and institutions in a religious and sacralizing idiom. In a broad and less formal array of literary settings, electrical politics helped perform the same fusion of secular and sacred. Rather than reduce the sacred to the secular, Patriots meant to invest the secular with the power of the sacred. If they tended rhetorically to dissolve God into the power of nature, it was not because they wished to eliminate the divine, but because they hoped to universalize His agency through the natural world. For this reason, it is perhaps clumsy but appropriate to characterize the agent at work in American electrical politics as nature *and* God. As a discourse of revolution by revelation, unmediated by reason, electrical politics reveals essential aspects of the relationship between electrical and political enlightenment in America.[18]

Political Electricity: Positive

As politics became electrified, electricity became politicized. For those who wished to draw political lessons from experimental philosophy in revolutionary times, Franklin was the inescapable figure. Through his theory of the economy of electric charges, his demonstration of the identity of lightning and electricity, and his invention of the lightning rod, Philadelphia's master electrician had stood on its head the imperial relationship of colonial fact-gatherer to metropolitan philosopher. How had he done so? As we have seen, commentators in the colonial era did not equate these achievements with a rupture in the imperial order of knowledge: Franklin's accomplishments were seen as those of a provincial cosmopolitan and loyal servant of empire that redounded to the glory of the greater British nation. The push for American independence changed all that, however, recasting Franklin's natural philosophy as a revolutionary reversal of the imperial division of intellectual labor, and inestimably raising the stakes of making Franklin's genius embody the suddenly rising glory of the American nation.

What kind of genius was Franklin?[19] One recurrent theme in revolutionary-era comment was the relationship between independence of mind and the American's lack of formal schooling or high social status. Franklin's genius was "irradiated scarcely by the twilight of education," yet "his Mind burst upon the Worlds of Science and Philosophy like the Sun on the Universe from the Womb of uncreated Nothing," in the words of one anonymous writer. Franklin's genius was the spontaneous work of nature, rather than education or status: "At his birth, the electric Spark, Nature's first and most attenuated Agent, sympathising with the fine aether of his Intellect, vibrated and coalesced, and unfolding the Laboratory of her works, taught him whence her Lightnings flash, and why her Thunders roll." Franklin's experiments thus took their place alongside John Bartram's botanizing and David Rittenhouse's mechanical virtuosity in an American triumvirate of natural genius in the Enlightenment. If Franklin had been born of middling artisanal stock in London rather than Boston, one British writer speculated, "the world would probably never have heard his name either as a philosopher or politician." In the mother country, he would have been condemned to a life of physical labor, making impossible the leisure for "speculative science" and the cultivation of genius. He could never have "become a powerful engine to shake a great empire, and to erect a congeries of republics from its dismembered parts." As it was, this Romantic and revolutionary genius, this "philosopher without the rules," became "the principal agent to introduce a new aera into the history of mankind . . . by procuring a legislative power to the western hemisphere." Surprisingly, this lowly son of a tallow chandler in the colonies had proved "a greater enemy to England than even Philip II, or Louis XIV."[20]

Eager comparisons to Newton as a figure of hitherto unparalleled force of intellect were common. Even before the Revolution, in an anonymous verse of the 1750s, Franklin's entrance into the sacred circle of Anglo-American philosophy was proudly forecast: "There Bacon, Newton will our F—lin greet / And place him in his Electrisic seat." On the eve of revolution, he became the choicest symbol of American cultural maturation: Hugh Henry Brackenridge and Philip Freneau's *Rising Glory of America* (1772) invoked his "genius piercing as th'electric fire, / Bright as the light'nings flash explain'd so well / By him the rival of Britannia's sage." Contemporaries not only pointed to Franklin's sheer strength of intellect but also reflected more particularly on the virtues of his method. "His mind was cast in a mould which Nature seems rarely to have used before," declared William Smith, former provost of the College of Philadelphia and sometime political rival of Franklin's, in his eulogy before the American Philosophical Society in 1790. Franklin's was an "original and universal genius." The moral honesty of this genius, and the humility with which it approached nature, demanded specific attention, Smith urged.[21] Admittedly, his electrical papers contained numerous conjectures and speculations, but Franklin took care to distinguish "Discoveries under the humble appellation of Conjectures or Guesses" from matters of fact. This "modest and cautious method of communication" displayed a "winning modesty . . . which gave him a great advantage over those who followed a more dogmatic manner." The Scottish whig lawyer and critic Francis Jeffrey was one of many to echo these sentiments: Franklin's was a genius in which imagination and observation worked in exquisite harmony. Though Franklin was "by no means averse to give scope to his imagination in suggesting a variety of explanations of obscure and unmanageable phenomena," Jeffrey wrote while reviewing Franklin's *Complete Works* for the *Edinburgh Review*, "he never allowed himself to confound these vague and conjectural theories with the solid results of experience and observation."[22]

What made Franklin's genius truly distinctive in Smith's view was the public service to which it was ceaselessly directed. Yes, Franklin had transformed electricity from an "amusement" into a truly philosophical study; but he then applied his insights for the benefit of humanity. "*Franklin*, as a Philosopher, might have become better than a *Newton*; as a Lawgiver, a *Lycurgus*: But he was greater than either of them, by uniting the talents of both, in the practical Philosophy of doing good." The "sublime and astonishing idea of draining the Clouds of their *Fire*" was "useful and beneficial to his Fellow-Creatures," Smith declared, with unrestrained pride, a "great achievement, which had eluded the industry and abilities of a *Boyle* and a *Newton*, [and] was reserved for a *Franklin*." Franklin's body now leapt into action as a decisive symbol of his social engagement. Representing his body was integral to celebrating the practical utility of his experiments, and to dramatizing his social and political activism. Public portrayals show how depictions of the natural philosopher had changed since the previous century. Here was no cloistered Christian ascetic who scorned social engagement, but an eighteenth-century embodied genius and worldly philosopher: Franklin, bon vivant and practical experimenter, turned revolutionary. Because of the convergence of experimental science and revolutionary politics in his life, more than any other figure of his age, Franklin embodied enlightenment in action.[23]

Franklin's practical genius for manipulating natural power through artificial machines like the electrostatic generator and the lightning rod at once foretold and confirmed the rise of American political power. The figure of the machine now came rhetorically to the fore, for unlike electrical politics, political electricity foregrounded human agency to praise (or blame) the agents in question. One writer for the *Virginia Gazette,* for example, just after independence had been declared, labeled Franklin "the greatest physician, politician, mathematician, and philosopher, this day living," and warned his readers that he intended "shortly to produce an electrical machine, of such wonderful force, that . . . render men of the same nation strangers and enemies to each other." A sympathetic British view of Franklin, published in the American press just after Lexington and Concord, chimed a fateful note: "The Doctor, from his uncommon abilities, perhaps not a little whetted by the treatment he met with here, will no doubt, contribute in some small measure to the winding up the feelings of his oppressed countrymen to a proper electrical charge, for a grand explosion throughout the continent of North America." Franklin would generate the political and military power to resist British authority from among his fellow colonials, as if they were a electrical machine primed for the charging. His bodily dexterity as a worker of machines symbolized his experimental genius and also the latent electrical-political power of the American populace, recalling once more Priestley's warning that the English establishment should fear the bourgeois operators of air pumps and electrical machines. During his tenure as colonial agent in London during the 1760s, Franklin had associated with a group of liberal and radical whigs, men of middling rank like Priestley who were Dissenters, experimenters, and critics of Parliament, and who sympathized with American grievances after 1763. The Revolution, crystallized in the figure of Franklin, stood precisely to realize the possibility that those with access to the active powers of nature would claim active political powers as well.[24]

The cosmopolitan celebrations surrounding Franklin now broke down, dividing into parallel discourses of patriotic nationalism and (as we shall see below) imperial treason. "A Subject of the Crown of Great Britain," President Macclesfield of the Royal Society had grandly declared Franklin in 1753. The image of Franklin as a British gentleman-philosopher was indeed repeatedly emphasized in London portraits of the 1760s. In Edward Fisher's 1763 mezzotint (after Mason Chamberlain's oil), for instance, a periwigged Franklin stands safely in his book-lined study while a violent storm rages outside (Figure 7). As late as 1772, commentators like the Venetian Jesuit Giambattista Toderini praised the Philadelphian for his "*grand esprit, propre de la Nation Angloise.*" With the Revolution, however, Franklinist electricity denoted American genius (even if the hallmarks of that genius—being practical and civic-minded—were in fact no different from the characteristics of the vaunted genius of enlightened Britain). Suddenly, in the early years of resistance, Americans claimed the lightning rod as one of their own inventions. In 1771, the *Pennsylvania Gazette* proudly published a list of "American Inventions" in which a prominent place was reserved for "Electrical point Rods." "From the current of the Gazettes, it is obvious to remark, of what wonderful efficacy in pulling down tyranny, a Committee of Correspondence is likely to be," another *Gazette* writer remarked in 1780. Now the Irish and the English were both following suit "after the example of the

Americans, who first taught [its] use." He faulted the English, however, for not showing due gratitude to their ingenious cousins across the Atlantic. "The glory of the invention is certainly ours," he declared, "as that of electrical rods, Hadley's Quadrant, or inoculation for the smallpox." Americans were now exemplary innovators in scientific and political technologies alike. "Discoveries in science have but slowly beamed upon the world," noted the *Royal Pennsylvania Gazette* in 1778. Genius was the sine qua non of great civilizations, old and new. "England boasts her Harvey . . . and her Newton, as the grand expounder of the laws of the planetary system." America, too, now "steps forward, as the sole discoverer of the identity of lightning and electrical fire."[25]

The Revolution also transformed the body politics of visually representing Franklin. After 1776, the British gentleman-philosopher was made to spring from the protected detachment of his study into electrical nature itself, now its heroic interlocutor and commander. Fragonard's allegory of the lightning rod, *Au Génie de Franklin* (1779), stages this scene with unabashed neoclassical grandiloquence: Franklin appears as Jupiter, monumental of body atop a celestial throne, meting out thundery justice to Britain (figured as Tyranny and Avarice)—the British gentleman turned American avenger (Figure 1). Benjamin West revisited this (now) revolutionary moment in the early nineteenth century, although in a more homespun idiom. In *Benjamin Franklin Drawing Electricity from the Sky*, he takes the viewer back to Franklin's demonstration of the identity of lightning and electricity, a crucial moment of sacred bodily knowledge, where Franklin draws a spark from the kite string with his finger (Figure 6). As with Fragonard, where Franklin defends the American republic (incarnate to his left), West's postrevolutionary allegory situates Franklin specifically as an American: witness the putti with Native American headdress in the background. But unlike Fragonard's deification of reason through Franklin, West's Franklin is very much a man rather than a god (and a rather worn one at that, owing perhaps to the fact that West did not paint Franklin from life). In both cases, however, the American Franklin is a man of corporeal virtuosity whose mastery of nature transcends mere (read, European) book learning.[26]

Representations of Franklin as a physically heroic genius incarnating a new relationship between knowledge and power became ubiquitous. While others were "falling martyrs to their experiments" (a dig at poor Richmann), Franklin defied all such dangers. He held "th'electric tube red glimm'ring in his hands," "seized" [the lightning] glimm'ring from the darken'd sky," and, recalling Turgot's epithet, "from tyrants snatch[ed] the rod." Franklin's example glittered for all of America's "unborn sages," insisted the diplomat and later bellettrist Joel Barlow: "Bid hovering clouds the threatening blast expire / Curb the fierce stream and hold th'imprison'd fire!" Inevitably, politicizations of electricity looped back into electrifications of politics. In a poem called "Liberty Fire" (1792), one "Populus" connected the two rhetorical poles of revolutionary politics and electricity, synchronizing Franklinist genius with universal natural power. "Latent long, and undetected, / Lay this heav'nly fire electric: / *Franklin* drew it from the skies, / Flashing Freedom in our eyes." "All nations [were] "now excited," Populus declared, as sparks flew from "minds *ignited*." These rivers of propagandistic ink made it impossible to separate Franklin's science from his politics. Franklin was the "Heaven-instructed Sage, skill'd in the bolt and balance of Jove," waxed one writer for the magazine *The Nightingale*, who framed "Revolutions on the Axis of Liberty and Law." As ever, Erasmus Darwin was not to be outdone. "Immortal FRANKLIN," a "young Sage" around whose head nymphs had "wreath'd the crown electric," had snatched Jove's lightning, bent his bolts over his knee, and "stabb'd the struggling Vampires" of British "Tyrant-Power" with his iron points. The rhetoric went on and on. Franklin was the icon of enlightenment in science and politics, an exemplar of world-historical significance. As we have seen, this was no automatic consequence of his experimental accomplishments; it was, rather, the belated result of an explosion of national self-definition.[27]

Political Electricity: Negative

The Revolution brought not only political and military conflict, but also a literary war of representation and counterrepresentation. The voices raised within the patriotic public sphere did not go unchallenged. Loyalists to British authority in America radically opposed the naturalizing and sacralizing narratives articulated by pro-independence writers. Opponents of independence also stressed Franklin's agency in science and politics, and his physical virtuosity. In doing so, however, they used questions of body to challenge his intellectual and moral status. To the enemies of independence, American Patriots' spontaneous knowledge of liberty-as-electricity was no more than a delusion: a deliberate work of interested, conspiratorial, enthusiastic art.[28]

Because Franklin embodied a revolutionary relationship between science and politics, rejecting his republicanism meant dismissing his natural philosophy as well. In 1777, when he was American minister in Paris, an anonymous pamphlet appeared in London examining his "pretensions to the title of natural philosopher." "Ever since the American disputes engaged so much of the public attention," its author began, "we have had our ears stunned ... with the surprising discoveries of so great a Philosopher as you have been represented." According to this critic, however, Franklin was not really a natural philosopher at all. Issues of class, body, and mind were brought together at once: Franklin possessed no rational genius, only the knack of a low mechanic. He was a man of body, not mind. The image of Franklin as backwoodsman sage, so winning among America's French allies, was hypocrisy, a tool to gain a "prodigious" reputation for what were in truth very modest accomplishments. His central failure, in contrast to

Newton's achievement with respect to gravity in the *Principia* (1687), was not to have reduced electricity to mathematics. Since Franklinism did not allow one to measure quantities of electrical force, it did not deserve the label "philosophy": "if a man cannot compute the effects, all his reasonings from them are but mere conjectures, and his finest conjectures only sports of the imagination." This metropolitan rejoinder turned the "sage of Philadelphia" back into a natural historian who collected experimental data and (precisely contrary to the praise Smith and Jeffrey would later offer) passed off conjectures as facts. In sum, Franklin was guilty of that most heinous eighteenth-century intellectual crime: system-building. In ringing failures of imagination management and philosophical modesty, Franklin had imposed his pet theories on natural phenomena, rather than allowing nature to speak for itself. He was like the gentleman who, having examined worms through a microscope, applied his "Theory of Worms" to everything, his mind "so much elated, and his imagination so possessed with his late discoveries, that everything must be converted into Worms." "Your *Seas of Globes and Cushions* (as you express yourself) are to be found every where, like the abovementioned gentleman's Worms, rubbing against each other, electrifying all nature, and producing every phaenomenon." He was seduced by his own overheated imagination into abandoning inductive philosophy for "phantoms" and pranks. How Newton "would lament to see this nation gazing, with admiration, at Electricians feasting on turkeys killed by electrical shocks, and styling themselves Natural Philosophers, and their feasts philosophical banquets!"[29]

Franklin's physical virtuosity now evidenced his philosophical and political fraud. Too many practitioners had stepped "out of their road to turn Philosophers," his anonymous critic regretted. Denigrating Franklin's physical prowess as a worker of machines expressed an acute class anxiety about power falling into grubby, dangerous hands. His dexterity was invoked to disqualify him, with a Platonic scorn for all things corporeal, as a true philosopher. Such men of body were "fit to be employed as hewers of wood and drawers of water for the service of the temple, tho' by no means proper to be admitted to minister at the altar." Franklin's mastery of objects and apparatus was part and parcel of the "modern method of *handling* Natural Philosophy," which nonetheless fell short of proper causal explanation. Such "modern Philosophers," "illiterate" men attended by "wagon[s] loaded with things," made a pitiful spectacle. Their penchant for "blowing up bladders in an air pump, or for drawing sparks from an electrical-machine" did not make them men of science. Body work should support the reasoning mind, like a scaffolding to a building, but should never stand alone: "were it to attend it constantly, it would be a monstrous incumbrance, and prevent the discovery either of its beauty or proportion." The result was experimental enthusiasm: Franklin was one among "swarms of Philosophers," inflamed by a "rage for making experiments," yet lacking the capacity to move beyond sensory experience to "remote consequences." Such practitioners dwelt in the domain of things "of which our sense can judge immediately," but failed to penetrate "into the secrets of nature." In the end, Franklinism meant vulgar, mindless amusement. Just look at his "mechanical and vulgar language": it was "easy to discover the *Worker at the Press*," displaying his "low breeding" for all to see. The Anglican minister William Jones of Nayland, a student of the biblical natural philosophy of John Hutchinson who lamented annually on January 30 the misdeeds of a regicide ancestor, echoed these antidemocratic sentiments. Franklin's lightning experiments were "an ominous prelude to the business he was soon afterwards to do in the world, in drawing down the fire of civil war upon his country, and spreading the confusion of anarchy over the earth." The American personified the danger of science without political control. "Philosophical men became so fond of this new art, whereby a thunder-cloud was turned into an electrical machine, that the practice became common, I think, rather too common, in many places."[30]

This diatribe by no means represented a consensus view of what constituted natural philosophy; indeed, no such consensus existed in the Enlightenment. The distinction between experiment and philosophy that was invoked (supposing the former to be mere mechanical art, and the latter rational mathematization) was particular, polemical, and misleading, since experimental philosophy in the eighteenth century flourished as the coordinated labor of head and hand. As political character assassination, however, it served the important function of stripping the traitor of his claims to natural mastery—claims that might legitimate his assertions of political mastery. Such attacks were buttressed by charges that Franklin had stolen his electrical ideas from others. In his history of the Revolution (1797), the Loyalist minister Jonathan Boucher insisted that Franklin had "stolen from an Irish gentleman, of the name of *Kinnersley,* many of his useful discoveries respecting electricity." It was not Franklin's genius that was remarkable, only the reputation that the "plagiarist" and "servile imitator" had fabricated. Depictions of Franklin as a thief, both critical and admiring, competed in accounts of the famous "Cockpit Scene" of January 29, 1774, when Solicitor General Alexander Wedderburn examined him before the Privy Council, in connection with the publication of private correspondence concerning the reinforcement of imperial authority in Massachusetts after the Boston Tea Party. In one section of this notable tirade, Wedderburn scorned Franklin as a *"homo trium literarum":* a man of three letters (*fur,* Latin for thief) and rebuked him for violating gentlemanly codes of epistolary privacy. In the war of revolutionary representation, however, all such accusations were subject to rebuttal. Some time later in the London *Public Advertiser,* an anonymous writer (possibly Franklin himself) granted that the American was indeed a thief although, along with "his greatest Admirers on the Continent" and "all the Philosophers of Europe," a singularly virtuous one, who stole the flame of liberty to ignite a new republicanism. Franklin as *fur* easily became Franklin as *vir*—the other three-letter moniker that became attached to him, as in the legend underpinning Joseph-Siffred Duplessis's 1778 portrait, which also featured the now obligatory lightning bolt. Franklin's manliness symbolized his virtuous defense of liberty. "It is universally believed in France that [Franklin's] electric wand has accomplished all this revolution," quipped John Adams with his customary envy. An anonymous late-century French watercolor made the same point more emphatically (Figure 12). This remarkable design, possibly of Masonic provenance, portrays the sexual prowess of the experimental philosopher who ravishes "America," while what appears to be an electrical machine sits in the background: Franklin as philosophical, political, and sexual plenipotentiary.[31]

Bodies figured in defenses of Franklin as much as in attacks. In May 1774, the *Pennsylvania Gazette* reported a pointed local sequel to the Cockpit Scene played out in the streets of Philadelphia. Resentment boiled at Wedderburn's mockery of "the venerable Dr. Franklin['s] Knowledge in Philosophy, universal Benevolence, just Sentiments of Liberty," and attempts at transatlantic reconciliation. In reply, a group of townspeople (possibly led by Kinnersley) made effigies of "the base born solicitor," as well as Governor Hutchinson (whose role in suppressing Massachusetts Franklin had exposed), parading them through the streets in a cart for several hours, recalling the Stamp Act protests of 1765. The effigies were brought to a gallows near the London Coffee House, where tickets for Kinnersley's electrical performances had been sold until just the previous year. Here they were hung and finally "burnt by ELECTRIC FIRE."[32] Franklin's critics were hardly deterred. Rather than deny his genius, some emphasized its corruption. In his sparkling invective in *The Origin and Progress of the American Rebellion*, written during the war for independence, the Massachusetts Loyalist Peter Oliver complained that "Dr. *Franklin* (pardon the Expression) . . . was a Man of Genius, but of so unprincipled an Heart, that the Merit of all his political and philosophical Disquisitions can never atone for the Mischiefs which he plunged Society into, by the Perversion of his Genius." Oliver quoted at length a poem by the prominent Loyalist Jonathan Odell, physician and Anglican minis-

12. Franklin as Plenipotentiary

Electricity and sex enjoyed intimate relations in the Enlightenment. Public demonstrations featured "electric kisses" from the "Venus electrificata," electric eels inspired erotic poems, and natural philosophers theorized sexual activity as electrical. This French watercolor, a vision of libertine enlightenment and an example of revolutionary-era propaganda, shows a bespectacled yet manful Franklin "ravishing" a female figure representing America, with what appears to be an electrostatic generator in the background. The image plays on Franklin's sexual reputation to depict him as the progenitor of the American republic. The symbols in the frame may signal freemasonic provenance and connotations.

ter, published in the *Gentleman's Magazine* for 1777. In Odell's view, the "Spark" that kindled the "Blaze of Sedition," lit by Franklin, came from Lucifer himself in the form of "a degrading Ambition." Ambition had substituted personal interest for the common imperial good. This son of a tallow chandler, who had secured the position of colonial postmaster (worth £500, Oliver noted), was devoted only to his own advancement. He was, in Wedderburn's view, "the true incendiary" of American affairs, using a "secret cabal" to "blow up the province [of Massachusetts] into a flame." It was precisely by masking his self-interest in the guise of a virtuous republicanism that Franklin had given "such a Shock to Government, and brought on such Convulsions, as the English Constitution will not be cured of in one Century, if ever."[33]

The charge of treason affected British debates about the trustworthiness of Franklin's science. In the early 1770s, a royal commission examined the question of whether lightning rods to protect royal palaces and military establishments should terminate in points (Franklin's design), or knobs (the view of Franklin's rival, Benjamin Wilson). After heated debate, Franklin's design was retained against Wilson's claim that points caused lightning to strike where it otherwise would not. The matter lay dormant until 1777, but in that year, Franklin rods failed to protect the Board House at Purfleet Ordnance on the River Thames. Wilson seized the opportunity, redoubling, and now politicizing, his insistence that knobs were safer than points. The dispute became a party affair between rebels and Loyalists; Wilson compared the physical danger of points to the political threat of America. Points, he wrote, had been "greedily adopted in England, at the time when Mr. Franklin was an Englishman." But since "he had ceased to be so," and was now "one of the Chiefs of the Revolution," Britons should discard "the discovery of an enemy" and switch their points for knobs. The confused sequel to this controversy seems to have led George III to order this substitution, although the order was not carried out. Once again, pro-American wits demanded the last word. A poem circulating in the British and American press supplied the philosophical quarrel with a political moral: "While you great George for knowledge hunt; / And sharp conductors change for blunt, / The nation's out of joint. / Franklin the wiser course pursues, / And all your thunder harmless views / By sticking to the point."[34]

Franklin's body also became a key site where evidence of the American conspiracy would be uncovered—unsurprising, since Franklin's body was a crucial instrument of his genius. As we have seen, Franklin was adept at manipulating his own body, as though it were a machine. But to what end? According to critics, his body was a machine for disguising interest as virtue and, even worse, formed part of a hidden circuit of political-electrical power. Franklin, Wedderburn raged, was "the first mover and prime conductor of this whole contrivance against his Majesty's two Governors" in Massachusetts: the political power of the American conspiracy flowed through his body like an invisible electrical charge, threatening a shocking explosion. Franklin supposedly represented the interests of Massachusetts in Westminster in the years before independence. But in reality he had "made the assembly *his* Agents in carrying on his own secret designs." Franklin's calm during his Cockpit examination—in this best known of many public silences, he declined to speak in his own defense—veiled "the coolest and most deliberate malevolence." The "bloodthirstiness" of the African was here "surpassed by the coolness and apathy of the wily American." Such "cool" inscrutability was a source of profound anxiety: how could Englishmen protect themselves from being deceived into a confidence that this Creole degenerate might exploit for his own selfish ends, such as in the publication of the Hutchinson letters? Peter Oliver resorted to physiognomy, the late-century vogue for reading moral character in facial characteristics. The "Features of his soul" were so marked on his face that "a Gentleman, whose Acumen was so great as to strike out a Character from a very slight view of a Face," later concluded "that he was calculated to set a whole Kingdom in a Flame." (The leading physiognomist Johann Caspar Lavater interpreted the same face as "the model of a Thinker of singular sagacity and penetration," however.) Wedderburn's solution was more sanguine: "Mark (and brand) the man," that he might never again enter polite society "with an unembarrassed face."[35]

But the British could not control Franklin's body. To Franklin's supporters, his protean body was a source of admiration and power. In the public gallery at the Cockpit in 1774, besides such literati as Priestley, Burke, and Jeremy Bentham, sat Edward Bancroft, the Massachusetts physician, electrician, and naturalist who under Franklin's sponsorship had secured a fellowship at the Royal Society and become an American political operative in London. According to Bancroft's meticulous recollection, Franklin "stood *conspicuously erect,* without the smallest movement of any part of his body. The muscles of his face had been previously composed, so as to afford a placid tranquil expression of countenance, and he did not suffer the slightest alteration of it to appear during the continuance of the speech in which he was so harshly and improperly treated . . . he kept his countenance as immovable as if his features had been made of *wood*." To weather the storm, Franklin turned himself into a lightning rod: "the Doctor seemed to receive the thunder of [Wedderburn's] Eloquence," reported the *Public Advertiser,* "with Philosophic Tranquility and sovereign Contempt." If only, the *Advertiser* lamented, "the American *Prometheus* could have call'd Fire from Heaven to blast the unmanner'd Railer." Franklin himself explained that the heat he felt was cooled by the justice of his cause. "What I feel on my own account," he wrote to a friend, "is half lost in what I feel for the publick." Unity in the cause of virtue protected all those grounded in the republican cause. "An Union of the Colonies, like an Electric Rod," averred one Philadelphian, "will render harmless the Storms of British Vengeance and Tyranny." "Virtue . . . Unanimity and Perseverance," he asserted, "are INVINCIBLE."[36]

Carrying off a full-scale revolution, however, ultimately necessitated manipulating the bodies of others as well. Where electrical politics omitted the figure of the machine to dissolve human agency into nature and the divine, the machine came to the fore in political electricity (especially its negative mode), to emphasize that what lay behind the image of immediatism and universalism in American resistance were interested social agents who produced rebellion through the art of manipulation. The link between machines and illusions was evident in skeptical accounts of evangelical awakening during the Enlightenment. "There is a very extraordinary and uncommon power" in conversions, Valentine Rathbun asserted in 1781. "A strange power begins to come on, and takes place in the body, or human frame, which sets the person a gaping and stretching; and soon sets him a twitching, as though his nerves were all in a convulsion. I can compare it to nothing nearer its feeling than the operations of an electerising *machine:* the person *believes* it is the power of God, and therefore dare not resist, but wholly gives way to it." The presence of the machine in Rathbun's account suggested that "belief" in profound spiritual experience might simply be the mistaking of artificial effects for spontaneous ones. Revelation might be nothing more than enthusiasm.[37]

Mechanical delusion also emerged as a political theme for Loyalists. In a counternarrative of the rebellion that has been all but ignored by nationalist literary histories of the Revolution, American resistance to British authority was portrayed as springing from the passions of the mob, wound up to the heights of enthusiasm by their new masters.[38] Well before Burke's antirevolutionary *Reflections on the Revolution in France* (1790), Loyalists during the American Revolution were the first to recast the problem of enthusiasm in secular political terms.[39] The Anglican dean of Gloucester Josiah Tucker blamed *"Mock-Patriots"* and "Republican-Zealots" who opposed the new taxes of the 1760s for "[mis]representing the poor Stamp-Act as a Monster . . . destructive to the human Race." To Peter Oliver, the Stamp Act riots were a "political Enthusiasm," a terrifying "Frenzy of Anarchy" that divided neighbors and poisoned even the pious with a thirst for violence. The author of the London pamphlet *The Delusive and Dangerous Principles of the Minority* (1778) decried the "political madness" of the Americans.

American republicanism was not the spontaneous work of nature and God, but a delusion, deliberately manufactured by demagogic mobocrats using the "innocent well-meaning farmers" of America as machines to generate their own independent power base. The colonists had been led awry by "false American prophets," "designing and turbulent men, [who] in order to promote their own ambitious views and party purposes" created "real from imaginary danger." Oliver painted the most vivid portrait of this monstrous hydra of conspiracy and enthusiasm. The American people "were like the Mobility of all Countries, perfect machines, wound up by any Hand who might first take the Winch." The conspiratorial vanguard in Massachusetts deceived the clergy, who deceived the People, "weak, and unversed in the Arts of Deception." Thus "the Wheel of Enthusiasm was set on going, and its constant Rotation set the Peoples Brains on Whirling; and by a certain centrifugal Force, all the Understanding which the People had was whirled away." These machines generated political electricity of sufficient moment "so as to disunite kingdoms," thanks to the concealed double agency of the machinator Franklin. He was, declared Wedderburn, "the actor and secret spring, by which all the Assembly's motions were directed, the inventor and first planner of the whole contrivance."[40]

The remarkable London broadsheet "Political Electricity; or, An Historical and Prophetical Print in the Year 1770," visualized the conspiracy narrative on the eve of revolution, in a series of thirty-one separate designs representing Britannia's distresses (Figure 13). Ministers gamble away public monies; physicians examine the corpse of a man

---

13. Electricity and Revolution

The American Revolution politicized Franklinist electricity as radical, republican, and dangerous. As a language of power, electricity provided an ideal metaphor for political power, especially hidden power. This exquisitely detailed broadsheet circulated in the London press in 1770, featuring over thirty scenes depicting the corruption and decline of Britain and its empire, and predicting their destruction—witness the carving of the British "Lion." Notice too the "electrical chain" of conspiracy that connects several of the panels.

ELECTRICAL POLITICS AND POLITICAL ELECTRICITY

Political Electricity, or An Historical & Prophetical Print in the Year 1770.

killed during riots over the election of John Wilkes, scourge of parliamentary corruption; Wilkes himself, scion of British liberty in the age of George III, stews helplessly in King's Bench Prison; the ports of London lie in disuse; the Royal Exchange is turned into a wilderness; and the Great British Lion is about to be carved up, while Boston is renamed "London"—a prophecy of the passing of liberty, industry, and prosperity to America. The separate designs do not represent separate events, however: they are linked by what the print's key describes as an "Electrical Chain." Returning us to the Jacobite joke about resistance disguised as loyalty with which this chapter began, the chain invokes electricity as the occult medium of imperial politics, a metaphor for the concatenation of hidden influences and conspiratorial designs that will blow the empire apart. Like laboratory electricity, political electricity flowed invisibly from body to body, visible only in communication. From the very highest figure of state, King George III (at the top of the image), the chain passes "over the water" to George's favorite, the Scottish Tory minister Lord Bute, widely reviled as the incarnation of ministerial corruption in the 1760s. Often depicted by satirists as a boot, Bute here appears in the singular form of an electrical machine, his head a glass cylinder being cranked by Britain's avowed enemies: "His Body ye Electrical Machine shaking hands with ye Principal Nobles in France" (Figure 14). (Below the image of Bute, meanwhile, a miniscule figure innocuously flies a kite between the British and French coasts—probably the first ever such image of Franklin.) Having passed through the bodies of ministers, the electrical chain terminates in the barrel of a musket, which it discharges, killing a protester outside King's Bench Prison, with Wilkes looking on from his cell window. The electrical chain *is* the conspiracy narrative: it reveals for one devastating instant the explosive connection between corruption, violence, the loss of liberty, and the destruction of empire.[41]

The metaphorical relations between electricity and politics generated by the Revolution were meaningful because, rather than being arbi-

14. THE CONSPIRATORS

During the Revolution, American patriots created a language of electrical politics in which they likened republican feeling to spontaneous experiences of bodily electrification. But opponents of American independence denied any higher agency was at work: they claimed that self-interested human agents led by Franklin were manipulating the passions of the American populace as if operating so many electrical machines. This detail from "Political Electricity" shows a Franklinesque figure flying a kite off the coast of France; Lord Bute, a potent symbol of British ministerial corruption, is depicted as an electrical machine being cranked by French noblemen. The ship symbolizes the international circulation of dangerous political-electrical power.

trary or infinitely plastic, they expressed distinct yet coherent ideological visions. Both electrical politics and political electricity used bodily electrification as a way of talking about hidden mechanisms of political power. Electrical politics used an idealized reimagining of electrification to celebrate republican resistance to Britain as the spontaneous and wonderful work of divine natural powers. It was not its association with the art of experimental science that lent this language its power; it was useful, rather, as an articulation of unmediated synchronicity between natural or divine will and the actions of human beings. By contrast, in its positive incarnation, political electricity stressed a monumental human agency: Franklin's prowess as rational philosopher and revolutionary new man. Emphasizing individual agency, however, was a strategy pursued with equal zeal by the opponents of independence. Loyalists inverted the logic of electrical politics: bodies were not transparent sites where the motive power of Providence was felt and displayed, but remained troublingly opaque, concealing hidden designs and subject to artful manipulations. To Loyalists, the American rebellion revealed the immense power, and danger, of wonder. The rebellion was no more than an artificial contrivance to secure power to a narrowly factional interest, organized by conspiracy, and founded on enthusiasm. Electrical politics, in other words, was really only a form of political electricity.

CHAPTER FIVE

How to Handle an Electric Eel

We became, in truth, convinced of the difficulty of handling this fish.

ALEXANDER VON HUMBOLDT AND AIMÉ BONPLAND, 1800

DEMONSTRATIONS OF electricity had been performed in British America since the 1740s. But while looking through the *South Carolina Gazette* in June 1774, Alexander Garden, a Scottish naturalist and physician resident in Charleston, discovered a notice for an altogether novel form of electrical spectacle:

> To be Seen at the house of Thomas Adamson, In Meeting Street, at the Sign of the Horse Mask, opposite Edward Rutledge, Esqrs. at any Hour of the Day. The Wonderful Electrical Fishes. They are natives of the Southernmost part of North America, and have never been seen before that we know of. These fish have the surprising power of darting the Shock thro' a Circle formed by any number of persons, the same as the Electrical phial doth in the Leyden Experiment . . . Many Ingenious Experiments have been made by a Committee appointed for that Purpose, by a Philosophical Society. Gentlemen and Ladies who choose to gratify their Curiosity by viewing this extraordinary Production of Nature, at the small expense of one dollar each, are desired to be speedy, as the Proprietor intends to stay but a few Days in this Place.[1]

As a recent fellow of the Royal Society (Linnaeus named the *Gardenia* for him), the aptly named Garden was eager to vindicate his induction with an article in the *Philosophical Transactions*. Despite a bout of illness, therefore, he forsook the comforts of bed to witness this new electrical wonder (perhaps he also hoped to find relief thereby: the advertisement mentioned that handling the fish was a "remarkable Cure for the Palsy or Weakness of the Nerves"). The result was a letter to the London naturalist John Ellis, one of Garden's metropolitan correspondents, published the following year. Garden's "Account of the Gymnotus Electricus, or Electrical Eel" offered a careful physical description but, like the *Gazette* advertisement, emphasized the creature's extraordinary behavior. "This fish hath the amazing power of giving so sudden and so violent a shock to any person who touches it," Garden reported, "that there is, I think, an absolute impossibility of ever examining accurately a living specimen." As something to be experienced rather than merely observed, the creature literally eluded the grasp of a natural historian whose aim was precise anatomical description. All who encountered the eel were fascinated by it. But how to handle this remarkable fish?[2]

The *Gazette* advertisement raises several tantalizing questions. These fish were "natives of the Southernmost part of North America," but where exactly did they come from? Who was transporting them, and to where? And what of their "astonishing" property of electrifying human bodies—how could an organic creature discharge its own electricity, something only hitherto possible through artificial generators? The solutions to these puzzles in turn raise a larger geographical issue: where did the history of "American" electricity happen? Despite Franklin's iconic identification with Philadelphia, we have already seen how the American electrical Enlightenment was defined by entrepreneurialism and itinerant circulation, with travelers, technologies, and techniques moving among European capitals and provincial American cities and seaports, from Newport, Rhode Island, to Antigua and Jamaica in the West Indies. By taking us to South America, the story of the electric eel confirms and extends this pattern, inviting us to see how the circulation of early modern colonial knowledge was often the result of movement across national lines, rather than through strictly national networks by which centers extracted knowledge from the periphery. Moreover, electrical fish turned out to be a vitally important discovery: they played a decisive role in the development of theories of bodily electricity ("animal electricity"), as well as the construction of the first current-generating batteries. These stories are usually told from the point of view of metropolitan science: we know much more about the European torpedo fish than the American eel. By contrast, this chapter connects narratives of metropolitan physical science (typically European narratives about laboratories) to social worlds beyond Europe and colonial natural history. In short, the colonial history of eels and experiments in the Americas is part of the history of science and enlightenment in Europe.

The Nature of Guiana

The electric eel first received sustained attention from Anglophone natural philosophers through the publication of *An Essay on the Natural History of Guiana, in South America* in London in 1769. The author was a physician named Edward Bancroft who had lived in Guiana roughly between 1763 and 1767. Bancroft's experimental history of the eel must therefore be understood in the social context of Dutch Guiana, and the intellectual context of early modern natural history (which in this case included not only Guiana's topography and natural productions, but also its indigenous peoples and slaves). Guiana had long possessed a mythic status in the European colonial imagination, particularly for the English, owing to the late-sixteenth-century efforts of Sir Walter Ralegh to reach the mythical El Dorado—a city made entirely of gold—and establish an English beachhead in the Iberian-dominated Americas. Although the English, French, and Dutch were all trading in Guiana shortly afterward, it was the Dutch who established an enduring presence in the seventeenth century, under the auspices of the Dutch West India Company, founded in 1621. Despite the relatively small size of the Netherlands, the Dutch became prime commercial movers in the seventeenth-century Atlantic, carrying English tobacco, breaking the Portuguese monopoly on the slave trade, and establishing colonies in the Caribbean and North America (and for a brief period in Brazil). Bordered to the east by the French colonies of Cayenne and French Guiana, to the west by New Spain, and to the south by Portuguese Brazil, Dutch Guiana was composed of four separate settlements: Demerara and Essequibo, which were run by the Company, and Berbice and Surinam (the latter founded by the English in 1651), which were effectively run by consortia of private Dutch interests in concert with the Company (Figure 15). Dutch Guiana's stability was precarious, however. Racked by raids from French privateers and the effects of the Anglo-Dutch wars in the mid-seventeenth century, Surinam was taken back by the English in 1667, only to be exchanged for New Amsterdam in the Peace of Breda (the Dutch settlements in Guiana would be officially ceded to the British in 1814).[3]

A highly contested colonial site, Dutch Guiana was also convulsed by the source of its own wealth: the "black gold" of slave labor. Sugar had been introduced in the Essequibo region in the 1630s, and together with coffee, cocoa, and cotton, provided a powerful incentive to plantation investment. The Dutch West India Company actively encouraged settlement, freely granting land to petitioners while ensuring that much of the produce benefited the mother country (for example, by restricting exports from Surinam and Berbice to the province of Zeeland in the Netherlands). The Dutch turned to African slaves to exploit Guiana's fertile coastal lands, producing a regime whose harshness resembled that of the sugar islands of the Caribbean. The immense profitability of sugar cultivation reoriented the economic life of the settlements away from interior commercial relations with Guiana's indigenous populations to plantation agriculture on the coast. The predictable consequence, however, was an implacable antagonism be-

15. An Electric El Dorado

Guiana had obsessed European travelers since the sixteenth century as the mythical home of a hidden city made entirely of gold. In the eighteenth century, it became a key site for experiments with electric eels that inhabited the Essequibo, Demerara, and other rivers in Dutch Guiana (now Suriname). Edward Bancroft of Massachusetts, the first to argue in English that numb-fish used electricity to stun their prey, worked in Guiana as a plantation doctor in the 1760s.

---

tween slave and master that produced violent uprisings and the formation of maroon communities of free blacks who encouraged further insurrections. One of the largest of these took place in Berbice in 1763, the year Bancroft landed in Essequibo. His *Natural History* describes a revolt of "thousands" who "massacred a considerable number of the white Inhabitants," with thousands more in the neighboring settlements of Demerara and Essequibo betraying a disturbingly "eager . . . disposition for revolting."[4]

The suppression of this rebellion depended on a set of intercolonial relationships that made possible both the prosperity of the Dutch and Bancroft's presence among them: cooperation with British colonists from the West Indies. Beginning in the 1730s, the governor of Essequibo, Laurens Storm van 's Gravesande, introduced tax exemptions to encourage British Caribbeans to invest their capital and expertise in sugar cultivation. Consequently, by midcentury British plantations and sugar mills had become integral to the wealth of the settlements. British Caribbeans came readily to defend their property when it was threatened by the uprising of 1763: before Dutch ships from St. Eustatius could arrive to bolster local Dutch defenses, British ships sent by Gedney Clarke, the customs collector at Barbados, arrived to protect British plantations in Demerara (though they did not see active service on that occasion). In the rebellion's aftermath, several hundred slaves were hanged or broken on the wheel. Economic and military interest, and racial solidarity in the face of a large black majority in an environment that encouraged slave resistance, thus opened the Dutch settlements to British colonial fortune-seekers.[5]

Edward Bancroft was one such fortune-seeker. A shadowy, restless, and elusive figure, he represents a fascinatingly ambiguous variation on Benjamin Franklin's model career of colonial self-making, especially given that Franklin was to become his friend and mentor. Bancroft's youth, like Franklin's, had been humble. Born in Westfield, Massachusetts, in 1744, he lost his father early on and was apprenticed to Dr. Thomas Williams of Lebanon, Connecticut, before becoming discontented and running off to sea at nineteen. "Insults received, a Haughty Disposition, and a Roving Fancy, conspired in effecting this adventure," he explained in a letter to Williams upon his arrival in Guiana in 1763 (after failing to find employment in Barbados). He later fictionalized this sequence of events in *The History of Charles*

Wentworth (1770), a novel (one of the very first published by an American) about an Englishman seeking his fortune in the West Indies and Guiana. Bancroft worked in Guiana as a general physician and surgeon, evidently on several plantations including that of a New Hampshire man named Paul Wentworth. Medical care for slaves was especially valued late in the century given the rising price of slaves and that humanitarian campaigners were calling for improved conditions. "I obtained in three days after my arrival the Employment of Surgeon to a gentleman of Fortune," Bancroft told Williams, "Owner of two large Plantations and Near two hundred Slaves in this River." When he arrived in London in 1767, he became apprenticed at St. Bartholomew's Hospital before taking a medical degree at Aberdeen, but it was Franklin's friendship—secured in part through a shared interest in electricity—that was decisive, allowing him to contribute articles to the *Monthly Review* and helping him to land a fellowship at the Royal Society (1773). In addition to the *Natural History* and *Charles Wentworth*, Bancroft published a pro-American political pamphlet in the early 1770s, and became secretary to the American delegation in Paris at the outbreak of the Revolutionary War. But the twist in this tale was dramatic: Bancroft betrayed Franklin and his associates, becoming a Loyalist and a spy who reported on American negotiations with the French during the Revolution.[6]

Bancroft is best known for the allegation (never proven) that he used his knowledge of Guiana's poisons to murder Silas Deane, the senior diplomat who had befriended him in his New England youth, and who may have been on the verge of exposing him in 1789. Indeed, scholars have neglected Bancroft's early colonial years and his career as a natural historian and experimenter, but this is precisely what interests us here. Enlightened natural history was distinctive for its systematic attempt at universal taxonomic classification. Several systems competed for adoption, but by far the most widely followed was that of the Swedish naturalist Carolus Linnaeus, described in his

Systema Naturae (1735), which organized flora according to sexual characteristics, employing a binomial method of linguistic identification (Linnaeus and others also sought to taxonomize human beings). Bancroft's natural history, however, demonstrates the persistence and vitality of older forms of early modern chorography through and beyond the Enlightenment. Chorographies were promiscuously inclusive natural and civil histories of domestic European localities that described everything from natural productions to social customs and political structures. The natural history of the colonial Americas was written in this heterogeneous vein, driven by a European desire to organize New World nature as a material resource for the metropolis and to understand the ethnography of native and enslaved peoples. While the simplicity of the Linnaean system stimulated both amateur and professional botanizing, and while Linnaeus actively coordinated the far-flung efforts of volunteers and traveling agents sent from Uppsala, natural history remained more than the will to classify; rather, it continued to accommodate and express a range of drives and purposes. Not the least of these was to transmute the experience of intercolonial travel into descriptive narratives for both learned and polite metropolitan audiences.[7]

Bancroft's was not only an era of classification; it was also one of imperial improvement. But he was not the agent of any specific European taxonomic or imperial project, and this is evident in the *Natural History*. In the late eighteenth century, natural history and botany became increasingly linked to British imperial interests under the leadership of Joseph Banks, president of the Royal Society from 1778. Banks not only sailed to the Pacific Islands with Captain James Cook; he also created a transoceanic network spanning the British empire that attempted to establish Linnaean economic botany as profitable national policy (Captain Bligh's attempt to move breadfruit trees from Tahiti to the West Indies is one example of the potential imperial utility of botany). In response to a series of grave crises including the Seven

Years' War, the American Revolution, Caribbean slave rebellions, and the Napoleonic wars, British imperialists used moralistic languages of agrarian improvement to justify their ongoing colonial projects in theaters like South Asia. Unlike the Linnaean or Banksian agents directed from Uppsala and London to collect, classify, and improve, however, Bancroft was a free if contingent agent: he was an American-born Creole who directed himself to the West Indies and Guiana, following opportunity wherever he found it. This relatively autonomous social position shaped both the structure and ideology of his *Natural History*. Since his intention was only to describe those plants, creatures, and peoples about which little was known, he identified Guiana's flora and fauna by reference to similar species described in the writings of John Ray, Hans Sloane, Mark Catesby, and Linnaeus, freely mixing Latin, English, and native names rather than adhering to a single taxonomic system. In fact, he explicitly claimed to have "inverted the order usually followed by Naturalists" in describing Guiana's topography, plants, and animals before its human population, an order he thought "more natural." To describe man without "premising the means by which his Creator has provided for supplying [his] wants," he reasoned, "appears to me unnatural." As a journeyman struggling to secure his fortune, an acute awareness of "wants and desires" shaped his view of both the social and natural economies of the colonial world.[8]

As part of its claim to speak authoritatively about the operations of nature, the discourse of experimental philosophy, within whose terms electricity typically fell, offered a "view from nowhere." Franklin, for instance, wrote about experiments performed in colonial Philadelphia, but his letters to Collinson described socially and geographically decontextualized phenomena, both assuming and advancing the notion that geographical and social provenance were irrelevant to their credibility. By contrast, Guiana's sociogeographical situation was the very stuff of Bancroft's descriptive natural history. He began by explaining the territory in terms of specific geopolitical contests that constrained the observer: he could only describe the Dutch settlements of Guiana, he pointed out, not the surrounding ones of Spain, Portugal or France, because those were "inaccessible to Foreigners." Decrying the "unpardonable indolence" of the profit-obsessed settlers, "few of whom have ever penetrated the woods farther than the confined limits of their plantations," he described his travel into the South American interior upriver along the Essequibo, leaving the coastal plantations behind him. His account would compel interest and credit as one of personal sensory experience that transcended mere eyewitnessing. Dismissing the testimony of others as "information" unfounded on experience, he explained to his reader, "I have spent many days . . . investigat[ing] the nature and qualities of these plants . . . by handling, smelling, tasting, &c." In a country rich in poisons, he presented himself as willing to taste nature to know it firsthand. After consuming the nuts of the *Ricinus Americanus* or Physick Nut Tree, for example, described to him as an emetic, he concluded: "I believe it from my own experience."[9]

But Bancroft openly acknowledged the fallibility of the senses in Guiana. His was not a decontextualized catalogue of flora and fauna, but a straightforward narrative that moved through tortuous social and epistemological terrain, a history of travels as travails (a "dangerous and almost fruitless endeavour," he called them). Nor was his a narrative of European mastery over colonial nature. To the contrary, Bancroft emphasized how Guiana overwhelmed the senses of the naturalist: "I have frequently found almost all the several senses, and their organs either disordered or violently affected, without being able to determine to which of the many subjects of my examination, I ought to attribute these uncommon effects." Disorder reigned in the *Natural History*. Unlike the botanical imperialists who succeeded Banks in the nineteenth century, Bancroft was not a prophet of improvement. Lacking the material support of a metropolitan sponsor or the ideology of orthodox Christian faith, he wrote instead of the precariousness of the human position within Guiana's natural economy—a precariousness informed by his own dependent status as a journeyman, the volatile character of Guiana society, and a pronounced ambivalence regarding the respective merits of European and Native American civilizations. Like other colonial American natural histories by Sloane, Catesby, and William Bartram, Bancroft's teemed with natural dangers, from the armies of ticks or "chiggers" that ravaged the feet of slaves, to the twenty-foot alligators whose skin was impervious to musket shot, to the fierce *Peri* fish who mutilated everything that went in the water, from ducks' feet to human genitalia.[10]

The wealth of snakes in Guiana staggered Bancroft, inspiring not the pious contemplation characteristic of natural theology, but a natural-historical modesty acutely expressive of human impotence. "Unhappily their immense number and variety constitute one of the principal inconveniencies of this country, and really endanger the safety of its inhabitants; and ought to humble the pride and arrogance of man, by convincing him, that all things are not made obedient to his will, nor created for his use." Mortal danger lurked in colonial nature. Natives, slaves, and settlers all slept in hammocks to avoid nocturnal visits from poisonous snakes, but they were still vulnerable to bats who "open[ed] the veins, and suck[ed] the blood" of victims, leaving them to "wake, and find themselves faint and wet with their own blood." Guiana's innumerable insects convinced Bancroft that much of the colony's nature actively defied improvement. "[They are] useless to the purposes of humanity," he concluded, "whilst the greater number are noxious to man, and consequently not created for his use." Bancroft saw the Dutch settlers not as masterful imperial improvers, but as deluded interlopers trespassing on a local natural economy to which they seemed irrelevant. "Every part of animated nature was created for its own happiness only," he wrote, and each part may have allowed an "adequate ... power of acquisition, or enjoyment" over nature, but not more. The "universal dominion" of humanity was a "flattering idea," but the notion "that all Terrestrial Beings are created for his use" was merely a "vain imagination." Personal experience of the colonies confounded projections of imperial dominion over American nature.[11]

Bancroft's freethinking vision of Guiana—that of a Creole critic of colonization—evoked a schizophrenic sublime, in which appreciation of the natural environment as a rational system (reminiscent of natural theology) mixed uneasily with sheer awe at its disorder, danger, yet exhilarating freedom. Quadrupeds bounded joyously about in "a life of rustic freedom and independence," while the ants, though regrettably lacking the "republican" organization and "provident industry" of their European brethren, enjoyed a "life of more ease and luxury" since Guiana's natural fertility required so much less labor (Bancroft drew a similar Rousseauvian contrast between the human inhabitants of Europe and the Americas). He revered the "Wisdom and Goodness" of the author of the "Order and Harmony of our material System"; even the "carnage and destruction" he witnessed in nature struck him as a functional economy. But he was also deeply impressed by what he took to be the natives' awe at the "deformities and convulsions of nature." It was convulsive nature, rather than "order, beauty, and regularity," he thought, that inspired their spiritual faith. And he, too, became captivated by the "lofty trees, whose elevated summits are obscured by the impending clouds" and the "Forests, where the liberal hand of indulgent Nature has ranged, in beautiful rustic disorder." In Romantic contrast to the artificiality of Europe, Guiana's was a moral social order based on nature, one whose dangers were outweighed by its "primaeval innocence and happiness."[12]

Experimental Natural History

One way to redress the heightened fallibility of the senses in tropical environments was to link them to norms of laboratory experiment. This is what Franklin had done in his lightning experiments, and Bancroft pursued the same strategy in his examination of the electric eel. "Numbfish" had been known since the ancient world. Artemidorus, the leading ancient authority on the interpretation of dreams, believed that dreaming of a torpedo swimming among other fish—an image that survives on Greek pottery—was a bad omen. Two thousand years before the Greeks, catfish with similar properties had been depicted in Nile River fishing scenes carved as friezes on Egyptian tombs, as at Sakkara, circa 2750 B.C. By the time of the Romans, such fish were also reputed for their medicinal value. One recipe called for torpedo prepared with coriander and sweet pickle. Pliny the Elder recommended placing the gall of a dead ray on the genitals to cool carnal passions; Galen urged torpedo bile taken orally to the same end. Pliny's contemporary Scribonius Largus was among the first to endorse the benefits specifically from the stunning power of live torpedoes placed under the feet, for the relief of such conditions as headache and gout. Such ideas, accepted by Dioscorides and others, appear to have remained in currency for centuries. And not only in the West: Ibn-Sidah, an eleventh-century Muslim physician in India, advocated placing live catfish on the brows of epileptics (how one went about this without getting shocked oneself is not clear). Jesuit missionaries in early modern Abyssinia described locals strapping patients to tabletops and shocking them with catfish as a method of expelling "Devils out of the human body."[13]

All observers since Pliny had noticed that torpedoes stunned their victims without direct physical contact. Before the seventeenth century, this ability was routinely explained by recourse to occult powers and magical virtues. In one grandiose tale, credited by Pliny, Antony's ship at Actium was stayed by the magical influence of a single *mora*. By the early modern period, such occult explanations were rapidly losing favor, however. "No one understands what power the torpedo has," lamented the natural historian Konrad Gesner in 1558. By the later 1600s, mechanical philosophers' corpuscular accounts of matter in motion made invocations of occult powers indefensible. Writing in 1654, Walter Charleton, the English Cartesian and follower of the French atomist Pierre Gassendi, attributed a microscopic "stupefactive emanation" to the torpedo as the purely mechanical means by which the fish physically affected its prey. This was also the thrust of accounts by the Italian natural philosophers Giovanni Borelli and Stefano Lorenzini, who argued along with Charleton that the torpedo struck by emitting a swift stream of "Corpuscles or *Effluviums.*" Anything, in an age of mechanical philosophy, to avoid "action at a distance"—the notion that matter could act where it was not, which troubled even Newton's account of gravity in the *Principia* (1687).[14]

Atlantic travel yielded European discoveries of new numbfish and ultimately, new electrical understandings of all such creatures. After Columbus's landfall in the West Indies, European travelers quickly noticed the existence of American numbfish. In his *History of a Voyage to the Land of Brazil* (1556), the French missionary Jean de Léry described the "dangerous and venomous" behavior of rayfish, when one was pulled into his boat and badly "stung the leg of a member of our company." The French astronomer Jean Richer wrote about a numb-eel he encountered in Cayenne on the northern coast of South America in 1671, but virtually no notice was taken back in Europe (more interesting to metropolitan philosophers were Richer's pendulum experiments on gravity). Seventy years later, the explorer Charles-Marie de la Condamine described a numb-eel he had encountered in Amazonia, but in noting the similarity of its physical effects with the torpedo's, he offered little more insight into their nature than had Richer, or even the ancients.[15] The first publicly to recognize the electricity of such fish appears to have been the French naturalist Michel Adanson, who compared the effect of the catfish he encountered in Senegal, West Africa, in 1751 to shocks given from the Leyden jar.[16] Adanson's was only a suggestion made in passing, however. It was the Dutch settlers in Guiana who were the first to conduct electrical experiments using live fish. Personal relationships spanning the Atlantic connected colonial reportage with metropolitan experiment. Hearing about the stunning effects of Guiana eels, Jean Allamand—one of the earliest Leyden jar experimenters of the 1740s—wrote to Governor Laurens Storm van 's Gravesande of Essequibo—nephew of the leading Dutch experimenter Willem Jacob van 's Gravesande—for more information about the *sidder-vis* (tremble fish). In his reply of November 1754, van 's Gravesande confirmed Allamand's suspicion based on his own personal experience: the eel "produces the same effect as the electricity, which I felt when I was with you, when holding in the hand a [Leyden] bottle fastened to the electrified tube by an iron wire." Within a few years, electrical views of the eel seem to have become common. Describing medical uses of the eel in Essequibo, a physician named Frans Van der Lott also wrote of the "shocks" of eels, comparing their effects to the Leyden jar's.[17]

Gentlemanly testimony about the sensational analogy between the effects of artificial and natural electric shocks suggested that like effects evidenced a like cause. This was the strategy Bancroft adopted in his 1769 account of the eel, the first English publication to argue definitively for its electricity. The "Torporific Eel" native to the Essequibo River was three feet long and twelve inches in circumference. It was "covered with a smooth skin, of a blueish lead colour, very much like that of sheet-lead which has been long exposed to the weather, being entirely destitute of scales." The head was flat on its upper and lower sides, with two small fins to the rear "which like the ears of an horse, are either elevated or depressed, as the Fish is pleased or displeased." The upper surface was "perforated with several holes, like those of a Lamprey Eel." The mouth was wide and toothless, and the tail was finless, although a fin three inches wide extended from the underside of the head along the belly. Bancroft broke off his natural history, however, to turn experimental philosopher, noting that the eel delivered a "shock perfectly resembling that of Electricity," either direct to the hand or through a metal rod, apparently at will. The colonial journeyman then dared to challenge the longstanding view of René-Antoine de Réaumur, the Parisian naturalist whose 1714 description of the eel ascribed its shock to swift mechanical contact. Bancroft displayed an improbable confidence: "all *Europe* have yielded an implicit assent" to Réaumur's mechanical hypothesis, but "it will be no ways difficult to demonstrate that the whole of M. Réaumur's pretended discovery is a perfect non-entity." Perhaps emboldened by Franklin's triumph over the mechanical theories of Nollet (Réaumur's pupil), Bancroft made an epistemological virtue of his peripheral status: in situ observations by travelers were more trustworthy than the dicta of metropolitan commentators. Fidelity to "great names," he insisted, led one into error and was a less sure defense against "the charms of novelty" than personal experience and the direct evidence of the senses (Réaumur had not evidently handled such a fish himself). "Whilst I have sense and faculties of my own, I am resolved to use them with that freedom for which they were given," he declared. "The shock of the Torporific Eel is not the *immediate* effect of *muscular motion*."[18]

Bancroft's experimental demonstrations of the eel's electricity closely followed those already conducted by the Dutch, from whom he had doubtless learned much. And like them, he transferred experimental techniques from the laboratory to the field, effectively substituting live eels for electrostatic machines to procure the same effects. The human body again provided the crucial venue for establishing this sensational analogy. Here is how Bancroft handled the eel: the first experiment involved circuit shocking, recalling the earliest electrical demonstrations of the 1740s and 1750s. When the eel is "touched with an iron rod, held in the hand of a person, whose other hand is joined to that of another, it communicates a violent shock to ten or a dozen persons thus joining hands, in a manner exactly similar to that of an electric machine." The second test involved a shock conducted through the water of the wooden troughs where live eels were kept: "a person holding his finger in the water, at the distance of eight or ten feet from the fish, receives a violent shock at the instant the fish is touched by another person." Bancroft's third observation strikingly related the intensity of the electrical discharge to the will of the creature. "This Eel, when enraged, upon elevating its head just above the surface of the water, if the hand of a person is within five or six inches therefrom, frequently communicates an unexpected shock, without being touched."[19]

Clearly, the fish was not an electrical machine: its discharges could not be automatically procured by physical stimulation alone but depended on the fish's conscious will (many reported the movements of its eyes in this regard): "No shock is perceived by holding the hand in the water, near the fish, when it is neither displeased nor touched." That the eel seemed to be an intelligent and self-directing agent was of great interest to philosophers with materialist inclinations, such as the London surgeon John Hunter (a dissector of electrical fish in the 1770s), who were eager to read such creatures as evidence of a vital universal force different from traditional notions of spirit or soul.[20] Yet to make claims about the kind of emission produced, Bancroft described the eel as though it *were* a machine that could be handled experimentally. According to this useful fiction, like corporeal effects demonstrated a like natural cause. The logic of Franklin's kite and sentry-box experiments again applied: the use of sensational analogy to interpret natural phenomena through laboratory experience. Based on his own physical interaction with the eel, Bancroft argued "that the shock is produced by an emission of torporific, or electric particles." "Either the mechanism and properties of the Torpedo and those of the Torporific Eel are widely different, or that Mons. *de Réaumur* has amused the world with an imaginary hypothesis." Auto-observation of bodily experience on the colonial periphery, when disciplined by protocols of the laboratory, could assert more compelling claims to knowledge than remote metropolitan conjecture. Electricity and the eel possessed an "affinity," Bancroft concluded, "not only in the sensations which they communicate, but in the medium through which they are conveyed."[21]

Printed in London, Bancroft's account of the trials conducted on the banks of the Essequibo circulated among Anglophone audiences on both sides of the Atlantic. Naturalists, physicians, and philosophers undertook to replicate his findings. Another American physician resident in Guiana, Dr. William Bryant of Trenton, New Jersey, described experiments with the eel in a paper to the American Philosophical Society in Philadelphia in February 1773. The "sudden and violent shock" delivered by the fish, Bryant confirmed, was "in all respects like that which is felt on touching the prime conductor, when charged with the electrical fluid from the globe." He too arranged a human circuit: the shock delivered convulsed all fourteen elbows "in the same manner as I remember to have been in the electrical experiment, when the several persons take hold of the wire and the equilibrium is restored by the fluids passing through their bodies." Adapting insulating techniques common in the laboratory, he found that if he held a sword blade by an end covered with sealing wax, or used a glass rod, he suffered no shock from the eel—further proof of the electrical analogy. Tangling with the eel suddenly became de rigueur for Guiana sojourners. The Scottish army officer John Gabriel Stedman cited Bancroft's experiments in his journal, destined to become the best-known account in English of eighteenth-century Guiana, but his own experience was less philosophical: "having turn'd up my shirt Sleeves, [I] tried above 20 different times to Grasp it with my *hand* but all without Effect, receiving just as many Electrical Shocks as I touched it." Stedman might literally be stunned by the eel but, by the 1770s, thanks to Bancroft (and the Dutch), the ancients' magical fish had become objects of rational experimental curiosity. If the nature of Guiana was itself wonderful, deranging, and awing to the senses, experiment provided a set of practices from a controlled environment by which to reestablish the cogency of the senses and the possibility of knowledge. Experiment was, in other words, a path out of exoticism and into enlightenment.[22]

Using Locals and Their Knowledge

Electricity in eighteenth-century America was not, on the whole, a science practiced among Native American or African American communities. British America was, of course, a society whose economy depended fundamentally on the profitability of slave labor and the triangular trade, and whose western expansion repeatedly pushed indigenous Americans off their ancestral lands. The growing wealth and leisure of British Americans, which enabled the pursuit of natural science, rested on these foundations. As we have seen, electricity was often regarded as a moral commodity for the improvement of the rational self, and in this sense electricity was largely a white urban science, for the cultivation of increasingly affluent white Creoles (though, as we have seen, there were times when it also played a role in racializing the difference between enlightened civility and unenlightened barbarism). In Guiana, the situation was different, and the role of non-European knowledge more fundamental to the business of conducting electrical experiments. Because of the need to access an extralaboratory specimen (the eel) in socially complex nonurban settings, experimenters necessarily depended on an array of assistants without whom work in this natural laboratory would have been impossible: Guiana's native and enslaved African populations.[23]

The importance of Guiana's indigenous population to the composition of the *Natural History* was fundamental. Bancroft did not present himself as a Romantic solitary traveler: he devoted a whole section to native customs, and throughout indicated how native expertise helped guide his movements and actions as a naturalist. As with his vision of Guiana's environment, he regarded these peoples with pronounced ambivalence, combining a Rousseauvian yearning for a natural social order with scorn for their "laziness" and "superstition," as well as anxiety about their potential corruption by Europeans. In his novel *Charles Wentworth,* Bancroft expressed this ambivalence through a fictional European exile named Gordon who explains to Wentworth both his admiration of Arawak customs (which lack the hypocrisy and immorality of European traditions), and his disdain for their superstitious belief in evil spirits called the "Yowahoo," which he tries to rectify (unsuccessfully) with demonstrations using electrical and chemical apparatus. In both *Charles Wentworth* and the *Natural History,* the natives are prisoners of their imaginations; their accounts of their own natural habitat, therefore, could not be taken on trust. Bancroft dismissed, for example, native accounts of the "Wild Man" (the Orang-Outang) as "being near five feet in height, maintaining an erect position, and having a human form, thinly covered with short black hair," and being given to ravishing human females. No nonnative testimony existed against which to test such an account, because Europeans preferred not to travel deep into the woods. Native eyewitnessing was untrustworthy because of the passionate nature of the observers: "I suspect that [the height of these creatures] has been augmented by the fears of the *Indians,* who greatly dread them, and instantly flee as soon as one is discovered." It was often not the fantastic features of the creature in question so much as the rational integrity of the observer that determined credibility. Bancroft doubted "fabulous" native descriptions of two-headed river snakes in Guiana, while he credited a similar account by a British soldier near Lake Champlain, New York. "This Serpent was a perfect monster, of whose existence I should strongly doubt," he wrote of the latter, "did I not think the veracity of the Gentleman, from whom I have this information, and by whom it was actually killed, unquestionable."[24]

But although Bancroft appeared to dismiss native testimony out of hand, he necessarily relied on native assistance as a colonial traveler and observer, and openly acknowledged this in the *Natural History*. Natives—Bancroft discussed the *"Carribbees, Accawaus, Worrows, [and] Arrowauks"* in turn, but often failed to specify which group he was talking about—led him into the woods to see apes, for example. Despite his wariness of their proneness to "vulgar error," he counted on them to "discover the names and properties" of snakes, which he collected. The very act of travel that brought him to the eel's habitat also depended on native guides, either Carib and/or Akawois (or Waikas), both of whom lived along the banks of the Essequibo River. Bancroft believed that since the time of Ralegh some two hundred years earlier, the Caribs had remained friendly to those who would help them fight off the Spanish. The assistance they lent the Dutch in putting down the Berbice slave rebellion was thus part of a broader commercial and diplomatic relationship involving strategic cross-cultural marriages and the exchange of canoes and wood for guns, metal tools, and beads. Since great stretches of Dutch Guiana lacked any form of public road, "the only method of traveling [was] by water, in Yatches, with convenient tents, elegantly ornamented, and six, eight, or ten oars, rowed by *Negroes,* or *Indians,*" with frequent stops at hospitable plantations."[25]

If both natives and slaves could lead Bancroft to the electric eel because they knew and navigated Guiana's waters, native prowess in fishing almost undoubtedly secured the live specimens for his experiments. "The [European] Inhabitants derive no small assistance from the *Indians,*" he explained to his readers, "particularly the *Arrowauks,* some of whom reside on almost every plantation, and are employed in various services, but especially in hunting and fishing . . . [and who] may be hired with a few baubles for several months." Bancroft drew especially on the expertise of the Akawois, who lived further inland at the source of the Berbice, Demerara, and Essequibo rivers, and whose medicine men possessed extensive knowledge of poisons, including one well suited to drugging fish. This poison, which came from the root of the Hiarree plant, was traded to the white settlers. Throwing one of these bruised two-foot roots "into a creek or river . . . when the water stagnates, is sufficient to inebriate all the fish within a considerable distance, so that, in a few minutes, they float motionless on the surface of the water, and are taken with ease." Almost all the fish taken in Guiana were captured by this method, Bancroft reported. Although some reported a strong fear of the eel among native peoples, fishing by poison made the eel a regular and "delicious" part of the native diet.[26]

Guiana's African slaves were no less important to the *Natural History*. "Agriculture, and all other labour, in these colonies, is almost wholly performed by Negroes," Bancroft noted, "as the white Inhabitants undertake no laborious employment"—exploration and naturalistic observation included. As a plantation doctor, Bancroft had considerable contact with slaves, affording him ample access to their labor, allowing him to employ them as auxiliary specimen collectors. The plantation economy and the economy of natural history were tightly related: in return for collecting dead snakes (he claimed his collection numbered some three hundred), Bancroft paid slaves in glasses of rum—the addictive product of their own enslaved toil. He also employed slaves as bird catchers. Like natives, slaves were regarded by Europeans as incapable of making knowledge in their own right. Yet Euro-American knowledge of places like Guiana could not be autonomous—its very foundation was the labor performed by local populations—and this included experiments with electric eels. Eighteenth-century accounts document the use of enslaved Africans both as experimental assistants and experimental instruments. Bryant of New Jersey, for example, noted the physical reaction of the "negro servant" who was badly shocked on emptying a tub of water previously electrified by its resident fish. The physician Philippe Fermin, who also published a description of Guiana in 1769, described the electrification of a human circuit comprised of himself and fourteen slaves: "I made everyone hold hands, and ordered the first negro to grasp the eel as firmly as possible . . . but hardly had he done so when he suffered such a violent shock to the arm that I, holding the last negro by the hand, also felt [it]."[27]

If there was an implied equality of bodily sensibility in Fermin's account—European and African feeling and communicating the shock equally—others pointed toward racial hierarchies of sensibility. Henry Collins Flagg, a South Carolina military officer in Guiana from 1782, noted that "if a number of persons join hands, and one touch the eel, they are all equally shocked, unless there should happen to be one of the number incapable of being affected by the eel." In the case in point, the insensible body belonged to a "lady" of Flagg's acquaintance, who could apparently "handle this fish at will." There was more than an air of the prodigious to this account: were ladies not the most sensible of creatures rather than the least? Flagg ventured that this surprising insensibility must have been the consequence of "something in [her] constitution." Maybe it was a temporary aberration caused by fever; he doubted she could "treat the fish with so much familiarity while in a perfect state of health." Coming from South Carolina, the American state with the largest black majority, Flagg was interested in racial variations in sensibility as well. He reported that "some Indians and negroes can [handle the fish at will]; whether by the assistance of any means to counteract the power of the eel, I know not." Perhaps some occult local trick might explain the Africans' uncommon endurance? "I have seen negroes take hold of it, at first very cautiously, receiving many light shocks, but presently have grasped it hard and taken it out of the water." This was not an unusual conclusion for a white commentator to draw: many promoted the idea of the bodily insensibility of Africans as a natural justification for plantation labor (in contrast to Native American delicacy). At other times, African prowess was simply reduced to a vain machismo. Recalling the scene in Aphra Behn's *Oroonoko* (1688) where Caesar the "royal slave" seizes a numb-eel in Guiana, believing it "impossible a man could lose his force at the touch of a fish" only to find out otherwise, Flagg recounted a story he had heard in which "a negro fellow formerly being bantered by his companions for his fear of this eel, determined to give proof of his resolution, and attempted to grasp it with both hands. The unhappy consequence was a confirmed paralysis of both arms."[28]

To Flagg, such paralysis was a prank played by nature on the foolhardy. Experimenters like Bancroft, of course, took more seriously the physical effects of handling the eel. These were philosophical shocks, not prodigious or merely entertaining blows. Nor were they a medical panacea. Having rejected Réaumur's mechanical interpretation, Bancroft threw out the Dutch physician Van der Lott's "fallacious" insistence that the eel's shocks were a natural form of electrotherapy. As the 1774 Charleston advertisement shows, the eel was quickly promoted as a therapeutic wonder, doubtless drawing on the ancient medical associations with numbfish. Bancroft knew better, however: "I have known the Eel frequently touched by paralytic patients, though I cannot say with much apparent advantage." He neglected to mention, however, that the Dutch had conducted a remarkable battery of tests concerning precisely this question, tests that reveal more of the interracial dynamics of experiment in Guiana. "Various people who to some degree had gouty pains," Governor van 's Gravesande informed Allamand in 1754, "and who touched the torpedo had been completely cured two or three minutes after contact." In these trials, Europeans acted as experimental subjects, deliberately putting themselves into contact with the eel to test its effects. Interactions between the eel and the bodies of natives and Africans possessed a different character, however. In one trial described by Van der Lott in 1761, a freshly caught eel was placed on the knees of a native said to be suffering "paralysis of the abdomen." The shock that resulted knocked over two assistants;

however, according to Van der Lott, after three repetitions "the patient who had had to be carried from the plantation where he was, walked back to the plantation without cane or crutches." Similar experiments using slave boys involved outright coercion and must have terrified their subjects: "Mr. Abraham van Doorn ex-councillor . . . *threw* [his slave] boy daily into a tub of water in which there was a large *Conger-aal* of the black variety." "The boy was so greatly shocked by this that he crawled out on his hands and knees"; again, Van der Lott insisted, he was thereby relieved of his disorder. Van Doorn also "*threw* a slave boy who had fever badly, similarly into a tub with a *Conger-aal*." This time the boy was so strongly shocked that van Doorn had to help him out. Yet once more, Van der Lott claimed the boy had been cured (he naturally quoted no testimony from the boy himself). Van Doorn's use of eels on his slaves was evidently habitual. Whenever "they complain of a bad headache, he puts one of their hands on their head and with the other they touch the fish, and immediately, without it having once failed they are cured." Where Europeans were experimental subjects who touched the fish of their own volition, native and slave bodies were experimental objects thrown into contact with live eels.[29]

Such trials with electrotherapy—a colonial echo of early medical trials with laboratory electricity like those of the Abbé Nollet in the 1740s—may in some cases have been an earnest attempt to relieve slaves of nervous disorders; just as likely, they may have provided a resource for masters to threaten reluctant laborers. In either case, we see yet again how interactions of body and electricity, ostensibly the same experience everywhere, could radically change character in different social contexts. In Guiana, the racial contrast between Kinnersley's moderate shocks for genteel audiences and the violent shocking of Caribbean slaves was replayed in the different ways Europeans experimented with the electricity of the American eel, using both their own bodies and those of natives and slaves.

The Atlantic Eel Trade

Sensational analogies between laboratory and natural electricity were a way of bridging distance, of connecting the metropolitan laboratory to the colonial field. They carried the claim that it was possible to take laboratory techniques into the field and use them to make natural phenomena intelligible and manipulable. Sensational analogies thus enabled electrical knowledge to circulate between the Americas and Europe. But circulation was, as ever, not without its problems. In December 1773, toward the end of his career, Ebenezer Kinnersley advertised a new electrical attraction in Philadelphia: "a curious representation of the astonishing electric Eel," modeled on the live specimen "lately seen in this city." The key to the attraction was the sensational analogy, only this time heading in the opposite direction, from the natural back to the artificial: "On touching [the model], while in the water, an electric shock may be as sensibly felt, as from a live one." But why exhibit a model eel? To be sure, experimental philosophers, electricians especially, were fascinated by the artificial (re)production of sensory experience; Kinnersley's performances, for example, thrived on the virtuosic recreation of natural effects. Exhibiting a model fish, however, was also a neat solution to the pressing constraints of transporting live specimens. In circulation, electric eels once again proved difficult to handle. Since interest in the eel centered on its ability to discharge electricity—an ability that depended on its vigor—keeping it healthy in captivity was an essential undertaking, but also an arduous one, given the logistics of eighteenth-century Atlantic travel.[30]

As the 1774 Charleston advertisement shows, the *Gymnotus* circulated in several simultaneous guises: as a philosophical curiosity (both experimental and natural-historical), a medical resource, and as a wonderful commodity in a flourishing Atlantic market for exotic marvels. For ornamental as much as intellectual reasons, not to mention the sheer difficulty of transporting them, live animals from distant lands made extremely prized possessions in the early modern world. Ship captains and merchants strove to exploit the commercial potential of these exotic commodities. Opportunistic entrepreneurialism, rather than any metropolitan organization of natural resources, drove attempts to ship electric eels out of Guiana. Bancroft's misgivings about European presence notwithstanding, his Guiana was the site of an intense commercial traffic in specimens. The "*Kishee-kishee*" bird, for example, sold for two gold coins a pair in Guiana, although attempts to ship the creature to the Netherlands had so far failed. Many chased butterflies, "catching and preserving these Insects for sale in *Europe*." Then there were the peacock pheasants, which Bancroft had seen both in Guiana and later (uncannily) "at an Exhibition of Birds in *Piccadilly*" shipped via Brazil. "Masters of Ships," he explained, "sailing to Foreign Countries, constantly purchase the most curious Birds, and transport them to *Europe*, by which means a considerable number of the Birds of *Guiana* have been already described by Naturalists, who never visited that Country." The difficulty of transportation was offset by the lure of profit. Bancroft himself was engaged in shipping large collections of snakes and poisons to London, although for philosophical (he claimed) rather than commercial motives.[31]

For decades, the Dutch had practiced what Bancroft called "colonization . . . by indulgence," allowing American merchants to ship Guiana rum to New England, ultimately for sale in Europe. Although this trade was restricted by the 1763 Sugar Act, by which the British tried to force the Americans to buy British sugar and rum, the trade routes remained open. And these rum routes temporarily became eel routes. Among the first to attempt to ship live eels out of Guiana was a British ship captain named George Baker, who sailed between Guiana, British America, and London. "The keeper of this fish informs me," Garden wrote upon meeting Baker in Charleston, "that he catched them in Surinam river, a great way up, beyond where the salt water reaches." While he provided live specimens for naturalists and philosophers to examine, Baker's motives were commercial—interacting with his eels came at a price. Kinnersley, David Rittenhouse, and others at the American Philosophical Society negotiated with "the Owner of the Torpedo" ("eel" and "torpedo" were often used interchangeably) "on terms to make a set of experiments, with a view to determine the nature of the shocks which it communicates . . . not exceeding 3 pounds for that privelidge." Baker's appearance in Charleston was merely one stop on a commercial itinerary that evidently promised higher prices at the final destination: London. "The person who owns [the eels] rates them at too high a price," Garden complained, "(not less than fifty guineas for the smallest) for me to get a dead specimen."[32]

In South America, the hot and humid climate inhibited close experimental work. Flagg ascribed his lack of progress on the eel to "the relaxation of the mental powers generally consequent upon the lassitude of body incident to the inhabitants of warm climates." The physician Fermin complained that the heat prevented him from performing the "perfect anatomical dissection which would have enabled me to locate the true motor of the impulsive motion" in the eel. Guiana was amenable to natural history, and perhaps some crude experiments, but it was too hot for precise causal investigations. Transporting eels out of Guiana to less sultry climates allowed experimenters a closer investigation of Bancroft's claims. But it also posed challenges to the sensational analogy between natural and artificial electricities. As the eel circulated between Guiana and London via British America, there emerged a series of troubling experimental inconsistencies, unreported by Bancroft, between the alleged electricity of the *Gymnotus* and that of the laboratory. Hugh Williamson, a Philadelphia natural philosopher, conducted a series of trials with eels shipped from Guiana in Philadelphia in 1773. In a letter printed by the Royal Society, Williamson once again rehearsed the fish-machine analogy, comparing the shock from the fish to the effect of "touching a charged conductor or charged phial." But how to explain the eel's wonderful capacity of generating its own electricity? Dr. Bryant back in Guiana for one had been stumped by this "unaccountable" faculty: he could not determine whether the eel collected electricity from its immediate aquatic environment, absorbed it from other animal bodies in the water, or somehow possessed an innate supply. Second, if the eel was electrical, how could it conserve its charge in water? Third, by the time it reached Philadelphia, the eel failed to demonstrate classic external signs of electrification. Restaging some of the most basic electrostatic trials with the fish, Williamson reported that cork balls suspended by silken threads over the eel's back barely stirred, whereas normally they would have repelled each other because of their common positive charge. After conducting a similar battery of tests the same year, Kinnersley, Rittenhouse, and company in Philadelphia also found that the eel failed either to electrify cork balls or (as Williamson, too, had noted) to display a spark visible to the naked eye.[33]

The problem of these disanalogies between natural and artificial electricity was related to another difficulty in making eels travel: keeping them healthy outside their native habitat. Live *Gymnoti* were transported in wooden troughs filled with water. To keep them fit for performances on tour, their owners fed them small fish that the *Gymnoti* stunned themselves, then ate. "Several attempts have been made to convey these Fish to *Europe*," noted Bancroft, "but the quantity of fresh water requisite to shift them as often as is necessary, together with the bruises which they must inevitably sustain from the motion of the ship, have hitherto rendered them unsuccessful." By the time eels reached Philadelphia, their vigor and electrical powers were diminished. "The eel being sickened by the change of climate," Williamson wrote, "its owner refused to let us take it out of the water, for the purpose of making experiments, on reasonable terms; and there were many experiments which I could not make on it in the water, to my own satisfaction." How was one to conduct the proper tests to resolve these disanalogies, when the specimen had lost the energy and will to discharge? Williamson noted how hard it became to provoke the eel: "I have frequently passed my hand along its back and sides from head to tail, and have lifted part of its body above the water; without tempting it to make any defence." This was a bad sign. Bryant reached a similar conclusion in Guiana: "Being sometime confined in a tub and wanting perhaps their natural food, they lose much of the strength of this extraordinary quality." (One could never be too sure, however: when Flagg handled a languid specimen in Guiana that had been out of the water for some time, he found that "applying my finger near the belly, the torporific power was very considerable, notwithstanding the fish was now almost dead.")[34]

Baker's ultimate destination was London, but Garden thought the mariner overly optimistic in trying to carve out his niche as an Atlantic eel trader. In the 1770s, eastward passage across the North Atlantic still took several weeks. Garden's skepticism was well founded: the eels had already taken a number of weeks just to reach British America from Guiana. So he advised Baker to "get a small cask of rum, with a large bung, into which he may put any of them that may die, and so preserve them for the inspection and examination of the curios when he arrives." Baker had already admitted that "when they were first caught, they could give a much stronger shock by a metalline conductor than they can do at present." Anxious to guarantee the customer's shock, he was careful to issue precise instructions for handling the specimen: "The person who is to receive the shock must take the fish with both hands, at some considerable distance asunder, so as to form the communication, otherwise he will not receive it." During the crossing to Britain, four of the five eels in Baker's care expired; the remaining eel, though it apparently discharged shocks at Falmouth in November 1774, perished en route to London. Baker's homage was to follow Garden's instructions. Having dutifully preserved the other four in spirits, he delivered them to John Hunter for dissection (Hunter reportedly "danced a jig" at their exquisite state), and was most likely compensated for his trouble at journey's end. Daniel Solander, Linnaeus's London associate, noted a plan to raise a subscription to send Baker out again, but no evidence exists that it came to fruition. Because the eel could not be made to travel as a live specimen, it ultimately reverted, as a dead one, from the domain of experiment (in the Americas) to the domain of natural history (in Europe). The plentiful supply of electrical torpedo fish in the Mediterranean and off France's Atlantic coast meant that Europeans needed neither to expend greater efforts to transport eels to Europe nor to travel to South America themselves to continue experimental work on electrical fish (although the Romantic natural philosopher Alexander von Humboldt, who "burned with desire to correct my ideas [on the eel] by direct experience" and to "subject [it] to experiments while it still possessed its full electric force," would do just that in 1799). Belying the tortuous history of handling that had characterized the eel's Atlantic journey, Hunter's precisely rendered anatomical diagrams (based on his still extant preserved specimens) offer mute and static testimony to this apparent conclusion to the eel's philosophical career (Figure 16).[35] It fell to London wags to capture the anticlimax of the American eel's rise and fall, returning us to the notion of sexualized active powers that we first encountered in the "Venus electrificata":

> That Eel which stood erect in beauty's pride,
> And nodded to and fro its coral head,
> Worship'd by untaught Indians far and wide,
> Like other creatures, is not stiff, tho' dead.
> Limber and lank the heaven-born charmer lies,
> From every virgin's hand with scorn 'tis hurl'd;
> No maid can make the poor Torpedo rise,
> Limp as a dish-clout—it forsakes the world . . .
> Here
> Rests—without further Hope of Resurrection,
> The Elastic

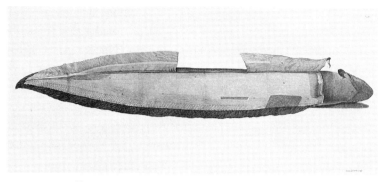

16. The Electric Eel as an Organic Machine

The remarkable stunning power of electric eels interested many observers, and in the 1770s several of these creatures were shipped from Dutch Guiana along the Eastern seaboard of North America to London. Although virtually all of them died en route, colonial experiments on the eels prompted research on European torpedo fish, which ultimately led to the construction of electric batteries modeled on the internal organs of such fish. This diagram of an American eel, dissected by the eminent London surgeon John Hunter, shows its electrical organs running along the length of its body. The remains of Hunter's eels are on display at the museum of the Royal College of Surgeons, London.

Body
Of
THE ELECTRICAL EEL, or GYMNOTUS ELECTRICUS.

Although the Marquis de Sade, in imitable fashion, would later hit on the ingenious notion of the electric eel as a vaginal torture, the difficulties of circulation meant that the potency of the American *Gymnotus* was indeed a fantasy, not a reality, in enlightened Europe.[36]

But the American eel's Atlantic career ended with a bang, not a whimper. Although the *Gymnotus* could not be transported as a live specimen, its circulation in print via Bancroft's *Natural History*, and the pages of several American and British journals, directly stimulated experiments on the European torpedo fish, which possessed similar properties to the eel.[37] In London during the 1770s, John Walsh—formerly Robert Clive's secretary in India and by this time an active experimenter in correspondence with Williamson—as well as the mathematician Henry Cavendish, worked to resolve the disanalogies between natural and artificial electricities. Although both worked on the torpedo rather than the eel, their concerns and their methods grew out of the experimental agenda already established in the Americas. Like his New World counterparts, Walsh shocked human chains using torpedoes caught off the French coast at La Rochelle, but his work went well beyond the colonial experiments, bringing the natural-artificial analogy in electricity to a resoundingly Franklinist conclusion. Using older yet precise anatomical descriptions of torpedoes, as well as Hunter's new diagrams, Walsh explained that torpedinal shocks were produced by the instantaneous charging and discharging of electricity within the fish's cylindrical internal organs (organs that ran the length of the fish in the case of the eel, as Hunter's other set of diagrams showed: see Figure 16). Through muscular contraction, the fish threw its organs into a radical disequilibrium of positive and negative charge, established a communication between these by means of an internal nervous fluid, and so completed the circuit to produce an equilibrium-restoring explosion. These fish were "animate [Leyden] phials" in Walsh's words: their natural electricity functioned according to Franklin's economy of charge, which had been developed to explain the behavior of artificial electricity in the Leyden jar. Because the charge and discharge of the fish was instantaneous, their electricity neither dissipated in water nor displayed external signs of electrification. Following Walsh's distinction between the "rarity" and "density" of electrical discharges, Cavendish developed the concept of electrical intensity to explain that no visible spark issued because of the low quantity of charge relative to surface area (the body of the fish)—this was the opposite of visible sparks from metal points (high quantities of electricity relative to a small surface area). Bancroft's experiments provided the original stimulus for Walsh's via Franklin, who was friends with both men. Aware of the eel experiments in Guiana, Franklin had encouraged Walsh in 1772 to repeat the same trials with the torpedo, and in his 1773 paper, Walsh duly asked Franklin to inform Bancroft, whose work he had cited, that he had "thus verified his suspicion concerning the Torpedo."[38]

In sustaining the analogy between natural and artificial electricities, Walsh and Cavendish advanced two distinct yet related experimental agendas. One would transform the pursuit of electricity in the late Enlightenment; the influence of the other would transform the entire nineteenth century. The first investigated the relationship between electricity and the animal economy, a development that also intensified as never before interest in exploiting electricity for medical purposes. The relationship between electricity and body would now become truly reciprocal: whereas philosophers and demonstrators had used the body to display the properties of electricity, physiologists, physicians, and therapists would now take center stage, exploiting electricity for what it could reveal about, and how it could improve, the workings of the body. The second experimental agenda involved using the natural organs of electric fish as models for the artificial generation of electricity, ultimately in the revolutionary form of a constant electric current. As early as 1775, Cavendish constructed an elaborate model torpedo made of pewter plates encased in leather and submerged in saltwater, which he used as an experimental apparatus for testing bodily response to electric discharges. By 1800, the Pavian natural philosopher Alessandro Volta described the construction of the first current-generating battery (the "Voltaic pile") by invoking the internal organs of electric fish as an organic model for his columns of copper and zinc submerged in brine. As the historian John Heilbron has written, "Volta fashion[ed] from the apparatus of natural electricity, as found in a fish, the prime organ of artificial electricity, the battery, with which the nineteenth century, and modern life, began." Early modern analogizing between nature and art had finally transformed electricity from a plastic cultural commodity into *the* material utility of the modern world.[39]

The history of the electric battery, and modernity, has traditionally been told as a story of metropolitan physics. But it should now be clear that the path to the battery, and the modern era, was first laid in a colonial setting, through the interactions of Europeans, Native Americans, and African slaves. The creation of artificial batteries depended on a prehistory of laboratory experiments that made it possible in the first place to recognize the electricity of exotic organic creatures. So we can complete Heilbron's sentence in the following way: "Experimental travelers, in particular the colonizers of Guiana, used sensational analogies to fashion from the apparatus of artificial electricity, as found in a machine, one of the prime organs of natural electricity, the electric fish. As a consequence, Volta could fashion from this apparatus of natural electricity, as found in a fish, the prime organ of artificial electricity, the battery, with which the nineteenth century, and modern life, began." If nature ultimately became a model for art in Volta's battery, it was because experimenters had learned to apprehend nature in terms of art. The colonial history of eels and experiments in the Americas is thus part of the history of science and enlightenment in Europe. By providing a natural model that could be subject to experiment, the American eel played its own role in the invention of a modern electrical world.

CHAPTER SIX

Electrical Humanitarianism

This perhaps... may with justice be termed the age of humanity.

THOMAS BARNARD, 1794

*A*s the eighteenth century drew to a close, electricity and the notion of "humanity" were drawn into ever tighter connections in what can be called programs of *electrical humanitarianism*. These programs combined physiology, physical therapy, and ideologies of social reform. Physiologists revised their view of the workings of the human body on the assumption that all animal life was electrical. In unprecedented numbers, experimental therapists began using electrical machines to heal diseased bodies. And moral reformers started promoting the use of electricity to reanimate failing bodies in the name of "humanity." Electricity thus became significant in the following questions: What was the physical basis of human life? By what artificial means could life be sustained or even recovered? And how best to manage such a technology in the interests of society?[1]

In the small town of Galway, New York, during the summer of 1796, a physician known to us simply as T. Gale was approached by a young man who was seeking help for his father. The young man, Gale later wrote, "informed me that his father was crazy." The father had settled in Galway some time earlier and purchased two farms, only to lose much of his property "by some deception." This "chagrinery" had evidently deranged the man's mind. His family was on the verge of putting him in chains because of his wild declarations: "he told them that the devil said he must kill a daughter of his." So violent was his raving that "they could not keep his clothing upon him." The young man had heard about Gale's use of electricity to relieve nervous disorders the previous year at the nearby spa town of Ballston Pool, and the father was brought to him. Could electricity relieve mental disorders? Could electricity restore reason itself? The agitated state of the patient made carrying out the therapy especially arduous: "I expected he would break [my machine] to pieces," Gale confessed. So he assembled a group of twenty assistants to subdue the man and bring him to his machine. Metal chains probably six to eight feet long were used to convey several massive electric shocks, from head to toe. "I charged the machine as high as I thought he could bear, and live through, for his arteries were in the highest state of action that ever I discovered," Gale noted. "I passed the shock upon him, which almost knocked him to the floor." The patient was repeatedly and violently shocked. Finally, Gale reported that the man was "sullen," so he sent him home. He appeared "more composed in his mind the next day." After a second session of electrification, Gale concluded that "the right use of his mind" was restored and he was "composed and well" again.[2]

Unnoticed by historians, Dr. T. Gale remains a man of mystery: besides the information provided in his own publications, almost no details of his life appear to have survived. His first name cannot be known for sure, while his surname does not appear even in local histories; nor can anything be established with certainty about his education or the origins of his interest in electricity. Such paucity of information is typical for itinerants—Gale circulated in upstate New York (and New York City) offering his services as an electrotherapist beginning, he recounts, in the 1780s. He was not alone: one historian has found evidence for over twenty such itinerants selling electrical therapies in New England and New York between the 1780s and early 1820s. Typically, such figures leave behind little more than a few scattered newspaper advertisements. As an itinerant, however, Gale is exceptional, because he published a three-hundred-page handbook on the theory and practice of medical electricity, the first such work by an American. *Electricity, or Ethereal Fire, Considered* (1802) was both a practical epitome of medical electricity and a philosophical account, discussing topics like "Animal and Vegetable Electricity" and "Astronomical Electricity." But Gale went further than this. In a remarkable passage entitled "Electricity Spiritualized," he confided the ultimate significance of his work to the reader: "I believe the Millennium is at the door; and that this ethereal fire will be as conspicuous a mean of purifying the body from disease in that day, as the fire of the spiritual kingdom will be, in purifying the souls of men; and that the publication of this medical treatise, is not without the intention of Heaven." In an epilogue entitled "Signs of the Times," the medico-electrician-turned-prophet offered a scriptural interpretation of Jefferson's election to the presidency in 1800. The vanguard of medico-electrical enlightenment in the early American republic was not merely evangelical, but millenarian.[3]

The relationship that existed between enlightenment and evangelicalism—between the agency of reason and the mystery of revelation—is nowhere more vividly exhibited in early America than by Dr. Gale. As we return to North America in this chapter, we find the social, political, and scientific landscapes altered, and working together in potent new ways, although still with discernibly colonial roots. As an itinerant and an entrepreneur, Gale was the heir of Kinnersley; as a philosopher, he owed fundamental debts to Franklin (and, for his fusion of philosophical and spiritual concerns, Newton himself). But Gale's electrified bodies, paradoxically made amenable to rational manipulation through the revelation of mysterious powers, were marked by different investments from Franklin's or Kinnersley's. Where the Philadelphians had used bodies as instruments for electrical experiments and demonstrations, largely shunning its medical potential, for Gale bodily electrification was no longer just a show; it was a miraculous form of healing. The relationship between body and electricity thus became at last truly reciprocal. Where bodies had displayed electricity, now electricity displayed and improved the workings of the body. The body electric became the electric body. In Gale's hands, electricity's status shifted decisively from a commodity for refining the colonial self to a material utility for restoring the republican body. This might seem a lurch toward materialism in an era of nationalist utilitarianism; in fact, it was an intensification of the religious impulse in electricity. As we shall see, medical electricity concerned not merely the self, but also the soul. In this respect, Gale was a quintessential enlightened electrician: a man of practical action, but also one whose head was filled with radical visions.

The concept of humanitarianism, which emerged in its modern form toward the end of the century, is crucial for seeing how issues of bodily management, morality, and social activism related to each other in the late Enlightenment. A variety of electrical humanitarianisms materialized at this time. Like the antislavery societies of the same era, humane societies, which sponsored the use of electricity for reanimation, pursued a collectivist approach to the relief of human suffering. Gale, by contrast, was a social immediatist: his ideal was individual intervention and the strenuous public dissemination of useful knowledge. Confidence in manipulating electric bodies was not universal, however; some had grave doubts, as shown by anxious discussions of spontaneous combustion. Where electrical humanitarians insisted that the human body could be reduced to and managed through Franklin's rational economy of charges, natural philosophers and Gothic allegorists explored the terrifying implications of exploding electric bodies, linking themes of natural inscrutability, enthusiasm, and the social disintegration of the United States. Radical faith in electricity contended with radical skepticism. Would electricity, that "latent, mysterious and powerful agent," help save the republic, or destroy it?[4]

Filling the Body with Fire

Before the invention of the Voltaic pile in 1800, the two central developments in electricity in the late eighteenth century were the practice of medical electricity and the theorization of what came to be called "animal electricity." But practice and theory were not automatically related. As we shall see more clearly in Chapter 7, the experiments performed on electric fish after the 1760s, followed by Luigi Galvani's observations of motor action in dead frogs, published in the early 1790s, fostered widespread belief in animal electricity (Galvani's phrase): the notion that electricity was inherent in the nervous system (humans included) and was its vital animating force. These theories often inspired the practice of medical electricity. Dr. Gale shows, however, that electrotherapists did not require galvanic inspiration, for Gale himself was not a Galvanist. Although he published *Electricity, or Ethereal Fire, Considered* in 1802, a decade after Galvani's work had become well known even in America, Gale neither invoked nor used galvanic concepts. This absence is surprising, but it reminds us not to take the circulation of scientific theories for granted, or to assume that local practices are necessarily driven by the "dominant" (metropolitan) theories of the day. Upstate New York was a peripheral locale when it came to science and medicine in this period. But in Gale's case being peripheral did not mean being intellectually subservient; it meant the freedom of a creative eclecticism. Taken individually, his electrical ideas and practices may not have been original, but their fusion with his local evangelical, political, and social visions shows how individuals could engage their own exuberant use of electricity to fashion personal cosmologies of passionate intensity. The power of the cosmos was accessible to nature's humblest witnesses. As Gale put it, electricity had revealed itself "to a babe."[5]

Medical electricity had in fact enjoyed a long if fitful career in the half-century before Gale, dating back at least to the 1740s. One celebrated early episode was the Abbé Nollet's 1748 Italian tour to investigate claims that electricity could conduct therapeutic effluvia (Peruvian balsam and opium) to patients. On this occasion, establishing a pattern of accusation destined to repeat itself over the years, Nollet (himself a sometime medical electrician) argued that it was the impressionability of lower-class experimental subjects, rather than electricity itself, that had effected apparent cures (a problem he tried to solve by using machines to gauge bodily response to electricity). Such concerns were to bedevil medical electricity (as well as cognate practices like Mesmerism) throughout the Enlightenment. Electricity's influence on the body remained open to charges of enthusiasm: it seemed impossible to know for sure whether its effects were produced by external causes or the workings of the imagination. This ambiguity did not deter repeated efforts to convert electrical generators and Leyden jars into enlightened medical technologies, however. In colonial America, the genteel social status of early practitioners, and skepticism about electricity's power to heal, inhibited attempts at systematic therapeutic application. In the late 1740s, Franklin had lent his apparatus to friends like the Philadelphia savant James Logan and the governor of New Jersey, Jonathan Belcher. With the blessing of the master electrician, both men experimented privately with shocks to treat the effects of palsy (stroke) and nervous disorders, respectively, but with no notable success. Undaunted, Franklin appears successfully to have used electricity to relieve the convulsions of a young woman in Philadelphia in 1752. He remained unconvinced by electrotherapy, however, ascribing such successes to the patient's desire for relief and the chicanery of the healer. Yet he also remained characteristically open-minded. In 1784, even as he headed the royal commission in

Paris that denounced Mesmerism as wonder-mongering, he privately admitted with casual heterodoxy to one French friend that "delusion may however in some cases be of use," and that people "may possibly find good Effects tho' they mistake the Cause." Here was Franklin again as modest (colonial) witness of nature, espousing an agnosticism that members of metropolitan establishments, both scientific and political, would be less likely to countenance. There was more in nature than was dreamed of by natural philosophy; enthusiasm could produce beneficial results. Franklin's liberal American economy of wonder allowed for the possibility that real good could come from mysterious sources, including medical electricity (and Mesmerism).[6]

By contrast with Franklin's genteel age of amateur private experiment, in the early republic medical electricity became a public, systematic, and commercial concern. The philosophical intensification of the electricity-body relationship was accompanied by a social transformation as well. If, in Kinnersley's performances, electrification was linked to ideals of genteel self-improvement, after the Revolution, medical electricity flowed through the less genteel hands of commercial therapists who sold electricity as a practical boon to health. In the republic, by contrast with the old colonial order, electricity itself was promoted as materially useful. Electrotherapy now turned up in an array of improvised venues, such as Kinnersley's own demonstrations, and at the stationery store of the Philadelphia bookseller and apothecary Samuel Dellap, who offered electrification with his "golden medical cephalic snuff" for two shillings and sixpence as a "certain efficacious remedy for almost all disorders of the eyes and ears."[7] The Boston surgeon-demonstrator Isaac Greenwood III offered "to electerise those who stand in need of that almost universal remedy, at 6s each time, at his House," and sold "electrical Machines with apparatus for experiments, and medical use." William King, an itinerant physiognotrace artist and ivory turner, made and sold his own machines and performed electrical cures at the Providence Bathhouse in

1800, while the Viennese Sigismund Niderburg (who claimed to have worked with Galvani) set up an electrical clinic in Manhattan in 1803. This commercial confidence in medical electricity was seconded by academic physiologists. In 1806, a medical student at the University of Pennsylvania named Richard Willmott Hall published his "Inaugural Essay on the Use of Electricity in Medicine," declaring, "We can direct the electric fluid through any part of the body with the greatest safety and convenience."[8]

Hall confined his analysis to practical therapeutic instructions. Gale, by contrast, was a philosopher who theorized about how medical electricity worked, weaving together the languages of humoralism and contemporary nervous physiology, and linking electricity's practical benefits to its singular ontological status. Although his book, like Hall's, was largely devoted to practical instructions, he was intensely interested in the question "What is electricity?" His answer was unequivocal: electricity was a form of "ethereal fire" akin to the "elementary fire" that emanated from the sun, animating the entire universe. The radiation of this fire across space to terrestrial bodies Gale described as "natural insolation"; applying artificially generated elecricity—"filling the body with fire"—simply extended this natural process. Making use of Franklin's demonstration of the identity of lightning and electricity, he observed that dissected bodies that had been struck by lightning revealed a massive distension of the pores and vessels. If electricity could be "temporized by art, administered in form and quantity," therefore, it could be made to reduce the excessive tension of bodily fibers contracted by inflammation or fever, replenishing their natural vitality. Shocks or "supernatural insolation" (positive electrification) could restore the "elastic spring" of the fibers. Borrowing concepts recently developed by John Brown of the Edinburgh school of nervous physiology, which decisively influenced the curriculum at the University of Pennsylvania, the leading American medical school, he explained fever as a state of deficient nervous "excitement," or nervous "debility." Gale, however, spoke in several explanatory languages at once: he also described this same nervous problem as a "vitiation of the humours" (balance among the four "humours," fluids thought to circulate in the body, was traditionally supposed to determine good health). Both of these states, nervous debility and vitiated humours, were in his view ultimately due to deficient ethereal fire. "Involuntary motion," he wrote regarding a case of convulsions, "proceeded not so much from debility, simply considered, as from those sharp humours, as I conceived, to prick and irritate the nerves, and cause them to spring and move, without the volition of the will." The solution? Artificial stimulation with electricity. "Purging off the humours," he concluded after relieving a fever through electrically induced perspiration, "was the cure of the involuntary motion of the nervous system."[9]

These overlapping physiologies vividly illustrate how humoral medicine, derived from the ancient writings of Galen, could be blended with the "medical science" of the Enlightenment. Creative borrowing from multiple traditions was the norm, especially beyond the controlled environment of medical institutions. As the bleeding of yellow fever victims in the 1790s by the premier American physician Benjamin Rush suggests, new medical sciences did not seem incompatible with the practice of "heroic medicine." Physicians' strategies for electrical healing incorporated elements of folk knowledge, too. In 1764, observing the failure of bleeding, vomiting, and purges to relieve the convulsions of an epileptic slave, Dr. James Greenhill of Stony Creek, Virginia, "resolved to give him a shock from the two Glass Spheres fixed to an Electric Machine." Evidently of an astrological turn of mind, Greenhill noted that these fits occurred when the moon was in the house of Capricorn. As an experimental art undertaken by a host of independents, medical electricity naturally converged with a range of medical traditions, ancient and modern. For an active physician like Gale, however, therapeutic practice was a more driving concern than arriving at a single definitive physiological theorization. Consequently, he encouraged experiments combining electrification with a variety of other strategies, including emetics, cathartics, and tonics, the taking of spa water (he took advantage of this fashion by soliciting clients at the resort town of Ballston Pool), the use of native remedies, and even bleeding.[10]

The ability to practice medical electricity depended, of course, on access to electrical apparatus. "Any man may use a machine," Gale promised, "but he must be well instructed, that doth it properly." He described a machine of standard eighteenth-century design, complete with Leyden jar and, to measure the quantity of charge, a simple electrometer, made of threads terminating in cork balls that rose from a position of rest to one of mutual horizontal repulsion when electrified. The apparatus could be used for shocking, for positive electrification (patients perched on an insulating stool for this, sometimes known as the "electric bath"), for the extraction of electric sparks, and for an optical treatment called "aural diffusion." All these techniques had been practiced for a number of years by European electricians, such as the London experimenter Tiberius Cavallo, whom Gale cited both as a technical authority and as a source for the writings of England's leading religious practitioners of medical electricity: Richard Lovett and John Wesley. Gale's practical instructions were resolutely systematic. For bilious fever and excitement of the stomach, for example, patients should take both an emetic and a cathartic, whose effects should then be accelerated by shocking. These shocks should be administered in an oblique pattern, from the left hip to the right foot, then from the right hip to the left foot, while patients held a chain or wire across themselves to facilitate the communication of electricity throughout the system. The shocks would relieve delirium by expanding the vessels, opening the pores, and stimulating the circulation of the blood. Hot, moist cloths could also be used to encourage perspiration, as could herbal tea. After the first round of electrification, the subject should wait fifteen or twenty minutes before receiving a further eight to ten shocks from the right side of the body to the left, strong enough to be felt in the breast and shoulders—light shocks to start with, gradually increasing in force. With minute variation, these simple instructions were repeated for a bewildering array of conditions: St. Vitus's Dance (a complication of rheumatic fever in which the face and extremities twitch), smallpox, vertigo, headache, menstrual blockages, deafness, blindness, ulcers, rickets, lockjaw, measles, agues—the list went on and on.[11]

Because electrification might be perceived as frightening and unpleasant, Gale stressed the importance of making it as humane as possible. He exhorted sympathetic healing, in which the doctor was sensitive to the bodily resistance of the individual patient. Shocks should be light, and continued for as long as necessary (sometimes weeks or even months), and the subject kept warm after electrification. Heavy shocks should be avoided, as should incomplete treatments and leaving the patient exposed to cold after electrification. "Trembling and faintness will sometimes occur," he warned, but "by consulting your patient's sensation of the shock," the requisite amount of charge for any given body could safely be administered. Crucially, shocks could be moderated according to the strength of the patient by using electrometers. "The operator must," he warned, "be acquainted with his own machine; He must observe the different degrees of the charge by the electrometer—must notice how well people are affected by them—must notice what degree will be felt well in a well man." Above all, humane electrotherapy required controlling the amount of electricity administered to the body.[12]

Relieving involuntary bodily motions provided a strategy for reining in involuntary mental motions as well. "There is a general utility in electrifying for all kinds of fits," Gale assured his readers. "Hysterics," for example, resulted from "spasms" caused by physical "compression and pressure," and could be relieved by shocking from the neck down to the feet. In hundreds of such cases, he claimed, a course of twenty shocks carefully applied in this manner had rendered "one uniform effect" in removing "rude fits." In the event of epileptic fits, the result of a deficiency of sensation and volition, "passing a few shocks from hand to hand, through the breast" would "bring any person out of a paroxysm immediately." Since "insanity" was the result of a "compression and irritation of the brain," which deranged its "exquisite sensibility," light shocks could "preserve the mental faculties" by acting "directly upon the brain and seat of the nerves" to relieve such pressure. (Gale's contemporary Edward Cutbush, another University of Pennsylvania medical student, went even further, theorizing insanity purely as a Franklinist economy: "All our motions, either of body or mind, depend on electric or electroid fluid; this has been emphatically called 'the enlivening spirit and soul of nature' . . . And may not the periodical attacks of Insanity be owing to a want or to an accumulation of this electricity in the brain"?) Executing the therapy was often far from straightforward, however. One of Gale's patients had decided to take "advantage of the absence of his family in the evening, [and] cut his throat across, with a case knife," only to be saved when a neighbor intervened. Made even "more insane" by this frustration, there was "no flattering of him to any thing, [so] they forced him to the machine, like a bullock to the slaughter." Subtler cases allowed for a more artful administration: the operator, Gale counseled, might to advantage use "soothing, entertaining, enlivening objects and funny topics of discourse" to gain the trust of an unwilling subject. In one sinister passage, Gale told the story of a young married woman from Saratoga County who had become unaccountably "terrified at the sight of her husband, with whom she had lived in perfect cordiality, until she became insane" and "emaciated almost to a skeleton." Attributing her affliction to nervous debility brought on by nursing her newborn baby, Gale revived the "Venus electrificata," cunningly passing his electrotherapy off as mere entertainment:

> Her husband came to the house—I observed her terror—I laid hold of this opportunity to gain upon her feelings; I would not suffer him to come nigh her, pretendedly so: It had the intended effect. To be brief, it was not long before I was able to persuade her to take a little wine; after this, under the appearance of entertainment, we got her to the machine, when I passed some very light shocks in all the before mentioned directions. We regaled her with all the lively appearances in our power . . . I alternated light shocks, with wine, diluted brandy, &c. and as her mind began to be caught with lively appearances, we endeavoured to furnish all the variety that was possible . . . she gradually assumed the appearance of cheerfulness . . . [In] four or five weeks she was able to unite with her husband again in keeping house.

This gallant medical electricity was even immortalized by Thomas Law in a poem tenderly entitled "Upon Curing Miss—*By My Electrical Box*": "Alas! With thoughtless zeal I sped, / Pleas'd to apply the electric art; / The aches are gone from Emma's head, / But I am suffering at the heart." Gale's success was no joke, however ("what a preservation of life has been ignorantly prostituted to mere amusement!" he scolded). That his deception played so well confirms how medical electricity used essentially the same techniques as electrical demonstrations: so indistinguishable were their gestures that the patient noticed no difference. In reusing methods like electric shock, practitioners like Gale revolutionized their significance. Where shocks had provided a source of playful sensory disorientation in commercial entertainments, they now became signal techniques for the restoration of reason itself.[13]

The Fire of the Spiritual Kingdom

Gale's attention to the intricate arts of electrotherapy might make him seem the quintessential utilitarian, but in fact his practical instructions were carefully situated within an extended discussion of electricity's status as spiritual fire. Rational practical instructions and a deeply spiritual cosmology presented no contradiction in Gale's mind. His spiritual concerns are in fact the key to his actions. Publishing on medical electricity was a way to engage in the salvation of God's flock: salvation of the body physical through electricity, and salvation of the soul through divine redemption. What does it mean that the first American handbook on medical electricity was also a millenarian tract? America was not unique in spawning theological interpretations of electricity, but Gale's fusion of evangelicalism and electricity shows two important things. First, electricity's status as spiritually useful knowledge coexisted with a new republican emphasis on medical utility. Second, electricity mattered perhaps most of all to those of zealous Christian faith, and the American public sphere in which they spoke and wrote was peculiarly receptive to unorthodox religious voices, owing to the lack of an established church, a centralist political tradition, or powerful scientific institutions.[14]

Gale's electrical philosophy drew on the familiar longstanding conception of the ether as a universal subtle medium believed to mediate between the divine and the material worlds. Although he did not cite Newton directly, his own musings followed a path of speculation laid down in successive editions of the *Opticks* (first published in 1704) that light, fire, and electricity were modifications of the same ethereal substance, and the source of animation in the universe. Gale explained that after God, the first cause of motion in the universe, electricity or "ethereal fire" constituted the "secondary cause of motion [able] to produce and support life throughout all nature." Ethereal fire was a highly spiritualized active power that animated the cosmos. "The electrical effluvia," Gale wrote, endorsing essentially the same view that Kinnersley had advanced in his demonstrations, "is far more subtile than air, is diffused through all space, surrounds the earth, and pervades every part of it." Gravity and motion, he continued, "have their origin in the various states or degrees of density of this ethereal electricity, if I may so call it." Filling "infinite space," electricity resembled "an ocean, into which the Author of nature has launched all worlds," and was directly imbibed by all plants and animals from the atmosphere. Ethereal fire was absorbed by "the lungs [which] serve as an electrical machine to all animals": in functional terms, the human body, indeed all animal bodies, were electrical machines. Electricity replenished the bodily stimuli necessary to sustain human life, encouraging fermentation and a "quickening of the circulations." And so medical electricity was merely an extension of this natural electrification. In Gale's words, "Such is the extreme fineness, velocity and expansiveness of this active principle, that all other matter seems to be only the body, and this the soul of the universe."[15]

Gale's ethereal account of electricity was distinguished by his insistence on its spiritual status. Such properties were not discovered by human reason but revealed by divine benevolence. What was electricity, fundamentally? Yet again, the answer was best conveyed through an analogy, though this time not between different natural domains, or between nature and society, but between the material and spiritual worlds. God, the sun, and electricity all partook of a universal "spiritual fire." "There seems to be a striking analogy existing between the natural and spiritual world," Gale intimated. Analogous natural and spiritual suns powered the cosmos. The natural sun sustained the solar system, its ethereal fire imparting warmth and motion to the universe and all its creatures, while the spiritual sun, God, "imparts spiritual nutrition" to all true believers in the form of His love. In the spiritual sun that is God, Gale continued, "there is a participation of the same element as the natural sun diffused through all the natural world." The realms of nature and spirit were separate but analogous, and ultimately connected, because each participated in the ethereal fire. For Gale, electricity as an ethereal fluid was an ontologically real entity, emphatically physical and spiritual, the means by which God gave physical and spiritual motion to his creation.[16]

In an extraordinary moment in the history of Franklinism, and American electricity, Gale put Franklin's phenomenal account of electricity to almost mystical purposes. Borrowing the notion of a rational economy of charge, Gale hypothesized that the sun repelled comets electrically, recalling the astronomical displays of Kinnersley and Hiller "as one plus electrified ball in an electrical machine will repel the atmosphere of another, and the lesser atmosphere gives place to the greater." At times, he appeared to strike a sober physico-theological tone, marveling at electricity's ability to inspire wonder at the rational and lawful structure of the Creation. "Behold the wisdom of the great Architect," he exhorted, his "innumerable laws and harmonies throughout the animal and vegetable world—all stampt with divinity—all paying homage to a God infinitely wise, who is seen in the things that are made." Electricity should lead faithful human beings "to admire that wonderful harmony, order and design which are so apparent in creation," and help to "refute many objections of Atheists to the existence of Infinite Intelligence and Providence." The "utmost glories" of electricity were "consummated in the sun," which was envisioned by Gale as an "infinite fountain of ethereal fire." "Behold him as the envoy of heaven, sitting regent, as on a throne of state, and by tacit thunders of his laws, bids his system roll in orders intricate and multiform." Yet Gale's sense of the universe's orderliness commingled with a mystical awareness of its sheer power. The sun was naturally the object of an instinctive awful reverence: "Was it not a mysterious instinct, that hath led so many of the untutored tribes of the earth, to pay divine honours to the sun, and even to fire, in some instances, as being of the same species?" Despite its rational guise, the power at the center of the universe remained shrouded in utmost mystery.[17]

The key to that mystery was the relationship among God, electricity, and the soul. What God did for the soul, electricity did for the body. Gale's discussion of solar electricity, the work of a primitive Christian, ultimately transcended rationalist cosmology to embrace "astonishing" sacred power through themes of animistic spiritual rebirth and (specifically through medical electricity) the Christian millennium.

Ethereal fire represented both the natural and spiritual medium by which living things were sustained and "subjected to [God's] laws," and by which they might physically and spiritually regenerate. According to Gale, therefore, spiritual regeneration was part of the lawful course of nature, and was executed through the subtle agency of electricity. "There is no life, no animation in the natural world, but by a participation of the ethereal fire, which constitutes the sun of nature; and by which participation also, the sun of nature becomes a quickening power in the subject of that participation; so it is in the spiritual world, 'except a man be born again, or except ye be baptised with the Holy Ghost and with fire (see the analogy) ye cannot see the kingdom of God.'" No creature could live, physically or spiritually, but by participating in the higher agencies of electricity and the divine. The nearness of Christ's return was signaled by the "allegorical allusion" that the healing powers of electricity bore to baptismal fire; the door to salvation stood tantalizingly ajar. Purification of the body betokened purification of the soul; and as we would embrace purification of the soul, Gale urged, we must embrace purification of the body through electricity:

> Can we suppose that nature hath endowed this element with so many astonishing powers, that all creation is nourished and supported by it, and that none of these invested powers can be made use of in curing diseases—that we must look for no aid in this respect, but from the fugitives of nature? or is this the interdicted fruit, and all mankind filially obsequious of late? or do we, as the Jews in another case, judge ourselves unworthy of the blessing? or are we disgusted at the allegorical allusion it bears to the spiritual world, to baptismal fire, that sanctifies the soul from spiritual disease; that we will not, by terrestrial fire, purge the body from its diseases? I will here confess, that I believe the Millennium is at the door; and that this ethereal fire will be as conspicuous a mean of purifying the body from disease in that day, as the fire of the spiritual kingdom will be, in purifying the souls of men; and that the publication of this medical treatise, is not without the intention of Heaven.

This electric fire was like the "pillar of fire": it would nurture the moral and devour the sinful. "The presence of the natural sun produceth different effects, according to the quality of the body present," he warned. "So is the spiritual Sun: His presence is a source of eternal transport to the moral soul; but to the immoral, *'devouring fire and everlasting burning.'*" Whether Gale saw this as a form of special intervention or not, he insisted that the means to judgment was the same ethereal element through which all life was "subjected to his laws." Law and mystery were not opposed, but conjoined: "His very presence gives law to his whole system."[18]

Where did all these ideas come from? Sometimes Gale explicitly cited Exodus and Revelation, but his sources were often implicit, making a precise genealogy for his intellectual itinerancy hard to trace. Possibilities abound. Discussions of solar electricity and ethereal fire can be found in numerous eighteenth-century works, such as the writings of British followers of the German mystic Jakob Böhme, the anti-Newtonian scriptural philosopher John Hutchinson, Bishop George Berkeley's treatise *Siris* (1744), the writings of Immanuel Swedenborg, and the works of Hermetic philosophers and Freemasons. Any or all of these may have fallen under Gale's eyes. His geographical movements suggest that he may well have moved in Masonic circles in Saratoga and Washington counties, New York, areas where lodges were well established by 1800, but we cannot know for sure. Although his denominational allegiance remains unclear, he may well have been a Baptist, since his publishers, the pro-Jeffersonians Moffitt and Lyon, published a number of Baptist and millennial tracts in the same period by writers like Benjamin Gorton and Elias Lee. In moving between the Ballston area and New York City, he anticipated the evangelical itineraries of visionaries like the Prophet Matthias and William Miller during the Second Great Awakening, which swept New York State in the early nineteenth century. Such a milieu would certainly be consistent with his fusion of millennialism and solar cosmology, the intellectual stomping ground a few years later for evangelicals in the Second Great Awakening. Gale, however, cited none of these traditions explicitly, and no single influence appears to have shaped his cosmology. Although systematic in intentions, he was undeniably eclectic in his sources.[19]

If Gale's intellectual origins elude us, his significance as part of an American tradition of zealous spiritual interpretations of nature is unmistakable. This tradition did not simply survive the Enlightenment, it drew strength from it and flourished long after it. The work of enlightened electricians with deistic tendencies, like Franklin, did not necessarily undermine zealous natural philosophy; indeed, it provided grist for spiritual mills. Recent studies of science in eighteenth-century Europe beyond the great metropolises of London, Paris, and Amsterdam have shown how important religion was to natural history and natural philosophy in provincial Europe. This work shows that the exceptionalist claim that the American Enlightenment, and American science, were unique for their close relationship to Protestantism (a claim made when the anticlericalism of the French Enlightenment was taken to be the norm) is misleading. We ought instead to regard American spiritual views of electricity as part of a transnational tradition of provincial enlightenment in which local religions played a highly animating role in the pursuit of science. The Enlightenment was a period in which German theologians, for example, readily helped themselves to the electrical research of their day, using it to recast Pentecostal fire as a form of "spiritual electerisation" and offering new electrical accounts of the Creation story in Genesis. Science was made to serve religion in a relationship that raised, rather than diminished, the importance of science.[20]

In British America, natural philosophers performed similar theological maneuvers. Samuel Johnson, the Anglican pastor and later first president of King's College in New York, provides a spectacular example of how the meaning of Franklin's science could be redefined by local religious concerns. Johnson embraced the ideas of Newton and Locke while a student at Yale, and converted to Anglicanism in the socalled Yale Apostasy of 1722. Johnson, however, came to fear the possibility of materialistic philosophies derived from the work of Newton. He was troubled, for example, by his friend Cadwallader Colden's insistence that matter was inherently active and self-directing, without divine assistance. Now in full retreat from Newton, he turned to the idealistic philosophy of Bishop Berkeley, Christian Neoplatonism, and the scriptural natural philosophy of the British divine John Hutchinson, author of works such as *Moses' Principia* (1724). In one remarkable passage, penned in the late 1760s, Johnson cast Franklin as a defender of scriptural truth against the menace of materialism. "It is remarkable that Bishop Berkeley in Ireland, Mr. Hutchinson in England, and the Abbe Pluch in France, the greatest men of the age, without any communication with each other should at the same time though by different media come into the same conclusion, namely that the Holy Scriptures teach the only true system of natural philosophy as well as the only true religion, and that Mr. Franklin in America should at the same time without any design by his electrical experiments greatly confirm it."[21]

How Franklin had confirmed the scriptures, Johnson did not make clear. But his conclusion was resounding. So much for Franklin the Newtonian. Franklinist electricity could mean different things, many things, in America. For Johnson, a prominent public figure in the eighteenth-century American Protestant community, Franklinism precisely did not mean the triumph of science over superstition (as it did to John Adams), but rather the defense of scriptural truths against "Newtonian" heresies. Gale likewise endorsed good Franklinist doctrines like the conservation of charge, turning them to his own evangelical ends. For Gale, Franklinism meant much more than an exquisite balance of positive and negative charge: it meant the mysterious, radical potencies of fire and soul. And he would doubtless have agreed with Johnson that scripture was the ideal guide for philosophy, since he valued electricity as a powerful weapon in the war of faith against infidelity.

The Age of Humanity

Gale's electricity was "the sister blessing of that grace, destined in due time to fill all hearts, and reign triumphant through our disordered world." "Adorable ELECTRICITY," he urged, was "the friend of human life"; its medical use must be embraced without delay for the sake of relieving a suffering humanity. But how was medical electricity to be organized to serve society?[22]

The body has long been recognized as a central target of the "humanitarian" campaigns that began around the turn of the nineteenth century, but what humanitarianism made of the body remains open to debate. According to Michel Foucault's influential account of penal reform in this era, bodies disappeared: the spectacle of public torture gave way to invisible incarcerations, and the internalization of discipline through surveillance, all in the name of greater humanity. Other humanitarianisms, however, gave suffering bodies a higher public profile than ever before. Advocates for the abolition of slavery mobilized graphic images of African bodies in pain, not as a form of social control, but to the contrary, as a means of social agitation to sway public opinion toward emancipation. In these literary and visual campaigns, slave bodies were a spectacle redeemed by the moral purpose of inspiring sympathy among audiences (even though, ironically, images of racial violence would later resurface as the objects of a detached voyeurism in nineteenth-century pornography).[23]

Electrical humanitarians were also committed to the relief of bodies in pain. In the 1750s, demonstrators like Joseph Hiller played around with reanimating model spiders made of cork. "We all think Raising the Dead one of the greatest of miracles," Hiller told his audience, "yet a Dead Body has all the necessary Organizations, the numberless Vessels; the solids & fluids are all properly Disposed, & only want to be put in motion." By the 1780s, reanimating human beings with electricity was becoming an experimental reality. Newly instituted humane societies employed a collectivist strategy in promulgating medical electricity to this end. The creation of the humane societies of Philadelphia (1780), Massachusetts (1784), and New York (1787) drew direct inspiration from European sister societies in Amsterdam (1769) and especially London (1774). The American societies were part of the flourishing of voluntary associations in the early republic, a mode of reconstituting social relations in a country whose traditional bonds of allegiance and organization had been disrupted by revolution. Protestant ministers mainly in the Northeast, driven to realize their vision of a Christian republic, were the moving force, although they enjoyed the moral and financial sponsorship of local patricians and prominent citizens like the Boston merchant James Bowdoin. In the campaigns conducted by these organizations, as it would be in the antislavery movement, the relation of humanitarianism to sensibility was fundamental. To humane society members, the spectacle of suffering provoked an irresistible rush of sympathy that produced social activism. "Spectatorial sympathy," a concept linked to the notion of an innate "moral sense" most notably in Adam Smith's *Theory of Moral Sentiments* (1759), was widely held to be a universal mechanism that compelled the individual to moral action when confronted by the suffering of others. "The mind, by contemplating those subjects," wrote the Massachusetts humanitarian Reverend Thomas Thacher, "which are connected with this institution, will be filled with noble and generous sentiments." The language of humanitarianism was saturated with that of moral sympathy. It was, suggested Reverend Thomas Barnard, minister of the North Church in Salem, "the age of humanity." Before the rise of antislavery, temperance, and other major reform movements of the nineteenth century, American humanitarians directed their efforts toward the reanimation of life. The very word took on unprecedented rhetorical and moral force. In 1800, in a speech delivered before the Humane Society of Massachusetts in Boston, Thacher gave a signal account of the zealous humanitarian sensibility, emphasizing its "respect for human life, and anxiety for its preservation." Whereas "among ferocious, unenlightened nations, animal life is but little regarded, and its loss but little deplored," cherishing life was the hallmark of only the most enlightened civilizations. Humanitarianism pleased God, moreover, being "perfectly agreeable to the system of our holy religion." Humanitarian action to eradicate suffering was, like Gale's medical electricity, a portent of millennial significance. "May we, by pious affections toward God, and by social virtues toward mankind," preached Thacher, "be prepared for the *second coming* of Jesus Christ, when he shall *wipe away all tears from all faces,* and banish natural evil and death from the creation." The Reverend John Lathrop, a trustee of the Massachusetts Humanitarian Society (and committed electrician), wrote of God as simply the "FATHER OF LIFE."[24]

Institutional humanitarians were driven by a vision of society as a collective of benevolent individuals, united by sympathetic "bonds of permanent friendship." "*Every man,*" Thacher wrote, "*is under a moral obligation, to perform not only all the good he can, as an individual; but all that may be effected by him in his collective capacity.*" Neither atheistic nor secular, humanitarians laid claim to a zealous but nonsectarian Protestant heritage in pursuing "the godlike ambition of preserving life." During an era of rapid commercialization and the emergence of national party politics in the formative years of the American republic, this was an attempt to capture a moral and ideological high ground by rejecting the interests of pocket and party in favor of the "common cause of humanity." Humane societies claimed not to weaken the political and religious unity of the republic, because the "party of humanity" served the common good, being neither a political organization nor a religious sect. Humanitarian organizations were therefore "perfectly agreeable to the principles of our mild and free government." Collective organization was necessary, moreover, since the "good purposes of humanity, are more likely to be secured by societies [rather] than by individuals." They would not threaten the integrity of the republic. "Here are no polemical party discussions on religion or civil government," Thacher declared, "no irritable emulations for literary fame; and consequently no stings of rage at disappointment."[25]

These societies embraced the relationship between electricity and the human body because electricity presented a direct avenue for humanitarian intervention: it could be used to reanimate persons apparently dead from drowning or lightning strikes. Around the turn of the century, electrical reanimation was not the irreligious Faustian adventure it would later seem by association with the fictional persona of Victor Frankenstein (1818), who galvanized an entirely new being from the parts of corpses. "The great cause of resuscitation" was, rather, a sacred charge, in which physicians investigating the phenomenon of suspended animation, and ministers promoting humanitarianism, made harmonious common cause. Vitalist metaphysics, practical therapeutics, and social benevolence all converged at the intersection of medical and animal electricities. "There is a great resemblance between the nervous and electric fluid," John Fleet told the Massachusetts Humane Society in 1797, citing an impressive array of British electricians: Charles Kite, who had been awarded a medal for his experiments with electrical resuscitation by the Royal Humane Society in 1778; the London surgeon John Hunter; Dr. John Fothergill; the Edinburgh physiologist William Cullen; and Erasmus Darwin, who had written an important treatise on organic life, *Zoonomia* (1794). "If a nerve be divided and a proper conductor of electricity be applied," Fleet explained, "from one divided end to the other, the motion in the part below will be restored, and some experimenters have been sure, that they have restored contraction to muscles, that had long lost their original life." Here again the Franklinist economy of charge was transposed analogically to the body, now to be read as electrical. Nervous disorders thus became reducible to excessive and deficient amounts of electricity in the body. Convulsions resulted from "explosions" of nervous-electric fluid, while "an exhaustion of nervous electricity, from want of friction in the muscular fibres if carried too far would end in death." According to Fleet, reimagining the animal economy through Franklinism and animal electricity—we can call this the emergence of an *animal-electrical economy*—made this economy eminently susceptible to rational management.[26]

Several works published in the 1790s pressed this case. For those whose "vital principle" had been suspended, leaving the heart or brain inactive (the target was most often sailors or dock workers on the Eastern seaboard, and other people struck by lightning), the New York physician David Hosack suggested "laying the body upon an *insulated stool,* and charging it *gradually* with the electric fire; and perhaps, after a little time drawing sparks from it." Humane societies circulated broadsides publicizing such techniques; individual humanitarians like the geographer Jedediah Morse endorsed them zealously in their sermons. If art achieved beauty by imitating nature, John Bartlett argued in a discourse on animation, reanimation was the greatest art of all in its imitation, and restoration, of life itself. Preventing premature burial became a minor obsession. Half a century before Edgar Allan Poe's protagonists lived in dread of such a fate, humanitarians attempted (unsuccessfully, it seems) to allay these fears by using electricity to illuminate "the boundary line between life and death." The Reverend Lathrop cited "numerous instances of recovery," suggesting that premature burials had been carried out because resuscitative techniques had simply not been tried. "We drop a tear over the grave of such of our fellow-creatures," he lamented, "as may have, alas! too soon, been joined to the congregation of the dead." In response, American humanitarians reprinted British instructions for reanimation in their own publications, encouraged acts of charity by publishing narratives of heroic interventions, and even offered prizes for performing such acts. A number of bodies, it appears, were successfully reanimated using these instructions and rewards given out.[27]

Gale was an electrical humanitarian, but of a different order. Medical electricity rather than reanimation was his main concern, though he was certainly aware of the "surprising recovery to life" effected by electrical resuscitation, and he took to repeating Franklin's instructions (then half a century old) for the preservation of life with conductors, including Kinnersley's insistence that using conductors did not "imply any distrust of Divine Providence." He also employed the moral language of humane intervention in his own promotions of electrotherapy: "In good conscience, and in obedience to God, and love to my fellow-creatures, [I] present them with a system of principles, and of practical rules, in a plain, familiar manner." But Gale's approach was significantly different from that of the humane societies. Gale was an entrepreneur and an individualist who worked independently of any institutional or collective organization. His was a humanitarianism that pursued a libertarian rather than corporate logic: rather than (re)form the republic through new social organizations and rely on social mechanisms like "spectatorial sympathy" to alleviate suffering, Gale addressed his public and patients directly through publications and healing, putting the resources for relief, both physical and spiritual, into their own hands.[28]

Medicine in the early United States entailed a variety of social projects for strengthening the republic, a form of government whose liberality made it notoriously prone, in the minds of classically educated eighteenth-century observers, to corruption and disintegration. No figure pursued republican unity with more single-mindedness than Benjamin Rush, the Edinburgh-trained leader of Philadelphia's medical enlightenment at the University of Pennsylvania. Rush's projects, from penal reform to tracing the effects of alcohol on social behavior (he constructed a chart that linked drinking habits to social destinies), reflected an obsession with finding a material basis for antisocial behavior and the medical means to correct it. The aim: preserve moral virtue, hence social unity, among the citizenry. Citizens belonged to the republic, that is, they were "republican machines" to be maintained by a technocratic elite of enlightened social mechanics—epitomized by Rush. Such grand projecting, however, had a limited practical reach. Early republican medicine was shaped by both a nascent medical system (centered in Philadelphia) and a variety of still-vital independent healing traditions. While Britain had more strict licensing standards and supervision for physicians and clearer class distinctions among physician, surgeon, and apothecary than did North America, both medical cultures fostered a flourishing market in unorthodox therapies. Auto-therapy was especially attractive given its emergence during an era of revolutionary anticentralism, proliferating nervous disorders, and medical practices in methodological flux.[29]

Rush's Foucauldian medicine was clearly an exercise in social control: enlightenment as elite management. But Rush's social mechanics was not the only game in town. Gale's medical electricity demonstrates an alternative social strategy in the early republic, one of medical enlightenment as self-emancipation from collectivist control. Where Rush dreamed of an enlightened republic of docile bodies, Gale's visions were of bodies activated and souls made ecstatic by electricity. Electricity, seemingly by its very nature, made for antinomian enlightenment: restless, fiery, spiritual, constantly circulating, explosive, animating, it was the stuff of liberation, not constraint. It fit beautifully with Gale's unilateralism and commitment to the ideal of personal autonomy. Auto-therapy meant emancipation. His touchstone was the English founder of Methodism, John Wesley, himself a practitioner of electrical auto-therapy, whose *Primitive Physic* (which advocated medical electricity) had been reprinted in America since the 1760s. Why did physicians (like Rush) seem to shun electricity? Gale found his answer in Wesley. "They must not disoblige their good friends, the apothecaries. Neither can it consist with their own interest to make (although not every man) so many men their own physicians." Gale's was a humanitarian cause directed against the antihumanitarians—physicians who, out of financial self-interest and even "malice," sought to "obliterate [electricity,] this friend of human life." For this reason, Gale wrote directly to the "man of common ingenuity." Quoting Wesley, he railed at the "gentlemen of the faculty," urging them not to "condemn what they know not." If only doctors would get out of the way. "Doctors in the city," he wrote of a trip to New York in the 1790s (where he claims to have treated yellow fever patients at Bellevue Hospital), "had the same [electrical] instruments, and did not use them scarcely in any case: I was from the country, and cut no popular figure [so my patient Mrs. Bower] did not know what to think of this new fashion, but was willing to own I had cured her very suddenly." As medical electricity's advance agent in America, Gale would disseminate the wisdom of Wesley, Lovett, and Cavallo across the Atlantic, pleading electricity's "cause before the august tribunal of the human race," case by case.[30]

If the key to therapeutic independence was auto-therapy, then the key to auto-therapy was access to an electrical machine. "A thorough knowledge of building the machinery, and keeping them in repair, is now what is wanting to consummate the whole business," Gale wrote, "and leave my reader master of the field." So he duly included detailed instructions for the construction of a classic electrical machine, at a cost of no more than two or three dollars, he claimed (cheaper than any apparatus that could be imported from Europe). Self-interest must bow to philanthropy so that "all mankind were in possession of the utmost knowledge and use of this precious mean of health." His "common construction" was no ornamental apparatus, but the simplest and cheapest machine imaginable, consisting of a single wooden frame with two posts five feet high, in which was set a wheel, four feet in diameter, which would rotate a "common decanter" to collect electric charge. Iron chains six to eight feet long for conveying shocks, a "square case bottle" for a Leyden jar ("that receiver which will be of the least cost"), and a cork-ball electrometer fixed to the prime conductor made up the remaining tools of the auto-therapist's home laboratory. "The machine already described is the cheapest that can be built to answer medical purposes," he proudly declared. Here was Gale as Poor Richard: the image of a homespun enlightenment in democratic action, through the "plain, familiar" dissemination of useful technology directly to "the man of common ingenuity."[31]

Yet Gale exaggerated his uniqueness: medical electricity was by no means as scorned as he made it out to be. When Moses Willard sent his manuscript treatise to the Massachusetts Humane Society in 1789, arguing for the reanimating and therapeutic potential of electricity, the physicians Joseph Warren and Aaron Dexter warmly encouraged his efforts in their reply, and offered official thanks from the trustees of the society. Perhaps surprisingly, Benjamin Rush, the preeminent early American physician, does not appear to have resorted very often to electricity in his practice, but even he was evidently not opposed to its use. When James Cunningham sought Rush's advice on treating muscle strain, Rush replied succinctly: "Advised—warm bath and Electricity—with frictions." Newspaper advertisements for auctions of doctors' effects often included electrical machines. Thomas Dancer's Caribbean *Medical Assistant* (1801) endorsed the view that "electricity is the most powerful of any of the external stimulants," encouraging the drawing of sparks in paralytic cases, and the use of shocks for sunstroke. By the turn of the nineteenth century, general reference works like the *New and Complete American Encyclopaedia* (1807) were describing a variety of medical applications for electrical techniques. In a decentralized medical culture in which eclecticism and experiment were the norm, electrotherapy flourished.[32]

The issues of cost and expertise in constructing electrical machines also deserve scrutiny. Despite his tirade against the self-interest of physicians, Gale himself sold electrical machines (sometimes to other physicians) and, like all itinerant electricians, he charged for the therapies he performed (two shillings for sixty shocks given in two hours).[33] The low cost of constructing an electrostatic generator for only two to three dollars may also be questionable. One of the most comprehensive extant price lists for electrical equipment in the early republic is that of John Prince, an artisan who lived and worked in Salem, Massachusetts. Prince supplied the College of Rhode Island with experimental apparatus at the turn of the nineteenth century, including an electrical machine with an eight-inch cylinder, at a cost of just over thirty-seven dollars. Isaac Greenwood priced his machine at one hundred and fifty dollars. How easy would it have been to build one's own apparatus without some prior experience in creating mechanical devices? Replication was bedeviled by technical controversies; reconstructing exactly the machines of others was notoriously tricky. It took Alexander Anderson, the Columbia medical student, for example, over one year to complete a machine he had begun in 1797, and he had assistance from local craftsmen. Gale's instructions for technological assembly were neither unique nor a guarantee of success. Yet Gale must also be viewed in his proper context. Although natural philosophers routinely published descriptions of their apparatus and experiments by the eighteenth century as part of a culture of public science, physicians (especially itinerants) tended jealously to guard their methods as profitable secrets of nature. As a commercial therapist, Gale was unusually open in describing his methods, since profit was apparently not his main objective.[34]

Gale's claims to transparency and universalism are further complicated, however, by his status as a machine therapist. Therapy performed with machines, like natural philosophy made through machines, involved an ambiguous relationship between the human operator and his technology. Philosophers who operated machines argued that it was their machines that produced natural knowledge, but they also boasted that they were uniquely adept operators of those machines. Similarly, Gale claimed that electrostatic generators effected natural therapies, while arguing that the therapies' success depended on the therapist's skill in managing both patient and machine. Although he promised that "any man of common ingenuity may perfectly understand [this] business," he insisted that the practice be undertaken only by professionals. Patients must trust "the operator's absolute judgment," Gale cautioned; "any man may use a machine; but he must be well instructed, that doth it properly." Gale's libertarian model of making every man his own electrotherapist thus ran into an internal problem of political authority: correct electrical practice should be overseen, but by whom? Gale suggested that physicians fulfill this function—presumably not the "gentlemen of the faculty," but a class of independent practitioners. Without fully articulating a plan for the social organization of electrotherapy, he outlined how each county in the Union might acquire its own electrical machine for medical purposes, to be operated and supervised by a qualified operator. How these operators would have been selected, and how the practice and expense of medical electricity would have been managed, remained unclear. Gale's strategy for realizing his vision of an autotherapeutic republic seemed to stumble because of the contradiction inherent in attempting to achieve individual medical emancipation while insisting that physicians must still "govern the regimen" of the nation's most valuable medical resource.[35]

The Jeffersonian traces in this vision are not coincidental. Where the humane societies strove to establish themselves as apolitical, avoiding any challenge to the republic's unity, Gale's unilateralism left him free to use electricity as a public platform for partisan political advocacy. The final section of his book, like many millennial pronouncements made during and after the revolutionary upheavals of the 1790s, was entitled "Signs of the Times." The provincial Gale once again turned universalist, outlining a bold vision of transatlantic fraternity in which the French Revolution and the election of Jefferson to the presidency were further signs of the millennium. Gale faulted the United States, particularly John Adams's recent Federalist administration, for not offering greater support to America's "sister republic" in her hour of travail. Americans had missed the significance of the Revolution in France: like their own, it was a sign that the "two witnesses" of the millennium, incarnated as religious and civil liberty, were at hand. Now in America in 1800, these freedoms were reunited in the person of the new president, the Democratic-Republican Jefferson, whose election heralded a republic of yeoman farmers and, reversing the Hamiltonian 1790s, a return to freedom from centralized power. To Gale, the election of "our greatly-beloved Jefferson" foretold, as did the healing powers of electricity, the day when "the kingdoms of this world [would] become the Kingdom of our Lord."[36]

Paradoxically, Gale's intense individualism, and here his endorsement of a political party, were driven by faith in the higher agency of God. While adopting a partisan stance as a Jeffersonian (Moffitt and Lyon, his publishers, were pro-Jefferson), Gale believed mundane agencies like political parties to be redeemed and subsumed by their role in realizing divine will. This was the revolutionary logic of electrical politics still at work in the republic: there was no individual agency—all were instruments of His agency. "No matter who is the instrument, God is the cause," Gale affirmed, and God would make his moral judgments known through the course of political revolutions now in motion. We must not resist "the immediate agency of heaven among the nations," which was driving out of office those who, like Adams and the Federalists, were "unfriendly to the equal rights of man." Jefferson's election confirmed that "this is the day, long looked for, in which politics involve prophecies, and prophecies politics." The new President would not act alone, moreover, but "co-operate with the Prince of Peace." A full half-century after Franklin's lightning rod had silenced the skies, supposedly ushering in a modern disenchanted view of nature, Gale—who saw himself as one of Franklin's true heirs and disciples—could publicly fuse medical electricity, Jeffersonian politics, and the Second Coming in a compelling vision of reason and mystery realizing God's will on earth. "Feel the dignity of acting in concert with the King Immortal," he exulted, "expecting, ere long, to see the mystery of God finished; when, by the seventh trumpet, every mountain of monarchy, or island of usurpation, shall be moved out of its place, and the kingdoms of this world become the Kingdom of our Lord, and of his Christ." The agencies of God, electricity, and man harmonized irresistibly as one.[37]

The Exploding Body Electric

In his instructions for constructing an electrical apparatus, Gale had been careful to include guidelines for building and using an electrometer, "an artificial measurement of the quantity of the charge, in any particular receiver." This was a decisive inclusion. Electrotherapy relied on the harnessing of potentially harmful natural forces; shocks, Gale emphasized, were after all simply miniature lightning bolts, "improved by art, or artfully adapted to medical purposes." In other words, fundamental to the project of electrical humanitarianism in the Enlightenment was faith in the ability to master the quantity of electricity supplied to the human body. But could bodily quantities of electricity be reliably managed?[38]

Those who rejected the possibility of rationally manipulating electric bodies raised their voices at the same time as others confidently proposed programs of electrical humanitarianism. If the dream of enlightened electrotherapy was a balanced animal-electrical economy, its alter ego, the exploding electric body (the unbalanced economy), found arch expression in the puzzling phenomenon of spontaneous combustion, and its signal representation in Gothic fiction. Both learned journals like the Royal Society's *Philosophical Transactions* and genteel magazines like the *New England Magazine of Knowledge and Pleasure* reported horrifying and mysterious human combustions. In 1745, the *Transactions* reported some of the earliest American observations of emanations of bodily electricity. The Presbyterian divine Henry Miles recounted an episode in Maryland in 1683, when the naturalist John Clayton of Virginia reported to Robert Boyle that one November evening the clothes of Mrs. Susanna Sewall had appeared with "a strange Flashing of Sparks (seem'd to be of Fire)," and that "when they were shaken, it would fly out in sparks." More startling were the accounts of full-blown combustion, like that of the mysterious death of the Countess Cornelia Zangari, as recounted by Giovanni Bianchini, the Veronese natural philosopher. Combustions like these were perplexing. They appeared to be the result of fire, but the nature and source of this fire could not be determined since there were no external signs of conflagration. The fire must be internal, it was commonly surmised, either electrical or owing to some uncontrollable action of animal heat. Bianchini's account was reprinted by the Philadelphia-based agriculturalist Charles Vancouver, who published a *General Compendium* on natural philosophy in 1785. Placing his "horrible" tale in a footnote to spare the delicacy of female readers, Vancouver noted that Zangari had been found one morning "in the middle of the room, reduced to ashes, all except her skull, face, legs, and three fingers," and that "the stockings and shoes she had on were not burnt in the least." What had caused her death? The mystery surrounding the origins and effects of combustion provided ample room for conjecture. "Some attribute the effect to a mine of sulphur under the house; others, to a miracle; while others suspect that art or villainy had a hand in it."[39]

Healing and annihilation were merely two points on the electrical continuum, distinguished only by degree. In 1785, Dr. John Brockenburry of Virginia was busily electrifying Mrs. Susanna Wiatt when she was fortuitously cured by "a moment's application of atmospheric electricity." At that very moment, however, she was by sheer chance struck by a lightning bolt that entered her house through her chimney. Electricity could either kill or cure depending on the quantity of charge. Spontaneous combustion threatened to engulf the humanitarianism of enlightened electricians with forces of natural devastation beyond comprehension or intervention. In "the interests of humanity," Vancouver lamented, "we wish [the cause] could be derived from something external to the human body: for if to the calamities of human life already known, we superadd a suspicion that we may unexpectedly and without the least warning, be consumed by an internal fire, the thought is too dreadful to be borne." Reports of spontaneous combustion were accounts of the unaccountable. It was precisely because philosophical speculations did not embrace supernatural interpretations of such phenomena that such accounts took on an aura of uncanny dread. Eighteenth-century sermons about fatal lightning strikes were in this sense far less mysterious than philosophical accounts of lightning or combustion, since they could at least explain such occurrences. Natural philosophers, disavowing any overarching spiritual interpretation of combustion, floundered among an array of possible reasons. One writer on the combustion of one Cornelia Bandi (possibly a modified account of the Zangari episode) recited a litany of speculation on the cause. "Common fire" it could not be, for the room had not been burnt; one Mondini attributed the cause to lightning; a Verona philosopher argued for "inflammable matters"; Bianchini for "internal fire"; Maffei for "lightening generated in her own body." Yet the cause might not always be lightning, the author cautioned, citing the effects of "destroying internal fires," "fiery evaporations," "hot liquors," and "ignes fatui" (a light thought to be caused by the combustion of gases), and electricity, which appeared often in flashes from human bodies, especially women's (yet another nod to the link between female passion and electrical volatility).[40]

One account published in 1792 in the *American Museum*, a "letter respecting an Italian priest, killed by an electric commotion, the cause of which resided in his own body," ventured an interaction between animal heat and "a flame coming from a highly electric atmosphere." In demonstrating the identity of lightning and electricity, Franklin's experiments had closed one field of electrical speculation, but inadvertently opened another: the question of atmospheric electricity's effect on the human body. By the 1790s, some commentators were confidently asserting the atmosphere's electrical effect on the nervous system, albeit without proving their assertions experimentally. In his *Brief History of Epidemic and Pestilential Diseases* (1799), the lexicographer Noah Webster adapted Franklin to describe the communication of positive charge from electrical thunderclouds to terrestrial bodies. Webster asserted that since "the fibres of living animals are the most perfect conductors of electricity," it followed that "in all the motions or operations of electricity in the atmosphere, the nerves must be the principal subjects of its influence." Excessive atmospheric charge mechanically stimulated the body, producing heightened states of nervous "excitement" or "irritability," leading ultimately to "debility." This debility, Webster speculated, helped to account for the yellow fever epidemic that struck the United States at the end of the century.[41]

Electrical combustions might therefore be triggered by external atmospheric conditions—and the *American Museum*'s account of the "electric commotion" became one of cultural moment as it passed before the eyes of one of the early republic's most ambitious young writers, Charles Brockden Brown, just as Brown was composing *Wieland; or, The Transformation* (1798). Brown's political persona was virtually the opposite of Gale's. While Gale was an evangelical and a universalist who saw civil and religious freedom proceeding in tandem through the transatlantic revolutions, Brown was a staunch advocate of American commerce, a Francophobic nationalist, and a secular cultural guardian of the republic. He wrote strenuously in defense of U.S. commercial interests abroad (his father and brothers were merchants), urging President Jefferson to increase naval protection for American shipping and publishing pamphlets in 1803 highly critical of the President's near "loss" of the Louisiana Territory and the Mississippi River to France and Spain, respectively. Fearful of Jacobin infiltration in the aftermath of the French Revolution, he rejected the divisiveness of party politics, undertaking to serve his country as a man of letters. To this end, he sought to promote national integrity through cultural unification, crafting an American literary style based on what he took to be the progressive rationalism of the classical republican tradition.[42]

Brown's rationalism was highly chastened, however, in both the form and content of *Wieland*, his self-conscious attempt at an American Gothic novel. Brown, an expelled Quaker, styled himself a "rhapsodist," and as such "equally remote from the giddy raptures of enthusiasm and the sober didactic strains of dull philosophy." In *Wieland*, he dramatized not the triumph of reason but its precariousness as it is beset by menacing unknown forces. The events of the novel—ventriloquism, insanity, mistaken identity, murder, and spontaneous combustion—expressed deep-seated fears that religious enthusiasm and the secular imagination would disintegrate the republic. Confusions of cause and effect provided the central theme. Based on an episode reported in New York, one that eerily recalls the anecdote at the start of this chapter, the plot begins with the story of Wieland's father, a German mystic (evidently modeled on the figure of Johannes Kelpius) who settles on the Pennsylvania frontier and is driven to murder his family by what he believes to be the voice of God. Subject to strange "fits" and convulsions, Wieland's father constructs a "temple" in the countryside supported by "twelve Tuscan columns." One day, while at prayer and envisioning the approach of "a person bearing a lamp," "a very bright spark was seen to light upon his clothes [and] in a moment the whole was reduced to ashes." As narrated in the novel, the scene shows a striking similarity to Loammi Baldwin's account of being overwhelmed by an electrical atmosphere (Brown could certainly have seen Baldwin's article, published just a few years before). A mysterious "fiery cloud" engulfs Wieland's father in his temple, an event that the narrator, in a footnote, links to recent accounts of spontaneous combustion. The scene, one of great allegorical moment in the early history of the United States, expresses profound doubt that the conditions of American life could support a rational republican culture when beset by the terrors of inscrutable nature:

> A blazing light was clearly discernible between the columns of the temple . . . Within the columns [my uncle] beheld what he could no better describe, than by saying that it resembled a cloud impregnated with light. It had the brightness of flame, but was without its upward motion . . . Fear and wonder rendered him powerless . . . the purity and cloudlessness of the atmosphere rended it impossible that lightning was the cause; what are the conclusions that we must form? . . . Is it a fresh proof that the Divine Ruler interferes in human affairs[?] . . . Or, was it merely the irregular expansion of the fluid that imparts warmth to our heart and our blood, caused by the fatigue of the preceding day, or flowing, by established laws, from the condition of his thoughts?

Electrical accounts of spontaneous combustion radically questioned the optimism of electrical humanitarianism. While physicians envisioned the application of electricity to the body as transparent and powerfully restorative, skeptics emphasized the terrifying opacity of that body, and the potential for catastrophic interactions of internal and external forces. Ironically, millenarian devotion like Gale's bred faith in manipulating electrical bodies, but trust in reason alone seemed to undermine confidence that there existed a sure basis for such manipulations—and the enlightened projects they supported. To Brown, the life-giving inner light or spiritual fire invoked by radical mystics like Gale (and Kelpius) was a terrifyingly unknowable destructive force.[43]

The idea that America was a continent of intense electrical power might have formed the basis for patriotic valorizations of the new republic, to bury the old European prejudice of natural American inferiority. Knowledgeable travelers suggested as much: Volney observed that the American atmosphere was more electric than Europe's; von Humboldt thought that American eels were ten times more powerful than European torpedoes. But in the United States the reverse was the case. In the Revolution, spontaneous electrical experience was enabling; but in the early republic, the electric powers perceptible to Webster and Brown seemed dangerously uncontrollable. What had happened to the prophets of glory who had used fables of Franklinist natural mastery to foretell the rise of America? The Revolution had changed the cultural geography of the Atlantic world less than it had its political geography. In the short term, independence left the United States a fledgling nation-state in a world still dominated by the naval and economic might of Britain and France. Material weakness fostered cultural underconfidence: fears of being overwhelmed by natural powers were an expression of the isolation and fragility Americans felt now that they had left the empire and were looking, in a sense for the first time, exclusively to their own continental situation. Though the United States had become a political center in its own right, Americans still looked, through provincial eyes, to the centers of cultural life in Europe. The centers of science remained European as well, with Americans on the margins, a situation that would only fully change with the economic development and political consolidation of the mid-nineteenth century and the Civil War. Until that time, American scientific culture, nationalist rhetoric notwithstanding, continued to be shaped largely by the geopolitics of the colonial order of knowledge, remaining descriptive and local rather than philosophical and universal. The power of American electricity thus could not ground the nascent American identity, but in fact threatened it.[44]

CHAPTER SEVEN

Electricity as Common Sense

We shall [soon] be able to make some return to our transatlantic brethren, for the rich stores of useful knowledge, which they have been pouring upon us for two centuries.

SAMUEL MILLER, 1803

IN the same years that Dr. Gale was electrifying the inhabitants of New York, another itinerant named Elisha Perkins, of Plainfield in neighboring Connecticut, was developing a unique new form of electrotherapy, as well as a distinctive way of promoting it. On July 14, 1796, Dr. Perkins received a letter from John Vaughan, a physician trained at the College of Philadelphia and member of the Medical Society of Delaware, informing him of the progress of his recently patented "Metallic tractors." "I have operated in a few cases only with your instruments," Vaughan explained, one of "nervous head-ache," the other of "odantalgia" (toothache). The patients were apparently cured in both instances. A third case involved treating a young man who had fallen ten or fifteen feet "and received a considerable contusion in the umbilical region, with tumefaction." Vaughan bled him, but to no effect. After "two minutes operation with the instruments," however, the man had "exclaimed, in extasy, 'I am well—I am well!—my pain is gone.'" Reprinted in one of several promotional pamphlets published in the United States and Great Britain in the late 1790s, Vaughan's account exemplified Perkins's marketing strategy for the tractors: use compelling firsthand testimony of their effectiveness, but ignore the question of how they worked.[1]

Perkins's tractors were disarmingly simple things. A set consisted of two three-inch metallic rods made of brass and iron, and they sold for twenty-five continental dollars in North America, five guineas in Britain (Figure 17). At the close of the eighteenth century, they were so popular that they became in some quarters a form of currency. Dr. Benjamin Parker recalled that "a gentleman in Virginia sold a plantation and took the pay for it in tractors. Nothing was more common," he noted, "than to sell horses and carriages to buy them." In addition to his own itinerant promotions, Perkins employed physicians and apothecaries as local agents to sell tractors from New England to South Carolina. In Britain, meanwhile, tractors also became high therapeutic fashion. In 1797, Perkins sent his son Benjamin Douglas Perkins across the Atlantic to be his London agent. "Mr. P. imports his tractors from America in parcels of two hundred sets, valued by him at one thousand guineas," observed the skeptical Irish satirist John Corry, equaling "fifty-two thousand guineas annually for this *base metal!*" Setting up a clinic in Leicester Square, in the former residence of the surgeon John Hunter, Benjamin Perkins's career paralleled his father's: he sold tractors through agents in Bath, Liverpool, and elsewhere in provincial Britain. Despite repeated accusations of quackery, damaging comparisons with the controversial Mesmerist movement of the 1780s (which had involved hypnotic experiences of "magnetic sleep"), as well as the untimely death of Elisha Perkins in 1799, "Perkinism" proved durable enough that a Perkinean Society of London was founded in 1803. When Benjamin Perkins finally returned to the United States in 1807, he is thought to have supplied or treated possibly as many as 1.5 million patients and made £10,000.[2]

Around a dozen promotional pamphlets for the tractors appeared between 1796 and 1804. These imageless and adjectiveless anthologies of successful case histories repeated many times a set of key terms to reinforce the idea of their effectiveness: *effect, fact, influence, experiment, observation, sense, evidence, certificate, operation,* and *perfor-*

17. Perkins's Tractors: Electricity or Imagination?
During the late eighteenth century, Americans were fascinated by the promise of electrotherapies such as "Perkins's tractors." Although their promoters preferred not to explain how they worked, they nevertheless encouraged consumers to believe that the tractors, which were to be passed over various parts of the body, worked by electricity. Critics, however, insisted that their apparent success was really due to the power of the imagination.

---

mance. These written testimonials were a peculiar form of "virtual witnessing," that is, they allowed readers to imagine the performance of experiments conducted elsewhere, and through which natural philosophers made claims about the causal workings of nature. Perkinist virtual witnessing was peculiar in two respects, however. Readers could not "see" the successful trials performed with tractors because although their beneficial effects were described, they were never explained. Tractors simply worked.[3]

What made these unexplained metal rods credible as therapeutic instruments? During the late Enlightenment, practices emerged that were similar to medical electricity—most notably Mesmerism during the 1780s. Tractoration was another. Previous studies have stressed Perkinism's reliance on the patient's imagination, and presumed its fall to be inevitable; or, they have emphasized its connection to electricity as the key to its (temporary) believability. To assume that Perkinism would fail because it was quackery would be grossly teleological; but attempting to explain its credibility solely through its connection to electricity would tell only part of the story. Electrical ideas and practices did provide an important source of authority for the business of tractoration. But to understand why Perkinism flourished, we must grasp the interplay among Perkinism, electricity, and the varied sources of social and ideological authority on which it drew.[4]

"Common sense" and "useful knowledge" were Perkinism's core strategies. In the late eighteenth century, these slogans did not simply denote particular schools of making and using knowledge; they were powerful ideologies of knowledge. That is, they were descriptions of the way knowledge ought to be made and used, flowing from the egalitarian politics of the republican revolutions, and made to flow, in turn, out of the tractors. Perkinists worked hard to make the tractors matters of natural fact independent of both wonder and reason, and founded instead on the self-evidence of individual experience. In seeking harder distinctions between facts and wonders, Perkinism departed from the complex cognitive festivals engaged in by earlier electricians (the contrast with the millenarian Gale, who was Perkins's contemporary, is especially striking). By the seventeenth century, a recognizable modern language of fact had emerged in the disciplines of law and natural science; with some exceptions, natural philosophers favored a probabilistic conception, to temper dogmatic truth claims and to reduce the possibility of civil discord or authoritarianism. The career of the tractors, however, is part of a later phase in the history of the fact, one that looks forward to the positivism of the nineteenth century, in which the probabilistic sensibility had begun to yield to an antirationalistic form of self-evidence.[5]

Battle was joined, however, over precisely this opposition between facts and wonders, returning us once again to the problem of enthusiasm: was it electricity or the power of imagination that lay behind the tractors' success? In breaking also with the theatrical tradition in electricity, Perkinism made a virtue of plain speaking and of defending the public's ability to make their own judgments about natural knowledge by relying on evidence supplied by their senses. Thus Perkinism cast itself as the antiwonderful epitome of enlightened science: accessible, useful, and based solely on experience. Critics, however, insisted that wonders had not ceased: this language of plain common sense was in fact a new theater of wonder, erected to appeal to an enlightened citizenry. That this theater played equally well to audiences on both sides of the Atlantic poses interesting questions about American scientific culture at the end of the Enlightenment, and how it stood in relation to Europe.

The Delusive Tendency of Theories

Useful knowledge and common sense are slogans traditionally taken to express some of the Enlightenment's core aspirations for natural knowledge. In the *Lettres philosophiques* (1733), Voltaire blasted what he regarded as the pretty but useless academic philosophy of France under the ancien régime, comparing it unfavorably with the empirical, improving tradition of Baconian Britain. American invocations of useful knowledge have been traced to the English Quaker Thomas Bray, who published *An Essay towards Promoting All Necessary and Useful Knowledge, Both Divine and Human* (1697) and who founded the Society for the Propagation of Christian Knowledge in 1699. Bray's useful knowledge was both religious and secular: texts in history, classics, the sciences, and theology all commingled in the more than thirty thousand volumes he had shipped across the Atlantic. A similar fusion of Christian and secular concerns—improving the lot of human beings as a way to please God—has often been found in historical genealogies of modern science *as* useful knowledge, particularly in the writings of Francis Bacon. Within programs for enlightenment, the notion of utility could accommodate a variety of meanings. In the industrializing and internationally competitive era of the late eighteenth century, however, useful knowledge increasingly came to be viewed in material rather than spiritual terms. The complementary relationship between American and British views is well symbolized by the role played by Benjamin Thompson of Massachusetts, later Count Rumford in his European incarnation, in founding the Royal Institution in London during 1799–1800. According to its charter, it was to be "a public institution for diffusing the knowledge and facilitating the general introduction of useful mechanical inventions and improvements," promoting "the application of science to the common purposes of life." This was the same language Benjamin Franklin had used half a century earlier in Philadelphia to call for the organization of the American Philosophical Society "for promoting useful knowledge among the British Plantations of America." As we have seen, revolutionaries emphasized the lightning rod above all to celebrate the practical character of American genius.[6]

But material utilitarianism also created obstacles to the pursuit of science in the early republic. In the first place, the United States lacked the powerful aristocratic and state patrons who had been integral to the development of science in Europe; instead, America nurtured antipathy toward aristocracy and all forms of political centralization. In the late eighteenth century, while France and Britain were industrializing and forging ever closer links among science, state, and empire, the leading citizens of the United States were debating whether the republic should even aspire to be a manufacturing power. Jefferson, president of the United States and the American Philosophical Society after 1801, best embodies the political paradox of early republican science: the American philosophe was committed to a decentralized republic of agrarian yeomen, and opposed to centralization, credit economies, urbanization, and large-scale manufacturing. In an era when whig polemics of liberty and genius cast Newton's achievements as born of a peculiarly English crucible of Protestantism and constitutional government, David Hume could confidently assert that *"the only proper Nursery of [such] noble plants [is] a free state . . . and that a republic is most favourable to the growth of the sciences."* But revolutionary republicanism posed serious challenges to American science by denying it traditional bases of support. Local natural history thus became the most recognizable scientific idiom in the republic, rather than a natural philosophy that aspired to universal causal explanation—although by the Jacksonian period if not earlier, the states were rife with applications of Newtonian mechanics to a variety of local engineering projects. As Alexis de Tocqueville would observe in the 1830s, science without obvious application appeared to strike the mass of Americans as worthless.[7]

Science was also ideologically suspect in the late Enlightenment. The French Revolution, in particular the events of the Terror, were decisive in this respect. Anglophone critics led by Edmund Burke charged that reason itself had been perverted into a dangerous enthusiasm, with science fostering a dangerously instrumental cast of mind, one that was unchained from conventional morality or religion and that produced radicalism, atheism, and violence. Conservative commentators in the United States followed Burke's lead. Joseph Dennie, editor of the Federalist *Port Folio,* cast aspersions on the intellectualism of the new Democratic-Republican president, labeling Jefferson a "cool-headed philosopher." (Rather than the abstract rationalist political enemies liked to depict, Jefferson was actually an arch empiricist whose faith in the senses moved him to modify Descartes' "I think, therefore I am" to the sensationalist "I feel, therefore I exist.") The notion that gentlemanly civility provided a foundation for natural philosophers to speak authoritatively about matters of fact does not automatically apply to the early United States. As Dennie's attacks show, American gentlemen like Jefferson were exposed to political attacks in the form of anti-intellectualism, licensed equally by American republicanism and the fear of French Jacobinism. "Although there are no nobles in America," wrote the French minister Louis Otto in 1806, "there is a class of men denominated 'gentlemen,' who by reason of their wealth, their talents, their education, their families, or the offices they hold, aspire to a preeminence which the people refuse to grant them." Genius itself was suspicious. To Benjamin Latrobe, the British engineer who moved to Pennsylvania in 1796, the United States was an antiscientific culture. "The want of learning and of science in the majority is one of those things which strikes foreigners who visit us very forceably," Latrobe wrote. "Superior talents actually excite distrust, and the experience of the world perhaps does not encourage the people to trust men of genius." In an era of incipient specialization, gentlemen of learning like Samuel Latham Mitchill of New York, editor of the *Medical Repository,* were lampooned for the virtuosic breadth of their interests. Dangerously Jacobinical or simply lacking social relevance, science appeared by turns suspect and ridiculous in North America.[8]

As with useful knowledge, the notion of common sense carried a variety of meanings. In the Anglophone eighteenth century, it figured prominently in three distinct though not unrelated discourses: social, political, and epistemological. The exemplar of social common sense was Anthony Ashley Cooper, third earl of Shaftesbury. In *Sensus Communis: An Essay on the Freedom of Wit and Humour* (1709), a highly influential work in the age of genteel clubbing, Shaftesbury used common sense to describe the sympathetic feelings that united gentlemen as a community of sociable beings, bound together by the free enjoyment of wit and raillery. Shaftsburyean common sense emphasized the naturalness of social affections and civil commonality among men, a theme that was later important in the "moral sense" tradition articulated in the Scottish Enlightenment by Francis Hutcheson, Adam Smith, and others—and that became a prominent feature of moderate Protestant sermons and college curricula in America during the second half of the century. In the words of the indomitable clubber and historian of the Tuesday Club of Annapolis, Dr. Alexander Hamilton, "There exists a certain affection or fellow feeling, between all bodies in nature, by which they have a strong tendency, to approach, one towards another, to Join."[9]

Shaftesbury also discussed common sense's second connotation, that of the moral reliability of lay practical judgment, even though he was keenly aware of its limits. "If," he wrote, "by the word 'sense,' we were to understand opinion and judgment and, by the word 'common,' the generality or any considerable part of mankind, it would be hard . . . to discover where the subject of common sense could lie. For that which was according to the sense of one part of mankind was against the sense of another. And if the majority were to determine common sense, it would change as often as men changed. That which was according to common sense today would be the contrary tomorrow, or soon after."[10] The political implications of invoking the reliability of public judgments were made clear during the American Revolution, when the practical judgment of "the people" served as a moral foundation for rejecting British authority and for creating the republic. In the *novus ordo saeclorum* envisioned by the Declaration of Independence, legitimate governments now derived their authority not from custom but from "the consent of the governed," and were called on to submit the "facts" justifying their existence to "a candid world." In *Common Sense* (1776), the British radical Tom Paine deliberately fashioned his prose to embody the political self-evidence of the new republic. Arguing for freedom from the anachronistic opacities of imperial government, he explained, "I offer nothing more than simple facts, plain arguments, and common sense." Regard for lay judgment and plain speaking was not, of course, an American innovation. And the public to whom Paine and the other Founders appealed was neither democratic nor universal, since the political nation was carefully limited to propertied gentlemen. Nevertheless, in comparison with ancien régime Europe, written constitutions subject to public scrutiny did emphasize an unprecedented role for lay judgment in American public life. "Equality," Alexis de Tocqueville later reflected, "stimulates each man to want to judge every thing for himself." Democratic peoples "mistrust systems and like to stick very close to the facts, and study them for themselves."[11]

Third, common sense described a school of epistemology linked to reaffirmations of the capacity of ordinary human beings to make reliable judgments about nature in the face of radical skepticism.[12] Since Aristotle, philosophers had spoken of the *sensorium commune* as the faculty of mind that integrated the evidence of the five senses into a pattern intelligible to human beings. By the eighteenth century, leading physiologists such as Albrecht von Haller and Charles Bonnet believed that the operation of such a faculty might be traced to a specific part of the brain where nerves from the five senses conjoined. To these practitioners, the integrative faculty of common sense became the key to locating and understanding human consciousness; to the less orthodox, such as the controversial healer Franz Anton Mesmer, integrative common sense was a sixth sense connecting the individual to the external influence of macrocosmic forces. "Common sense realism," meanwhile, was a distinctive late-century response to the philosophical skepticism of David Hume, and original to the Scottish Enlightenment, but once again widely taught in American colleges, particularly in the decades after the Revolution. This school, led by the Aberdeen philosopher Thomas Reid, responded to Hume's attack on the empirical demonstrability of cause and effect, and defended the human capacity to make reliable judgments about nature (and religion) based on the evidence of the senses. Sensory evidence enjoyed the status of self-evidence: "No man seeks a reason for believing what he sees or feels," Reid asserted. "There is no reason," he went on, "why the opinion of a philosopher should have any more authority than that of another man of common sense." Yet he did not overstate his case. Reliance on the senses was not fallacious, but that did not mean the senses were infallible. From intuition, Reid argued, human beings justifiably believed in regular and efficient causation, but it was above the power of natural philosophy to demonstrate this. "We are very much in the dark with regard to the real agents or causes which produce the phenomena of nature," Reid conceded. In an educational culture inclined toward Protestant piety and useful knowledge, as the American republic was, commonsense realism provided a serviceable alternative to skepticism and a justification for the continued pursuit of Baconian science.[13]

As we shall see, Perkinism did not draw on useful knowledge and common sense as static and unchanging traditions, but intervened in them, making its own relationship between them. Its central, controversial claim was that tractors were credible not despite their lack of elite scientific explanation, but because their effects were evident without such explanation. Perkinism thus made reliable knowledge of the useful facts of tractoration reside in the common sense of the public alone, relegating the opinions (and questions) of academic physicians or natural philosophers to irrelevance. That is, it used common sense to establish the utility of tractors by drawing on the republican animus against elite knowledge. As Shaftesbury sagely hinted, however, one man's common sense often turned out to be another's delusion.

As with Gale, Elisha Perkins's origins are obscure. Little is known about his youth, education, or religious persuasion. Originally from Plainfield, Connecticut, Perkins was the son of Dr. Joseph Perkins of

Norwich, and possibly attended Yale College before establishing his own medical practice. By the mid-1790s, Elisha (who was apparently the father of ten) was heavily in debt and had taken to mule trading to improve his finances. The decisive turn in his fortunes came in 1795. In that year, he noticed that his penknife appeared to have certain unusual curative effects on his patients, and he hit upon the use of what he initially called metallic "transfers," owing to his early belief that they cured nervous disorders by drawing off "surcharge[s] of the electric fluid" responsible for "the greatest part of our pains." Through trial and error, Perkins worked out a method of treatment superficially similar to the practice of "stroking," a controversial therapy employed by the seventeenth-century English healer Valentine Greatrakes. According to the *Directions for Performing the Metallic Operation with Perkins's Patent Tractors* later issued by Benjamin Perkins with each set of tractors sold in Britain, a large number of rheumatisms, convulsions, inflammations, and aches could be relieved by "drawing the Points of the Tractors over the Parts affected, and continuing them along on the Skin to a considerable Distance from the Complaint, usually towards the Extremities." "Thus, if the Pain, is in the Shoulder or Elbow, operate over those Parts, and extend the Tractors along on the Skin, to the Hands; if in the Hip or Knee, from those Places to the Foot." In all cases, "Tractors should be used alternately, changing them in the Course of two or three Minutes." Guarantees of success were studiously avoided. The *Directions* stressed that some complaints were more susceptible to treatment than others; moisture or oil on the hands of the operator or the body of the patient, too, inhibited the therapy's effectiveness. The method described was the one "*generally* practised . . . though it is sometimes necessary to vary it as the Circumstances of the Case require." Additionally, it might be necessary to experiment with different operators, because the tractors "in a few instances operated more effectually in the Hands of some Persons than in those of others." These qualifications aside, the *Directions* declared that "many Complaints which have for Years baffled the Efforts of Medicine, have been perfectly cured."[14]

Unlike other eighteenth-century therapies, tractors were appealing because of their painless, practical simplicity. Even the most enlightened physicians still bled their patients, as shown by Benjamin Rush's response to the yellow fever epidemic (an outbreak that claimed Elisha Perkins in New York in 1799; his body is thought to be buried under Washington Square). Therapies were often as painful as the ailments they targeted. The same could be said of electric shocks: as Gale shows, patients sometimes had to be forced to submit to them. Tractors, however, required minimal expertise and involved a painless treatment that required only twelve to thirty minutes on average to complete. This austerity also contrasted with the practice of Mesmerism, the therapeutic movement that claimed to restore nervous health and that became a spectacular Parisian craze of the 1780s until French medical authorities publicly criticized it as the work of the imagination. Mesmerist sessions lasted at least two hours and were conspicuous performances, involving the direct physical manipulation of participants' bodies; the use of mysterious tubs containing glass, water, and protruding metal tubes; atmospheric devices like the playing of musical instruments, notably the glass harmonica; and the organization of "crisis" rooms with mattresses onto which patients could comfortably swoon. Mesmerism also made bold explanatory claims, describing itself as the channeling of "animal magnetic" fluids between body and cosmos.[15]

Next to the "ridiculous" practice of Mesmerism, Perkinism was far less performative, billing itself as an external treatment only, in which small metal rods rather than human hands intervened in the animal economy. Moreover, no claims were made about why the procedure worked; instead skeptics were invited to simply witness the visible effects on the body. "The knowledge, which may authorise us to speak on effects produced by an external application," wrote Benjamin

Perkins, "is very unlike that which would be necessary, where internal remedies are used." "Good eyes, in a character uninfluenced by prejudice, or *interest*, may give testimony respecting the change *observable* in an inflammation or a tumefaction, on an application of the Tractors, which is more satisfactory than all the medical knowledge in the universe, without those requisites." Useful facts derived from the senses formed the "solid basis" of Perkinism. "I flatter myself," Benjamin Perkins confided to his readers, wearing his empiricism on his sleeve, that "I shall have the satisfaction, on being favoured with suitable subjects, of affording evidence, which to every person must be more satisfactory than any testimony, viz. the evidence of the *senses*." According to Dr. James Tilton, an associate of John Vaughan and president of the Delaware Medical Society who helped to market tractors in Wilmington, the crucial difference between Elisha Perkins and Mesmer was that the Austrian had *"obscured those simple facts, which should have been used for the benefit of Society, with a pile of empirical frauds."* Perkins, on the other hand, *"barely disclose[d] useful facts."* It was the "delusive tendency of theories" that had led to the disgrace of animal magnetism, and Perkins would not make the same mistake. Taking pains to distinguish itself from the cultish secrecy that surrounded Mesmerism, Perkinism claimed to promote the public interest through a transparency it described as "perfectly intelligible to every capacity," including even "the most unlettered in society." "The opinion I early imbibed of the inutility of theories, has induced me not to attempt them," wrote Benjamin Perkins, echoing his father. "I have thought it more worthy my attention to collect FACTS, articles more substantial in their nature, and useful in their tendency." Perkinism thus argued for a particular relationship among common sense, useful knowledge, and matters of fact: lay practical judgment, based on the evidence of the senses, demonstrated that the utility of the tractors was a fact that required no explanation. "The judicious physician at first leaves the flowery path of speculation, for the more arduous one of experiment," Benjamin quoted Elisha, "and builds his theory, so far as is possible, on the solid basis of facts." "These facts he establishes by the concurring testimony of his senses, accompanied with such critical and candid observations, as alone are competent to detect a fallacy, or support the truth." "The great object with Dr. Perkins," he emphasized, "was first to ascertain the truth relative to the *usefulness* as well as the *existence* of the discovery." "To accomplish this, it was not absolutely necessary to understand minutely the theory, but rather to observe critically the effects."[16]

Through their publications, both Elisha and Benjamin sold tractors directly to consumers, allowing them to "witness" (secondhand) the seemingly spontaneous testimony of the human body, uncomplicated by philosophical explanations. For example, in a letter to Elisha dated October 7, 1796, which was subsequently published in *Evidences of the Efficacy of Doctor Perkins's Patent Metallic Instruments* (1797), Pierpont Edwards, federal district attorney in New Haven, reported the case of a neighbor named Mrs. Beers who had been so distressed by her rheumatism that "she had not been able to walk across her room even with crutches." She "procured a set of your Metallic Substances, and in less than an hour after she had begun to use them, in the manner directed by you," she rose from her chair and walked about the house, and "on the next day she went abroad to her neighbors, having thrown aside her crutches." "She is," Edwards concluded, "in a fair way to be restored to perfect health." "It is a duty," he insisted, "you owe yourself and the world, to promulge this event." In a similar account from the same year, Reverend Elhanan Winchester testified to seeing the "*Metallic Tractors* tried with success, upon several patients in the Alms or Bettering house in Philadelphia, especially upon a man who was unable to lift his right-hand to his head . . . but who in a few minutes, was able to move it at pleasure." All those whom Winchester had seen receive the treatment agreed that they had been successfully cured, and he pledged his unswerving gratitude to the inventor. "From what I saw *with my own eyes, and what I heard,* I have great hopes that this discovery will greatly tend to alleviate the miseries of mankind."[17]

Josiah Meigs, professor of natural philosophy at Yale College, described an experiment with tractors on his eight-year-old son in 1797 in exemplary Perkinist fashion:

> Dr. Munson being again called, pronounced his case a hazardous one; after having prescribed what he thought proper, suggested a trial of your *Tractors*. This I immediately undertook, and in about half an hour he declared the pain was gone, turned himself without difficulty on his right side, and fell into a profound sleep . . . he awoke in perfect health, and has continued so to this day . . . I cannot tell why the waters of Jordan should be better than Abana and Pharpar, rivers of Damascus, yet since *experience* has proved them so, no reasoning can change the opinion. Indeed, the causes of all common facts are, *we think,* perfectly well known to us, and it is very probable that fifty or an hundred years hence, we shall as well know why your *Tractors* should in a few minutes remove violent pains, as we now know why cantharides and opium produce opposite effects; viz. we shall know but *very little* about either, excepting *facts.*

In his testimonial, Meigs swore to the effectiveness of the treatment without describing his method or any physical process to explain it. Perkinist prose allowed the reader to "witness" the fact of the body's responsiveness to tractoration, but little more. Tractoration was useful knowledge that, ironically, occasioned no actual claims to knowledge.[18]

In both America and Britain, Perkinists explicitly set the common sense of the public against the trustworthiness of licensed physicians. Charles Langworthy, a surgeon who acted as the Perkinist agent in Bath, expressly addressed his promotional writing "to the publick, and have of course chosen to write it in a popular way, without applying the technical language of the profession, where it was possible to avoid it." Langworthy went on: "It is not to display *myself* that I write, but in the language of simplicity to convey my subject 'home to the business and bosoms' of the illiterate and the afflicted, as well as to the philanthropist, the philosopher, and the physician." It did not matter that the cause remained unknown, since the good effects of the treatment were self-evident. The causes underlying the effects of magnetism, opium, and Peruvian bark were also obscure, but their efficacy was scarcely questioned. "None can pretend to say how the bark cures an intermittent fever; mercury the syphilis," Benjamin Perkins protested, "or how opium produces sleep, and yet they are not rejected on that account." Even Professor Peter Christian Abildgaard of the Danish Royal Academy of Science, who experimented with reanimating dead animals and worked at formulating a physical theory of tractoration in Copenhagen, thought it unwise to reject the practice just "because we cannot immediately see the connexion between the cause and effect." *Terrible Tractoration* (1803), a mock-heroic satire directed against critics of the tractors by Thomas Green Fessenden, the Vermont projector and London associate of Benjamin Perkins, warned opponents that by condemning the tractors "you make a sacrifice of *truth, decency* and *common sense.*" He rammed the point home in a lyric reportedly delivered at the opening of the Perkinean Society in London on July 15, 1803: "What, though the CAUSES may not be explain'd, / Since these EFFECTS are duly ascertain'd, / Let not self-interest, prejudice, or pride, / Induce mankind to set the means aside: / Means, which, though simple, are by Heaven design'd, / T'alleviate the woes of human kind."[19]

Metallic Tractors and Galvanic Circles

Despite this barrage of antiphilosophical rhetoric, the tractors' advocates actively encouraged the perception of a link between tractoration and animal electricity, or "Galvanism." Engaging with this compelling if controversial branch of natural philosophy might not produce a definitive explanation of tractoration, but it might deflect charges that the tractors were somehow not what they seemed.

By the late eighteenth century, the speculative writings of Newton, experiments with electrical fish, and investigations of the nervous system of frogs appeared to support the view that electricity was in fact the fundamental basis of all animal life. In the 1710s, Newton had influentially speculated that the "animal spirits" that communicated messages from the brain to the muscles might be related to the existence of a subtle ethereal or electrical fluid. By the 1770s, as we have seen, research conducted by the London experimenters John Walsh and Henry Cavendish had confirmed Edward Bancroft's suggestion that eels and torpedoes naturally generated their own electricity in their internal organs. The discovery of electricity in fish strongly encouraged the belief that electricity animated all organic creatures. Although taken up by natural philosophers like the Abbé Pierre Bertholon during the 1780s to explain the functioning of the nervous system, the concept of "animal electricity" attained a far more spectacular notoriety with the work of the Bolognese anatomist Luigi Galvani. In *De viribus electricitatis in motu musculari* (1791), Galvani described experiments in which he procured physical responses from the bodies of dead frogs using metal instruments. According to Galvani, this postmortem muscle action was produced by the continuing activity of a form of electricity that he believed was inherent in the nervous system, and thus the motive force for the animal economy itself. Animal electricity (or "nerveo-electric fluid") thus provided an alternative to the animal spirits as the conducting medium by which muscular motion was thought to be initiated.[20]

This striking claim was taken up by many well-known electricians such as Tiberius Cavallo and the Edinburgh-trained physiologist Richard Fowler. Yet Galvanism also became one of the most contested bodies of experimental knowledge in the Enlightenment. Alessandro Volta, who would later invent the first current-generating battery, questioned the existence of animal electricity, arguing that Galvani's "ambiguous frogs" convulsed because they served as a conductor for electricity between two charged pieces of metal, not because they discharged any electricity of their own. A war of interpretation raged throughout the 1790s among Galvani, Volta, and their supporters. Some commentators, meanwhile, conformed to neither camp. Fowler viewed Galvanism as an entirely "new influence," whereas Alexander von Humboldt believed the nervous fluid to be galvanic in nature, but unlike conventional electricity. Galvanism also became morally controversial when Galvani's nephew, Giovanni Aldini, tried to reanimate human corpses using electricity in London in 1802, a scandal that inspired the plot of Mary Shelley's *Frankenstein* (1818). If animal electricity ultimately proved to be an untenable doctrine for Volta and his allies, it nevertheless achieved spectacular public notoriety and stimulated unprecedented interest in the theory and practice of medical electricity (not to mention a run on frogs).[21]

Americans were exploring the relationship between electricity and life during the same period. In 1789, even before Galvani published his findings, Moses Willard of Worcester, Massachusetts, sent his "Treatise on Medical Electricity" to Benjamin Rush, in which he began by explaining that "the nerves are tubes which conduct the nervous, or Electrical fluid from the brain." During 1794–1795, the prominent Philadelphia physicians John Redman Coxe and Charles Caldwell offered reviews of European galvanic writings in their early publications, as did polite periodical literature like the *American Monthly Review*. Joseph Macrery's *Inaugural Dissertation on the Principle of Animation* (1802), written at the University of Pennsylvania, was one of several medical theses composed at American colleges advancing the view that "all the phenomena of life and motion, are owing to the energy of a subtile, active principle, called the electric fluid."[22] Animal electrical ideas also circulated beyond the academy. In 1803, while the *New York Herald* was publishing front-page reports of Aldini's galvanization of London cadavers, Sigismund Niderburg, an itinerant physician from Vienna who claimed to have been Galvani's assistant (and who later continued his practice in Havana, Cuba), published *Improved Galvanismus* and established his own electrical clinic in New York, observing that "the life of organized bodies consist in a continual Galvanic activity." The formation of a New York Galvanic Society was announced that same year, organized by one "M. Carreudessez," who intended to use the "Galvanic Pile" to explore animal electricity's "astonishing power over living things." "The Apparatus pours unfeelingly the principle of life," declared David Launy, yet another medical electrician, in 1804, "and restores health without the help of medicine." As we have also seen, humane societies were quick to adopt galvanic theories to justify their own reanimation programs. And even general works like Samuel Miller's *Brief Retrospect of the Eighteenth Century* (1803) and Thomas Dobson's *Encyclopaedia* (1803) made room for detailed summaries of galvanic phenomena.[23]

Perkinists made no attempt to resolve the ambiguities surrounding animal electricity; rather, they exploited them. From the very beginning of his use of "transfers" in 1795, Elisha Perkins had evidently believed that the therapy he devised was electrical in nature, a view he eagerly recorded in his diary. His account book also verifies that he possessed a copy of Franklin's writings on electricity. In public, however, Perkinism was a promotional strategy that marginalized the issue of causation. Elisha's early pamphlets, published around 1797 mostly in New England, barely mentioned electricity at all. It was only after 1797, in the British pamphlets published by Benjamin Perkins, that Perkinism engaged publicly with Galvanism. Suddenly, the relationship with electricity had always been there. In recounting his father's discovery of the tractors, Benjamin Perkins now connected this historic event to the experiments of Galvani. This chronology might suggest that the link to electricity was a feature of British Perkinism, rather than American. Indeed, the American patent under which tractors were sold (the first federal medical patent granted in the United States) was for "Removing Pain, etc., by Metallic Points" (1796), whereas the royal British patent explicitly referred to tractors as an "Application of Galvanism as a Curative Agent" (1798). But the link to electricity as a causal explanation was not unique to the British context. For one thing, the tractors' British career exceeded their American one for reasons of historical accident—Elisha's untimely death in 1799. And although it was the later British pamphlets that explored the relation between tractors and Galvanism most fully, the first person to publish at length on this subject was an American: John Vaughan.[24]

Vaughan was a well-established physician by this time. He knew the likes of Rush and Caldwell from his university days in Philadelphia, and was an associate of James Tilton, president of the Medical Society of Delaware and keen promoter of the tractors. Vaughan published his own independent examination of the galvanic connection as early as 1797, entitled *Observations on Animal Electricity, in Explanation of the Metallic Operation of Doctor Perkins*. In this work, he deployed the electroconductivity of metallic points, the concept of animal electricity, and the Franklinist economy of charge to explain how tractors worked. *"Electricity is already proved by philosophers, to be concerned in almost all natural phenomena,"* he proffered. He also worked out in this treatise an electrophysiological account of the will in which the brain set the nerves in motion by surcharging the muscles—organic Leyden jars—with electricity. All physical sensation depended on excess or deficiency of nervous energy, or what he called *"hominal electricity"*: the "different movements [of muscles] may very rationally be imputed to their different degrees of electricity." Pain, pleasure, and health, therefore, became simply a matter of equalizing positive and negative charges in the animal economy. This was what tractors did: "the metals being susceptible of this fluid, conduct the extra degree of energy to parts where it is diminished, or out of the system altogether, restoring the native law of electric equilibrium." Tractors worked as an intervention in the animal electrical economy. Sensibility was, in fact, electricity, with pain being "merely an accumulation of electricity, in a particular part." "The subsequent state of ease is obtained by abstracting the extra degree of sensibility." True, tractors produced no sensible shocks or visible sparks. But they could still be credible as electrical devices because by the 1790s experimenters were exploring the "weak electricity" that issued when different metals were brought into contact, the forerunner of the electric current that was later to flow from the Voltaic pile. This weak electricity was invisible and did not spark. In this context, the electricity of tractors was credible as an idea, even if it was not evident to the senses.[25]

Benjamin Perkins excerpted long sections of Vaughan's galvanic account of the tractors in his promotional pamphlets along with similar writings by other American supporters, such as Yale Medical School founder Eneas Munson's "Dissertation on Animal Electricity and Magnetism" and a treatise by the Reverend John Devotion of Old Saybrook. On one occasion, after tractorizing several patients, Devotion declared himself to be possessed by a *"universal relaxation and nausea"* akin to "the sensation of an electric shock." Unusually for a Perkinist, Devotion took this as an invitation to theorize. "Thus, sir," he triumphantly announced, "with experiments on my own leg opens a system of philosophic speculation." "I then reasoned thus," he continued, relying on Vaughan's theory: "I have drawn the electric fluid of three persons into my own body." His conclusion was unequivocal: "As the *Metallic Tractors* act upon an established law of nature, there is no fear they will want power so long as the electric rod will draw electric fire." Lengthier accounts by Matthew Yatman, a member of the Company of Apothecaries and codirector of the Perkinean Society in London, and Professor Abildgaard of Copenhagen (whose electrical theories Benjamin Perkins also reprinted) both linked the tractors to Franklinist electricity. The different metals in each rod, they argued, created a "galvanic circle" of positive and negative electric charges (this even though patients were instructed to use their tractors alternately); their pointed ends could draw off, redistribute, or supply electricity to afflicted areas of the body.[26]

For all this, Perkinists never made the concept of animal electricity integral to the credibility of the tractors. The ideas of Vaughan, Yatman, and Abildgaard remained undemonstrated hypotheses. Even after their theories had been published, the Perkinist agent in Bath, Charles Langworthy, publicly admitted that there was no "satisfactory theory on their phenomena; nor has Dr. PERKINS as yet published any theory upon which the effects produced by his discovery may be accounted for." Vaughan himself equivocated about the status of his interpretation. "Call it electricity," he insisted, "the nervous fluid, the galvanic fluid, or what you will." When the surgeons C. G. Herholdt and J. D. Rafn of the Royal Frederick Hospital in Copenhagen performed experiments with the tractors with favorable results (their reports were translated from German and edited for incorporation in the British pamphlets), the best they could do was to offer multiple possible explanations. "Perkinismus," as they called it, "must be explained on the principles, either of *mechanical stimulus, electricity, galvanism, or imagination.*" Similarly, Benjamin Perkins lumped together an array of electricians—some of whom, like Eusebio Valli and Volta, were in open philosophical conflict—as possible authorities for the tractors' electricity. These clumsy attempts at explanation, which recall the causal ambiguities of spontaneous combustion, achieved a less-than-prominent place in the pamphlet literature, being relegated in one case to a long footnote appearing after some three hundred pages of doggedly "factual" eyewitness testimonials. The footnote even included the opinion of Cavallo that the "surprising effects" of electricity were "generally inexplicable" and seemed "to admit of no theory sufficiently probable or satisfactory."[27]

Benjamin Perkins dismissed the relevance of the speculative interpretations he himself quoted, preferring to return his readers instead to the more solid ground of facts. "That the tractors act on the galvanic principle is very generally believed," he averred, "but I shall not consider it any disparagement to the Practice, if I admit that the present knowledge of the laws and properties of that principle is so limited, as not to allow of our giving of the *modus operandi* a complete and satisfactory explanation." Following a discussion of Abildgaard's theories, Perkins reminded readers that "electricity and Galvanism are now just enough involved in mystery to afford an ingenious theorist an ample field for expatiating on the principle of this metallic operation." But even while Perkinism shunned the language of wonders, it displayed its disdain for reason, pressing its reassuringly conservative epistemological allegiance to experience alone. "As theories, on almost every operation in nature, are generally as various as the individuals who give them, I shall at present confine myself to what I conceive of more importance, viz., *to establish the facts,* and leave it with the philosophic reader to gratify himself with *theorising* as much as he pleases." Perkinist utilitarianism made the very concept of electricity—indeed, any concept—irrelevant to successful therapeutic practice. Such philosophical innocence was proof of moral virtue, and solid Americanism to boot: "Whilst the Philosophers of Europe were engaged in their researches into the phenomena of Galvanism, or the effect of metals on the nerves and muscles of various animals, and thus ascertaining the extent of the metallic influence upon the inferior parts of creation, the late Dr. Perkins, an eminent physician of Connecticut, in North America, conceived the more noble purpose of directing his pursuit to the examination of their effects in the diseases of man."[28]

The Age of Quackery

Public healing traditionally relied on various forms of showmanship to attract customers; even enlightened medical electricians routinely promoted their services through theatrical display. William King, for example, was an itinerant physiognotrace artist and ivory-turner who performed electrical cures at venues like the Providence Bathhouse in 1800, cures that he "exhibited" as "amusements"; Sigismund Niderburg, too, sold admission tickets to his electrotherapeutic clinic in

New York, where "spectators [were] admitted from one to two o' clock every day." Conspicuous display in public healing, however, conjured up the specter of wonder-mongering and charlatanry. James Graham and Anthony Yeldal, British itinerants who toured North America as "oculists" and "aurists" in the early 1770s (before they turned with great controversy to electrical healing upon their return to Britain), evidently inspired such fear of imposture in New England that in October 1773 the Connecticut legislature passed an Act for Suppressing Mountebanks. The law, passed against those "publickly advertising and giving notice of their skill and ability to cure diseases," took seriously the power of the spectacular in medical fraud, banning public stages and prohibiting "any plays, tricks, juggling or unprofitable feats of uncommon dexterity and agility of body, [which] tends to draw together great numbers of people, to the corruption of manners, promoting of idleness, and the detriment of good order and religion." From the colonial period—when authorities legislated against the establishment of the theater, citing mainly religious and moral objections—to the emergence of an independent republican nation, theatricality bred suspicion in America.[29]

Perkinism did not seek to persuade its public through elaborate rhetoric or conspicuous theatrical display, the traditional strategies associated with unorthodox healers. Perkinists were neither charismatic "charlatans" (loquacious verbal deceivers) nor "mountebanks" (itinerant doctors who literally mounted stages publicly to display their wares) in their conventional senses. Rather, as the product of and a product for a culture committed to antitheatrical republican transparency, and hence the superior morality of common sense and useful facts, Perkinism employed a form of "natural theatricality": the art of making nature appear to speak for itself. Perkinism's credibility resided not in the social status of its supporters or any institutional affiliation, but rather in a depersonalized material thing (metal tractors) amenable to universal manipulation. Such a strategy resonated with the morality of depersonalizing the public sphere in the age of republican revolution. Whereas the "charlatan wanted to be seen," the Perkinist aimed "not to display [himself]," in Langworthy's words—it was, to paraphrase Paine's *Common Sense*, the instrument or doctrine rather than its advocate that deserved public attention. Rejecting the conspicuous performance and controversial philosophical engagement of Mesmerism, Perkinism aimed to display not itself but merely the bodily effects of the tractors. Charles Wilkinson, a hostile British critic, drew a particularly illuminating distinction between the two therapies in this respect: "In the Mesmerism doctrine there is something fascinating, something calculated to strike the minds of the multitude ... It is difficult to conceive what interest can be attached to the tractors, which are founded on no medical or philosophical principles." If there was anything mesmerizing about the tractors, it was paradoxically their very matter-of-factness. The distinctiveness, and success, of Perkinism's appeal to common sense and useful knowledge lay in its appearance as a form of republican nonperformance.[30]

In lieu of conspicuous performance or methodological secrecy, inscription became an important strategy for controlling the marketing and use of tractors. According to critics, tractors were simply expensive pieces of metal. Branding helped to fix the identity of these pieces of metal *as* tractors, and justify their price, which was higher than conventional electrotherapies. (Gale, as we saw, claimed that his readers could build their own electrical machines for only two to three dollars, following his instructions.) Legal inscription also guarded against counterfeit reproduction by drawing on the patent system then taking shape in the Industrial Revolution and thereby establishing claims to the tractors' legal originality. As a result of such inscription, however, Elisha Perkins was expelled from the Connecticut Medical Society in May 1797 for practicing (in the words of his colleagues) "delusive quackery." In addition to the opprobrious claim that Perkinism had been "gleaned up from the miserable remnants of animal magnetism," it was the charge of monopolizing his invention as a "nostrum" for personal gain that provoked the expulsion. Gentlemanly disinterestedness and profit were incompatible, the society insisted.³¹

But Elisha Perkins exploited the new industrial ideology of inventor compensation to maintain that no such conflict existed. Commenting publicly on his expulsion, Perkins cited the American federal patent signed by George Washington, Timothy Pickering, and Charles Lee, which granted tractors legal protection by giving the inventor "the full and exclusive right and liberty of making, constructing, using and vending [them] to others" for fourteen years. American patent law weighed equally the rights of inventors with the good of society. Perkins's patent served "to defray the expenses that have arisen in disseminating a knowledge of its importance" under "the sanction of our first constituted authorities"; nothing more. Perkins was merely "promoting the progress of useful arts," as the Constitution demanded, to which no loss of moral face or social credit was attached. Tractors were easy to counterfeit, however, so these claims to legal originality had ultimately to be inscribed on the tractors themselves. "INSTANCES *occurred during the last Year, where the Tractors suffered discredit through illegal and dishonourable attempts to circulate spurious imitations,*" Benjamin Perkins informed his British readers in 1801. It was essential, therefore, to mark the genuine articles from false. "*To guard against Impositions, Applicants will please to observe, that every* Genuine *set of Tractors is stamped with the words* 'PERKINS PATENT TRACTORS;' *and to the printed Directions accompanying them is subjoined a receipt for the Five Guineas, signed in the Hand-writing of the Patentee. To counterfeit this is Felony.*" Fetishization via inscription promised control over the identity and sale of "real" tractors by distinguishing them from mere counterfeited metal things.³²

The ideal of medical emancipation was also central to the tractors' appeal. "Did not the magnanimous Mr. Perkins, in open defiance of the winds and waves, traverse the vast Atlantic Ocean, to work miracles in this favoured isle?" sarcastically asked the Irish satirist John Corry. Universal benevolence was precisely the claim. "The writer has crossed the Atlantic," Benjamin Perkins declared, "and become a resident of London, that he may devote his time and attention to the diffusion of this important discovery, and its application to the relief of the miseries of mankind." By selling tractors directly to consumers, Perkinists were selling the ideal of autotherapy, of control of one's body free from the pricey interference of established physicians, to a "candid and enlightened public." This social crusade was of course no innovation. Autotherapy possessed a long history as an emancipatory, even antinomian project; enlightened medical electricity, as Gale vividly but by no means uniquely demonstrates, was emphatically linked to notions of individual freedom. It was precisely because academicians (allegedly) scorned unorthodox therapies like medical electricity that the common sense of a skeptical public would recognize their good effects (the revelation of which vested interests naturally opposed).[33]

Perkinism thus relied in many ways on promoting the idea that nothing mediated the public's apprehension of the facts of tractors' good effects. But this immediatism was itself the product of a series of careful mediations. Although presented as the spontaneous testimony of bodies, the credibility of Perkinist testimonials drew on the social authority of the testifier. In reality, Perkinists did not dismiss the authority of all gentlemanly physicians, only those who criticized the tractors. One factor that explains why Gale's influence remained local and provincial, while Perkins's spread throughout the United States and crossed the Atlantic, is that Perkins built a network of agents while Gale remained an individualist. Physicians like Tilton in Delaware and Langworthy in Bath constituted a web of entrepreneurial promoters that proved crucial to the tractors' geographical reach, from New England to the Carolinas in North America, from London to Liverpool and the West Country in Britain.[34]

Let us here reconsider Meigs's testimony about his son being cured by tractors. Meigs provided an eyewitness account that emphasized the evidence of the senses, but for readers, the stated identity of the author as a professor at Yale College was surely not unimportant. Recall also his mentioning that a physician had endorsed the tractors to him. In addition to those with a direct stake (Tilton and Langworthy) or a medical interest (Vaughan), American testimonials were overwhelmingly drawn from the professional classes, especially clergymen, judges and lawyers, military officers, and college professors. Recommended to the public not so much as men of high class as men of good character, those who testified nevertheless included some of the most prominent citizens of the United States, like the geographer and Federalist politician Jedediah Morse, Massachusetts congressman and envoy extraordinary to France Elbridge Gerry, and U.S. Supreme Court Chief Justices Oliver Ellsworth and John Marshall. Perkinists even claimed that the father of the nation, George Washington, owned a pair of tractors.[35]

In Britain, by contrast, where the Perkinean Society was sponsored by patrons like Lord Rivers, Sir William Barker, and William Franklin (Franklin's loyalist son and former governor of New Jersey), the social character of Perkinism was more aristocratic. Perkinists claimed that the tractors had been successfully used in institutional settings in both countries: in poorhouses and almshouses in Philadelphia (including the Pennsylvania Hospital), New York, Boston, and around New England as well as in the St. Pancras Poorhouse, Guy's Hospital, the Regimental Hospital of the Duke of York, and several locations in provincial Britain, the most important being Bath. It was in Britain, however, rather than America, that Perkinism gained an institutional footing in the form of the Perkinean Society, an "Institution for Relieving the Poor" sponsored by aristocratic patrons and offering free treatments for London's impoverished laborers. This process of institutionalizing social benevolence mimicked the formation of the humane societies and the creation of specialized medical establishments like the London Electrical Dispensary. The principles put forward in the first edition of the short-lived *Transactions of the Perkinean Society* (1803) declared its aim: to "afford relief to the diseases of the afflicted and industrious poor of the metropolis." In Britain, where urbanization and industrialization were more advanced than in America, Perkinists identified tractors with noblesse oblige and projects of urban regeneration. To this end, the Perkinean Society was to serve as a clearinghouse, vaguely styled after the Royal Society, in which disinterested persons might publish the results of experiments with tractors.[36]

Making American testimony travel across the Atlantic did not always convince skeptical Europeans, however. A specifically European challenge to Perkinism emerged over the issue of the testifier's social rank. Langworthy, the Perkinist agent in Bath, described Elisha Perkins to Britons as "a man of honour and integrity," reassuring his readers that the tractors enjoyed the support of "some of the most eminent physicians, and natural experimentalists of the United States." Perkinists, however, hedged their bets by using both high-status and socially anonymous testifiers, "not wish[ing] to rest the credit of the discovery merely on the authority of great names, however repeatable and influential." This approach encountered problems. According to one story, tractors had reached continental Europe when the wife of a Major Oxholm, animated by "a laudable desire of extending their utility to her suffering countrymen," had brought a pair back to Denmark. With pride, Benjamin Perkins subsequently boasted of a universal therapy, presenting to his readers "cases in America," "cases in England," and "cases at Copenhagen." But the response of a Danish professor named Tode indicated the limitations of using classless, nonprofessional testifiers for European audiences. "Who may this Mr. CALVIN GODDARD be that we should so implicitly give him credit," Tode inquired, deriding "the fashion with all those physicians, who trumpet forth their cures, to bring witnesses who have no other credentials than their names." Extraordinary cures, in particular, merited attention to "the *rank* and condition of the witness"; ideally, "he should have a certain *notoriety*." "How can a person in Europe trust to such witnesses," Tode persisted, "when they certify to things so extraordinary"? Perkins protested that "the success of the practice in America was substantiated by as full and respectable testimony as has ever been deemed necessary to establish any medical facts," claiming that the German translations of Perkinist tracts had accidentally omitted professional titles from the testimonials. But Tode was implacable: "They are from persons of no note or character, and consequently entitled to no credit." The evidence of Ebenezer Robinson, a Connecticut physician who was not identified as such in the German translation, "does not deserve any credit," concluded Tode, "as the witness is *not a professional man*." Although mistranslations might have occurred, as Perkins claimed, the occasional use of socially anonymous testimonials in the American and British pamphlets was deliberate, not accidental.[37]

More damaging were satirical and theatrical traditions that opponents of the tractors drew on to recast Perkinist common sense as a confidence trick. Attacks on the tractors emerged quickly in the United States. Even before Elisha Perkins had been expelled from the Connecticut Medical Society, local wits had lampooned tractoration in the *Connecticut Courant* in a "Patent Address" in 1796. In this belletristic salvo, which anticipated Fessenden's mock-critical *Terrible Tractoration,* a fictional Perkins sings of the tractors' ability to restore sight to the blind and speech to the dumb, while lining his pockets with dollars. Tractors would make the old young again ("Yet, notwithstanding it so strange appears, / I soon *stroke* sixty, down to twenty, years"); even redeem human sin ("Man shall no longer feel old Sin's controul— / The powers of sickness no more pain his soul / . . .You shall cure the maladies of all, / Old sores expunge—and wipe away the Fall"). In Philadelphia the next year, Dr. James Currie reported to the

American Philosophical Society that tractors worked by "the impression made on the Patient's mind," invoking the comparison with Mesmerism once again. But these criticisms failed to offer a clear demonstration that tractors were a fraud, and they did not reach a broad public (the anonymous satire was only published locally, and Currie never published his paper); as a result, these criticisms were easily outrun by the transatlantic itinerancy of the Perkinses.[38]

The more decisive attacks on the tractors took the form of published accounts of theatrical dramatizations of the power of the imagination staged in Britain. When unchecked by reason or the senses, the faculty of imagination in its pathological mode inspired trepidation among enlightened natural philosophers, incarnating the persistent threat of ungovernable powers of mind. By the 1790s, the American and French Revolutions had moved critics to link imagination explicitly with the specter of mobbish populaces (we have seen how electricity was linked to critiques of American enthusiasm in the 1770s). Natural philosophers understood this problem in terms of politicized passions and secular enthusiasm, in which artificially engineered effects were mistaken for having real natural causes. The use of placebo trials, used to effect by the 1784 Royal Commission headed by Franklin investigating Mesmerism, was therefore revived to debunk the tractors. The key was to expose Perkinist common sense as a theatrical trick of the imagination; that is, to bring back the language of wonders to show that tractoration was not self-evident experience, but an artfully produced effect of the mind.[39]

The lead was taken by the Bath physician John Haygarth. Concerned that "persons of rank and understanding" were lending their support to Perkinism in Bath (where Langworthy operated as the Perkinist agent), Haygarth asked publicly *"how the Patent Tractors produced such wonderful effects"* and determined to intervene, if necessary, to "correct public opinion." In *Of the Imagination as a Cause and as a Cure of Disorders of the Body* (1800), he described experiments he and others (primarily Richard Smith, a hospital-trained surgeon who worked at the Bristol Infirmary) had performed both with wooden and real tractors, declaring that "the whole effect undoubtedly depends upon the impression which can be made upon the patient's Imagination." Tractors were instruments of wonder, they claimed. By passing off a pair of false tractors as real, then secretly substituting the genuine article, Haygarth claimed to have secured virtually identical results. Of those patients treated with the false tractors at the General Hospital in Bath, he reported that one "could walk much better," another "was easier for nine hours," and a third "had a tingling sensation for two hours." Those treated with real tractors, meanwhile, "were in some measure, but not more relieved by the second application, except one." Haygarth assured his readers that "a fair opportunity [had been] offered to discover whether the metallick Tractors possessed any efficacy superior to the ligneous Tractors, or wooden pegs." How did tractors work? The answer was not electricity. It was "the wonderful force of the Imagination."[40]

Haygarth's trials dramatized a competition over the direction of experimental subjects and triggered a series of accusations and counter-accusations about witness manipulation. Perkinist pamphlets claimed that unsuccessful treatments resulted only "for want of a proper knowledge of the mode of using the tractors, and of the diseases subject to their influence." Benjamin Perkins denied any manipulation of his patients, insisting that successful tractoration resulted specifically from "the perfectly tranquil state of the patient's mind." His opponents were the true manipulators, he insisted, forcing fantastic stories on experimental subjects about the healing powers of the tractors before treating them with false ones. This was in fact true, and Haygarth gleefully admitted it. "It was often necessary to play the part of a necromancer," he confessed, "to describe circles, squares, triangles, and half figures of geometry, upon the part affected, with the small ends of the tractors." He now played up the association between tractoration and Galvanism as pure theater of the mind. "During all this time we conversed upon the discoveries of FRANKLIN and GALVANI, laying much stress upon the power of metallick points attracting even lightning, and conveying it to the earth harmless." Above all, the tests turned Perkinism into theater, and satire, subjecting its participants to ridicule and laughter. "To a more curious farce I never was witness; we were almost afraid to look each other in the face, lest an involuntary smile should remove the mask from our countenances, and dispel the charm. But to return to my patient:—In one minute, he felt a smarting in his loins." In identifying the imagination as the occult agent of tractoration (the same charge leveled by the Commission on Mesmerism a decade earlier), Haygarth's and Smith's performances lent the tractors that aspect of conspicuous theatricality that Perkinists had so scrupulously avoided. When put on display in this manner, the commonsensical facts about tractoration could be rewitnessed as "marvellous" and "astonishing" effects that owed more to the "wonderful and powerful influence of the passions of the mind upon the state and disorders of the body" than to any physical effect of the metallic rods.[41]

True to ideological form, Perkins attempted, in turn, to rewrite anti-Perkinism as an elitist attack on the public's ability to make judgments about useful knowledge. Where Haygarth had been concerned lest respectable citizens unwittingly endorse quacks, the satirist John Corry openly derided the affluence of Perkinism's hypochondriac London patrons, who imagined themselves "indisposed when only labouring under the torpor of indolence." Perkins, however, claimed for himself a transparency of practice that the trials with fictitious tractors lacked, conducted as they were behind closed doors. The anti-Perkinists were the true wonder-mongers. It was in these private settings that they had used "terror and awe" to contort "the minds of the credulous patients," some of whom they had "almost frightened to death." In asserting that tractors were a fraud, anti-Perkinists had insulted "the penetration of the public," since thousands believed they had been successfully treated. Indeed, how suddenly to dismiss the testimony of all those who were fully convinced they had been cured? How could real effects be imagined? "Can the imagination cure a *gout?*" Perkins implored. Children, even horses, had been cured "in the presence of many spectators." Surely the imagination could not explain "the cure of an animal of the brute creation"?[42]

This appeal to the public interest notwithstanding, Perkins refused to grant credibility to anti-Perkinist testimony because of the lowly social standing of the subjects involved. Ironically, social status now became an openly avowed criterion for credibility for Perkins, too. Trials conducted at public hospitals were "far less satisfactory than those on persons of respectability in private practice," he now declared, "where the character of the patient, as well as the disease, is better known," contradicting his own demand that all trials with the tractors should be public, not private. This unease over the tractors' public status induced Perkins finally to discuss the class and stage dynamics of such practices that he and his father had always sought to minimize. "No declaration of relief from the poor credulous paupers in a hospital, ought to be admitted as evidence in this practice, unless there is a *visible* proof to the by-stander of the alteration," he demanded. Suddenly, in order to regain control over testimony about real tractors, the mediation of the eyewitness became more reliable than the spontaneous testimony of the subject's body. "To persuade patients of this class to declare, that they are relieved, and even to think so, nothing more is necessary than to impress on their minds a favourable opinion of the remedy, to induce them to believe that thousands have been cured by the *wonderfully efficacious* means, and they will be very ready to acknowledge that they begin to feel what others have previously experienced." Here was the "trick which has been played off on *Fictitious Tractors,* and public hospitals have been sought as the best theatre where such experiments ought to be exhibited." Even at a linguistic level, the tractors' identity had become unstable, reduced to "pieces of stick," "ligneous tractors," and "wooden skewers." Smith admitted to using pieces of bone, slate pencils, and tobacco pipes at the Bristol Infirmary specifically "to render the trials the more ridiculous."[43]

The satirist Corry supplied the coup de grâce, cementing the image of Perkinism as a theater of wonder that for too long had made itself invisible in a cloak of enlightened common sense. It was no longer the Age of Reason, Corry observed, but the "age of Quackery," in which *"miracle-mongers"* and *"retailers of sanity"* preyed upon an inexhaustible "public credulity." *"Cheats can seldom stand long against laughter,"* however, and they would not stand against the defense of those "who cannot think or judge for themselves" through "well-intended Satire." In *Quack Doctors Dissected* (1810), he recounted the tale of "Wilkinson," a John Bullish "rural philosopher" whose custom it was to make commonsense examinations of all forms of "literary quackery" (including the moral effects of "the new philosophy," the doctrine of sexual egalitarianism, and "similar paradoxes of this enlightened age.") The fictional Wilkinson, mimicking Haygarth, employed fictitious tractors to embarrass a foppish female aristocrat named "Dame Thomson," who was made to flee in shame when, imagining herself cured, Wilkinson revealed that the *"genuine American metal"* had had no effect on his own dog. Wilkinson's "Tractors" were, in fact, "part of my old kitchen poker, which Ben Perkins, our blacksmith, took to the smith yesterday and hammered into skewers." Corry consigned Perkins to a pantheon of quackery styled the *"Grand Pantomimic-farcical-tragi-comical Drama."* "Their theatre," he wrote, "might be a temporary structure of wood, emblematic of the transitory nature of all earthly blessings ... The first scene should exhibit a number of old men and women hobbling in on crutches, and *groaning,* to the great delight of the hearers, while Mr. Perkins, like a kind magician, came forward and by touching the old women with his talismanic tractors, they should appear suddenly restored to health and ease." In the next scene, "masquerade" and "pantomimic gesticulation" dazzled the eyes of the spectators, before a Faustian denouement brought "Justice" down from heaven and "by one touch of her fiery sword the ground . . . opens beneath the feet of the beneficent advertising physicians and their satellites." Some years before, the American writer Fessenden had jokingly made his fictional anti-Perkinist Dr. Caustic warn that he was "preparing a most awful Tragedy for Drury Lane Theatre" entitled the "Dreadful Downfall of Terrible Tractorising Confounded Conjuration." For Perkinists, however, the farce was no laughing matter: with their common sense exposed as wonderful theatrical performance, tractors indeed passed from the Age of Reason to the Age of Quackery.[44]

The sale and use of tractors petered out on both sides of the Atlantic halfway through the first decade of the nineteenth century, although the metallic rods would be invoked for years as an emblem of quack medicine. Benjamin Perkins returned to Connecticut in 1807, where in his final years he continued to pursue commercial promotions of useful medicine, became involved in scientific publishing, and even associated with leading American natural philosophers like Benjamin Silliman of Yale, to whom he sold a mineralogical cabinet of some two thousand specimens he had collected while in Europe.[45]

The success of the tractors turned on their promoters' ability to turn generic things (metal rods) into specific objects ("tractors"). The tractors' history shows how simple material things could became loaded with extrinsic properties and meanings to establish (paradoxically) the claim that these properties and meanings were inherent, self-evident, and a matter of common sense. Rejecting equally the languages of wonder and reason in favor of facts based solely on experience was a strategy designed to exploit the ideological and epistemological conservatism that followed the American and French Revolutions, yet also appeal to a democratic ideal of individual medical emancipation. So it was that Perkinism broke with the culture of display in electricity and Mesmerism. The controversy over tractors turned on the same issue: critics insisted on reducing tractors back into mere matter, mere pieces of wood or metal. Ultimately, tractoration was a "marvelous fact" in both senses of the phrase: depending on one's point of view, it was either a process that produced genuine effects that defied explanation, or a delusion masquerading as a fact. It is a telling irony that such a "simple" practice had to rely on so many discourses to make it credible as such. The tale of the tractors is equally ironic as the final chapter of the electrical enlightenment in America. From the economy of charge and the lightning rod to Galvanism, electricity was so well established in the public mind by the end of the century that it could now be invoked as an idea that could compel attention even without empirical evidence of its presence. Electricity, in other words, was itself now a form of common sense; but therefore, more than ever, possibly only a wonderful illusion.[46]

Mesmerism, it has been argued, marked the end of the Enlightenment in France, representing the triumph of a form of pseudo-science that perverted the conventions of natural philosophy. But in adopting recognizable enlightened ideals and styles, Perkinism (and indeed Mesmerism) did not signal the end of the Enlightenment so much as its universalization, revealing in the process the difficulty of telling where enlightenment ended and imposture began. By appealing to common sense, utility, and facts, Perkinism demonstrated how the anticharlatan and antimountebank were becoming uniquely potent persuaders in the enlightened public sphere, both in America and Britain. What did this transatlantic success signify? Without question, important differences existed between the two contexts at the turn of the nineteenth century. British science was increasingly imperial in its ambitions, while American science remained local and provincial. But success in both scientific cultures signals that important commonalities were in play, too. At the beginning of the chapter, the Presbyterian intellectual Samuel Miller was quoted as expressing his hope that Americans would soon offer Europeans important new discoveries, repaying the two centuries of knowledge that they had received. The tractors were one such return, and they point to some enduring features of science and medicine in the eighteenth-century Anglophone Atlantic, in particular, the commercial circulation of techniques and languages for making and using natural knowledge. United by a thriving open market for unorthodox therapies, the North Atlantic formed a single itinerary for promoters like Benjamin Perkins. But equally important, Perkinism articulated ideals for natural knowledge—common sense, useful knowledge, and the value of empirical facts—that Americans and Britons held in common. Tractors sold well in both countries because their promoters spoke languages of enlightenment that these "transatlantic brethren" understood.[47]

CONCLUSION

What Is American Enlightenment?

*I*N the eighteenth century, electricity raised compelling questions about agency and power. When electricity made sparks and moved matter, what was really at work: the divine power of God or some secular force of nature? And could human beings, through the practice of experiment and the use of reason, control natural powers that at their most violent included the lethal force of lightning? There was no single answer to such questions, just as there was no definitive answer to Kant's question "What is enlightenment?" This book, however, has described the contours of American debates about what electrical enlightenment meant for the relationship between human agency and divine and natural powers. It is useful to think back, first, to the image with which we began: Franklin as a revolutionary Jupiter hurling thunderbolts at the enemies of America (Figure 1). This is where our understanding of American Enlightenment has always been, with Franklin's godlike power at the center of a glorious story about electricity and the emergence of American agency: an electrical 1776.

The broadsheet cartoon "Political Electricity," however, provides a counterimage that helps us to place the electrical revolutionary in context (Figure 13). We still have the figure of Franklin as the kite flyer, but he is now only one of many actors in a larger drama of enlightenment, defined here as the history of communication and revolution in an Atlantic world linked by oceangoing ships and electric chains, along which power and agency travel (albeit in a diffuse, even hidden, way). Franklin is here juxtaposed with the figure of the human electric machine (Figure 14): an icon not of agency but of human manipulation by natural powers (in this case through the art of experiment). We can now see the extraordinary conceit of Franklin as the American Jupiter for what it really was: a nationalist claim to natural mastery that suppressed experimental skepticism about the possibility of manipulating electricity, and that depended on, yet also reacted against, a history of Atlantic circulation.

Was there a distinctively American Enlightenment? Ways of knowing cannot ultimately be separated from political ways of being. In the early modern era, European cities became imperial capitals and centers of global science, obliging their peripheries to report experience of the natural world for interpretation at the center. Franklinist electricity was the first major rupture in this Atlantic hierarchy, generating new philosophies of nature at the periphery that commanded metropolitan assent, although their success did not immediately lead either to the creation of powerful scientific centers in North America or to a new theoretically emboldened American way of knowing. The emergence of T. Gale and Elisha Perkins at the end of the century as the most dynamic American electricians points to a social difference between European and American science at the end of the Enlightenment. While Europeans, many of them in recognized institutions and academies, sought to discipline electricity through strategies of disembodiment like the mathematical representation of forces or the increased mechanization of experiments, American electricity's most creative practitioners remained hands-on medical entrepreneurs and men on the margins.

Gale and Perkins also show that there were different paths to enlightenment in America. Gale was an individualistic spiritual natural philosopher for whom reason and revelation were conjoined in personal experience, a tradition that survives to this day despite secular narratives of scientific progress that made such emphases seem to disappear, and despite famous interpretations—above all, de Tocqueville's in *Democracy in America*—of the starkly utilitarian character of American science. Perkins, by contrast, was an antirationalist who built a network for utilitarian autotherapy that rejected both reason and wonder in favor of the "facts" of common experience. Perkinism was not, however, simply an essential American epistemology. The religious tradition, wild with ideas and visions as epitomized by Gale, shows that there were several American traditions, and that for many there was no contradiction between spiritually and materially useful knowledge: the so-called Age of Reason and Age of Enthusiasm did not always clash. Both Perkins and Gale, each in their different ways, expressed political hostility toward elite scientific knowledge, viewing it as a form of tyrannical authority. This is what American anti-intellectualism (or antirationalism) has always been: a form of resistance by the periphery to the very idea of a political center. Anti-intellectualism is not a timeless American mentality but a specific eighteenth-century product of the revolution against Atlantic hierarchies.[1]

As everyone knows, the leviathan of imperial control stalked eighteenth-century America but failed to subject it. British authority was destroyed in what became the United States precisely when it attempted to establish a true transatlantic empire exercising military, political, and economic dominion over its American colonies. The American Revolution did not immediately produce a strong American nation-state, however, but a country divided between centralizing and anticentralist urges, only later united through a bloody civil war in the mid-nineteenth century. This tension—born of the localist character of colonial American society, crystallized by the Revolution, and carried over into the United States—structured the American Enlightenment and early American science. Certainly, there were those in positions of power, epitomized by Benjamin Rush, for whom science and medicine promised elite control of the physical bodies and social behavior of republican citizens. There also were those, like John Adams, who applauded Franklin's experiments as a modernizing blow against superstition and enthusiasm. In the colonial era, mastery of electricity allowed Americans to align themselves culturally with their British empire, helping to identify white Creoles as enlightened political masters of uncomprehending native Americans and African slaves.

But the careers of electricians like Ebenezer Kinnersley, Gale, and Perkins show how American enlightenment about nature was animated by antinomian and ecstatic impulses more than any statist agenda for elite social management. Even Franklin, the bookkeeper of nature who sought to control electricity through economic management, marveled at its fiery tendency to liberty and violent separation. In early modern Europe, scientific institutions like the Royal Society began sanctioning specific ways of knowing and talking about nature in public, while identifying and disqualifying those rival epistemological traditions they deemed unenlightened. But in America, the practice of science was not disciplined by national institutions until the middle of the nineteenth century. In the eighteenth century, therefore, the science of electricity, which circulated as a wonderful commodity through a society that lacked metropolitan institutional controls, was made meaningful through personal experience. Indeed, the anticentralist impulse that animated so many American electricians replayed, time and again, the Revolution's rupture of the hierarchies that had structured the Atlantic world under the old regime. The electric fire was not a tool for the mechanical construction of a new American leviathan, but the vital moral instrument of individual experience in the happy absence of a strong state. And so it remains controversial to speak of "the American Enlightenment" because enlightenment continues to be associated with a (European) program of social control at odds with a democratic (American) revolutionary heritage.[2]

This anticentralism was not unique to eighteenth-century America, but an extension of the liberal structure of British political and commercial culture (in which early modern Americans participated). In terms of experimental practice, moreover, there was a striking and necessary continuity between American and European electricity: Americans like Franklin and Bancroft modeled their techniques on European precedents, precisely so that their experiments would be credible to metropolitan audiences. The art of experiment made possible a white bourgeois enlightenment that commodified wonders. But experiment did not completely strip electricity of its marvelous or spiritual aspects. This was in part because of the practical difficulty of subjecting electricity completely to rational laws, and in part due to the sheer religiosity of its bourgeois cultural consumers. American engagements with electricity reveal a pattern in which nature was approached with humility, with nature (and its Creator) revered for its wonderful ability to surprise the human body and astonish human reason.

Although the zeal and pluralism of the American public sphere, unstructured by any state-sponsored church, gave this pious modesty broad and powerful voice, Protestant emphasis on humility before nature's divine powers was not unique to North America. The Christian zeal that drove Joseph Priestley's English experimental career, and John Wesley's, prove the point (although it is telling that Priestley emigrated to Pennsylvania in the 1790s). Yet what we might call the American economy of wonder was more liberal than its metropolitan European equivalent in the eighteenth century. This was in part due to the absence of an established religion in America. But the American embrace of nonrational experience was also an effect of Atlantic geopolitics. Colonial Americans were necessarily modest witnesses of nature: their claims to knowledge would have to defer to those of their metropolitan betters on the other side of the Atlantic. American philosophical modesty was thus an expression of enforced political modesty. The freedom to wonder at nature was in fact a privilege of being subordinate.[3]

The twist in this tale of early American science, however, is that trust in modest, personal experience of the natural world was not rejected when political independence was achieved; instead, it became ideologically crucial for the preservation of republican freedom (not to mention Protestant piety) given the threatening possibility that European-style centers, both political and intellectual, might establish themselves in the United States. Whereas Europeans had insisted that colonial Americans *couldn't* theorize about nature, independent Americans from Elisha Perkins to Thomas Jefferson, taking their cue from Franklin, agreed they *shouldn't*, because to do so would betray common experience and reinstate the hierarchies against which the Revolution had been fought. In America, enlightenment lay in rejecting, rather than pressing, claims to rational mastery.

Notes

Abbreviations

BFP   Benjamin Franklin, *The Papers of Benjamin Franklin*, ed. Leonard Labaree et al. (New Haven, 1959–)
EO   Benjamin Franklin, *Experiments and Observations on Electricity, Made at Philadelphia in America*, 5th ed. (London, 1774)
PG   *Pennsylvania Gazette*
Phil. Trans.   *Philosophical Transactions of the Royal Society*
WMQ   *William and Mary Quarterly*, 3d series

Introduction

1. Loammi Baldwin, "An Account of a Very Curious Appearance of the Electrical Fluid, Produced by Raising an Electrical Kite in the Time of a Thunder-Shower," *Memoirs of the American Academy of Arts and Sciences* 1 (1785): 258–260. This is Loammi Baldwin senior; on the engineering careers of both Loammi senior and junior, see Daniel H. Calhoun, *The American Civil Engineer: Origins and Conflict* (Cambridge, Mass., 1960).

2. Kant heralded Franklin as the "Prometheus of the new age" *("dem Prometheus der neuer zeitern")*: Immanuel Kant, *Kants Werke*, 12 vols. (Berlin, 1902), 1:472. On Fragonard, see *Benjamin Franklin's Experiments: A New Edition of Franklin's Experiments and Observations on Electricity*, ed. I. Bernard Cohen (Cambridge, Mass., 1941), xxvii, and Mary D. Sheriff, "*Au Génie de Franklin:* An Allegory by J.-H. Fragonard," *Proceedings of the American Philosophical Society* 127 (June 1983): 180–193; on French views of Franklin, see Alfred O. Aldridge, *Franklin and His French Contemporaries* (New York, 1957); on liberty and power in eighteenth-century Anglo-American political thought, see Bernard Bailyn, *The Ideological Origins of the American Revolution*, rev. ed. (Cambridge, Mass., 1992), 57–58.

3. See, however, Sylvia Plath's disquieting evocation of the effects of electroconvulsive therapy on the patient in *The Bell-Jar* (New York, 1963); as well as Jürgen Martschukat, "'The Art of Killing by Electricity':

The Sublime and the Electric Chair," *Journal of American History* 89 (2002): 900–921; more generally, see David E. Nye, *Electrifying America: Social Meanings of a New Technology, 1880–1940* (Cambridge, Mass., 1992).

4. Recent textualist accounts of the American Enlightenment include Henry F. May, *The Enlightenment in America* (Oxford, 1976); Ned C. Landsman, *From Colonials to Provincials: American Thought and Culture, 1680–1760* (New York, 1997); and Robert A. Ferguson, *The American Enlightenment, 1750–1820* (Cambridge, Mass., 1997). David Jaffee's "The Village Enlightenment in New England, 1760–1820," *WMQ* 47 (July 1990): 327–346, is mainly concerned with textual circulation in provincial America, but his more recent "Curiosities Encountered: James Wilson and Provincial Cartography in the United States, 1790–1840," *Common-Place* 4 (2004), www.common-place.org, usefully embraces material culture as well. For works that question the impact of the Enlightenment in colonial America, see Herbert Leventhal, *In the Shadow of the Enlightenment: Occultism and Renaissance Science in Eighteenth-Century America* (New York, 1976), and Robert Blair St. George, *Conversing by Signs: Poetics of Implication in Colonial New England Culture* (Chapel Hill, 1998). For a literary approach to the social history of the American Enlightenment, see David S. Shields, *Civil Tongues and Polite Letters in British America* (Chapel Hill, 1997); for social histories of enlightened penology and freemasonry, respectively, see Michael Meranze, *Laboratories of Virtue: Punishment, Revolution, and Authority in Philadelphia, 1760–1835* (Chapel Hill, 1996), and Steven C. Bullock, *Revolutionary Brotherhood: Freemasonry and the Transformation of the American Social Order, 1730–1840* (Chapel Hill, 1996). The principal older studies of early American science are Brooke Hindle, *The Pursuit of Science in Revolutionary America, 1735–1789* (Chapel Hill, 1956), and Raymond P. Stearns, *Science in the British Colonies of America* (Urbana, Ill., 1970). There has been a wave of recent work, however. On the eighteenth century, see Pamela Regis, *Describing Early America: Bartram, Jefferson, Crèvecoeur, and the Influence of Natural History* (Philadelphia, 1992); Amy Meyers and Margaret Beck Pritchard, eds., *Empire's Nature: Mark Catesby's New World Vision* (Chapel Hill, 1998); Leigh Eric Schmidt, *Hearing Things: Religion, Illusion, and the American Enlightenment* (Cambridge, Mass., 2000); Nina Reid-Maroney, *Philadelphia's Enlightenment, 1740–1800: Kingdom of Christ, Empire of Reason* (Westport,

Conn., 2001); Susan Scott Parrish, *American Curiosity: Cultures of Natural History in the Colonial British Atlantic World* (Chapel Hill, 2006); Joyce E. Chaplin, *The First Scientific American: Benjamin Franklin and the Pursuit of Genius* (New York, 2006); Sara S. Gronim, "Ambiguous Empire: The Knowledge of the Natural World in British Colonial New York," Ph.D. diss., Rutgers University, 1999; and Andrew Lewis, "The Curious and the Learned: Natural History in the Early American Republic," Ph.D. diss., Yale University, 2001. On science and colonization in America before the Enlightenment, see Joyce E. Chaplin, *Subject Matter: Technology, the Body, and Science on the Anglo-American Frontier, 1500–1676* (Cambridge, Mass., 2001).

5. Epigraph to *Benjamin Franklin's Experiments;* Horace Walpole quoted in Simon Schaffer, "Natural Philosophy and Public Spectacle in the Eighteenth Century," *History of Science* 21 (Mar. 1983): 15; Ebenezer Kinnersley, advertisement in the *New-York Gazette, Revived in the Weekly Post-Boy,* June 1, 1752; Jacob Green and Erskine Hazard, *An Epitome of Electricity and Galvanism by Two Gentlemen of Philadelphia* (Philadelphia, 1809), xxxiv, xxxvi. See also I. Bernard Cohen, *Franklin and Newton: An Inquiry into Speculative Newtonian Experimental Science and Franklin's Work in Electricity as an Example Thereof* (Philadelphia, 1956), a book that possesses the unique distinction of being dedicated to both Perry Miller and Alexandre Koyré.

6. See Michel Foucault, *Discipline and Punish: The Birth of the Prison,* trans. Alan Sheridan (New York, 1977), and for Foucault's influence on recent histories of enlightened science, the introduction to William Clark, Jan Golinski, and Simon Schaffer, eds., *The Sciences in Enlightened Europe* (Chicago, 1999). On elite concern with the social danger of enthusiasm and imagination in relation to early modern science and society, see Schaffer, "Natural Philosophy and Public Spectacle"; Lorraine Daston and Katharine Park, *Wonders and the Order of Nature, 1150–1750* (New York, 1998), 14–19, 324, 334–364, esp. 361–362; Steven Shapin and Simon Schaffer, *Leviathan and the Air-Pump: Hobbes, Boyle, and the Experimental Life* (Princeton, 1985); and Michael Heyd, *"Be Sober and Reasonable": The Critique of Enthusiasm in the Seventeenth and Early Eighteenth Centuries* (Leyden, 1995). On electrical wonder, see Marcello Pera, *The Ambiguous Frog: The Galvani-Volta Controversy on Animal Electricity,* trans. Jonathan Mandelbaum (Princeton, 1992), 3–18; on wonders in colonial New England, see David D. Hall, *Worlds of Wonder, Days of*

Judgment: Popular Religious Belief in Early New England (New York, 1989). On empiricist responses to Cartesian dualism, see Jessica Riskin, *Science in the Age of Sensibility: The Sentimental Empiricists of the French Enlightenment* (Chicago, 2002), esp. 21–28, 47–51; and Roy Porter, *Flesh in the Age of Reason* (New York, 2003), part 1. Riskin's fine account of sensibilist science in eighteenth-century France complements, in my view, the history of American electricity offered here. Two overall differences are worth noting, however. Where Riskin focuses on sensibility as a moral and natural philosophy, I concentrate on the material history of the body in experiment. Second, religion plays a much larger role in my account than in Riskin's, as a contextual factor that conferred epistemological credit on corporeal experience in America. On idioms of experience in American Puritanism, see Jim Egan, *Authorizing Experience: Refigurations of the Body Politic in Seventeenth-Century New England Writing* (Princeton, 1999). The highly suggestive discussion of "ecstatic" experience in nineteenth-century travel and ethnography in Johannes Fabian, *Out of Our Minds: Reason and Madness in the Exploration of Central Africa* (Berkeley, 2000), is also relevant here.

7. Benjamin Franklin, "Observations and Suppositions, towards Forming a New Hypothesis, for Explaining the Several Phaenomena of Thunder-Gusts" (1749), *EO*, 48. Medical electricity was certainly a utilitarian project, but the claims of electrotherapists remained highly controversial throughout the eighteenth century.

1. Atlantic Circuits

1. *Procès-Verbaux de l'Académie des Sciences* 65 (1746): 6, quoted in John L. Heilbron, *Electricity in the Seventeenth and Eighteenth Centuries: A Study of Early Modern Physics*, rev. ed. (Mineola, N.Y., 1999), 313–314; on the different discoveries of the Leyden jar in this period, see 312–316.
2. William Caruthers to Thomas Jefferson, July 29, 1801, Thomas Jefferson Papers, Massachusetts Historical Society.
3. Francis Bacon, *New Atlantis* (London, 1626); Richard Ligon, *A True and Exact History of the Island of Barbadoes* (London, 1657). The classic statement of the center-periphery model in the history of science beyond Europe is George Basalla, "The Spread of Western Science," *Science* 156 (May 1967): 612–619; for recent directions in the history of science, colonialism, and empire, see Roy MacLeod, ed., *Nature and Empire: Science*

and the Colonial Enterprise, special issue of *Osiris* 15 (2000), and "Focus: Colonial Science," *Isis* 96 (Mar. 2005): 52–87; on Banksian empire, see David Philip Miller and Peter Hanns Reill, eds., *Visions of Empire: Voyages, Botany, and Representations of Nature* (Cambridge, 1996); John Gascoigne, *Science in the Service of Empire: Joseph Banks, the British State, and the Uses of Science in the Age of Revolution* (Cambridge, 1998); and Richard Drayton, *Nature's Government: Science, Imperial Britain, and the "Improvement" of the World* (New Haven, 2000). On economic botany, see, most recently, Londa Schiebinger, *Plants and Empire: Colonial Bioprospecting in the Atlantic World* (Cambridge, Mass., 2004), and Londa Schiebinger and Claudia Swan, eds., *Colonial Botany: Science, Commerce, and Politics in the Early Modern World* (Philadelphia, 2004).

4. David Hancock, *Citizens of the World: London Merchants and the Integration of the British Atlantic Community, 1735–1785* (Cambridge, 1995); Robin Blackburn, *The Making of New World Slavery: From the Baroque to the Modern, 1492–1800* (London, 1997), 1–27; Peter Linebaugh and Marcus Rediker, *The Many-Headed Hydra: Sailors, Slaves, Commoners and the Hidden History of the Revolutionary Atlantic* (Boston, 2000); Amy Meyers and Margaret Beck Pritchard, eds., *Empire's Nature: Mark Catesby's New World Vision* (Chapel Hill, 1998); Richard Drayton, "Knowledge and Empire," in P. J. Marshall, ed., *The Oxford History of the British Empire*, vol. 2: *The Eighteenth Century* (New York, 1998), 231–252; Larry Stewart, "Global Pillage: Science, Commerce, and Empire," in Roy Porter, ed., *The Cambridge History of Science*, vol. 4: *Eighteenth-Century Science* (Cambridge, 2003), 825–844; Ian K. Steele, *The English Atlantic, 1675–1740: An Exploration of Culture and Community* (New York, 1986), esp. parts 1–2. On information in colonial Massachusetts at the turn of the eighteenth century, see Richard D. Brown, *Knowledge Is Power: The Diffusion of Information in Early America, 1700–1865* (New York, 1989), 16–41.

5. Ralph Bauer, *The Cultural Geography of Colonial American Literatures: Empire, Travel, Modernity* (Cambridge, 2003), 179–199; Thomas Jefferson, *Notes on the State of Virginia* (1785), ed. William Peden (Chapel Hill, 1954); Antonello Gerbi, *The Dispute of the New World: The History of a Polemic, 1750–1900*, trans. Jeremy Moyle (Pittsburgh, 1973), esp. 3–34, 52–79; Jorge Cañizares-Esguerra, *How to Write the History of the New World: Historiographies, Epistemologies, and Identities in the Eighteenth-Century Atlantic World* (Stanford, 2001). On Creole consciousness in North American science, see Carla Mulford, "New Science and the

Question of Identity in Eighteenth-Century British America," in Carla Mulford and David S. Shields, eds., *Finding Colonial Americas: Essays Honoring J. A. Leo Lemay* (Newark, Del., 2001), 79–103.

6. Newton's description of Brattle is quoted in Raymond P. Stearns, *Science in the British Colonies of America* (Urbana, Ill., 1970), 153. On European wonder at America, see Stephen Greenblatt, *Marvelous Possessions: The Wonder of the New World* (Chicago, 1991). On Providential wonders in New England at the turn of the eighteenth century, see David D. Hall, *Worlds of Wonder, Days of Judgment: Popular Religious Belief in Early New England* (New York, 1989), esp. 71–116; and Michael P. Winship, *Seers of God: Puritan Providentialism in the Restoration and Early Enlightenment* (Baltimore, 1996); Carolyn Merchant, *Ecological Revolutions: Nature, Gender, and Science in New England* (Chapel Hill, 1988); and Joyce E. Chaplin, *Subject Matter: Technology, Science, and the Body on the Anglo-American Frontier, 1500–1676* (Cambridge, Mass., 2001), esp. chap. 8. The seminal thesis about the relationship between (English) Puritanism and science is Robert K. Merton's *Science, Technology and Society in Seventeenth-Century England* (New York, 1970), originally published in 1938. On cosmopolitanism, see Thomas Schlereth, *The Cosmopolitan Ideal in Enlightenment Thought: Its Form and Function in the Ideas of Franklin, Hume, and Voltaire, 1694–1790* (Notre Dame, Ind., 1977); Lorraine Daston, "The Ideal and the Reality of the Republic of Letters in the Enlightenment," *Science in Context* 4 (1991): 367–386; Christopher Iannini, "'The Itinerant Man': Crèvecoeur's Caribbean, Raynal's Revolution, and the Fate of Atlantic Cosmopolitanism," *WMQ* 61 (Apr. 2004): 201–234; see also Anne Goldgar, *Impolite Learning: Conduct and Community in the Republic of Letters, 1680–1750* (New Haven, 1995).

7. Benjamin Franklin, "A Proposal for Promoting Useful Knowledge among the British Plantations in America" (1743), *BFP,* 2:380; Amy R. Meyers and Margaret Pritchard, "Introduction: Toward an Understanding of Catesby," in Meyers and Pritchard, *Empire's Nature,* 1–20; on Bartram, see Stearns, *Science in the British Colonies,* 575–593; Alfred R. Hoermann, *Cadwallader Colden: A Figure of the American Enlightenment* (Westport, Conn., 2002), 19–21; on Garden, see Konstantin Dierks, "Letter Writing, Masculinity, and American Men of Science, 1750–1800," *Explorations in Early American Culture/Pennsylvania History* 65 (1998): 167–198; on provincialism, cosmopolitanism, and enlightenment, see David Jaffee, "The Village Enlightenment in New England, 1760–1820," *WMQ* 47 (July

1990): 327–346; on American provincial cosmopolitanism, see Ned C. Landsman, *From Colonials to Provincials: American Thought and Culture, 1680–1760* (New York, 1997), esp. 57–63; as well as Daniel Hulsebosch, "Imperia in Imperio: The Multiple Constitutions of Empire in New York, 1750–1777," *Law and History Review* 16 (1998): 319–379.
8. Charles Morton, *Compendium Physicae* (1687), *Publications of the Colonial Society of Massachusetts* 33 (Boston, 1940), 40, 76.
9. Isaac Greenwood, *An Experimental Course of Mechanical Philosophy* (Boston, 1726), title page and 2. On Greenwood, see Stearns, *Science in the British Colonies*, 446–455; on Desaguliers's Newtonianism, see Larry Stewart, *The Rise of Public Science: Rhetoric, Technology and Natural Philosophy in Newtonian Britain, 1660–1750* (Cambridge, 1992), chap. 7; on the variant meanings of Newtonianism, see Simon Schaffer, "Newtonianism," in R. C. Olby et al., eds., *Companion to the History of Modern Science* (New York, 1990), 610–626.
10. Heilbron, *Electricity in the Seventeenth and Eighteenth Centuries*, 169–192.
11. Roy Porter, *Flesh in the Age of Reason* (New York, 2003), part 1; on the imagination, see Lorraine Daston, "Fear and Loathing of the Imagination in Science," *Daedalus* 127 (Winter 1998): 73–95; Lorraine Daston and Katharine Park, *Wonders and the Order of Nature, 1150–1750* (New York, 1998), 339–341; Jessica Riskin, *Science in the Age of Sensibility: The Sentimental Empiricists of the French Enlightenment* (Chicago, 2002), 209–225; Jan Goldstein, "Enthusiasm or Imagination? Eighteenth-Century Smear Words in Comparative National Context," *Huntington Library Quarterly* 60 (1998): 29–49; and Simon Schaffer, "Self Evidence," in James Chandler, Arnold I. Davidson, and Harry Harootunian, eds., *Questions of Evidence: Proof, Practice, and Persuasion across the Disciplines* (Chicago, 1994), 68.
12. Heilbron, *Electricity in the Seventeenth and Eighteenth Centuries*, 229–239; Paola Bertucci, "The Electrical Body of Knowledge: Medical Electricity and Experimental Philosophy in the Mid-Eighteenth Century," in Bertucci and Giuliano Pancaldi, eds., *Electric Bodies: Episodes in the History of Medical Electricity* (Bologna, 2001), 43–68; Geoffrey V. Sutton, *Science for a Polite Society: Gender, Culture, and the Demonstration of Enlightenment* (Boulder, Colo., 1995), 304–305; Simon Schaffer, "Experimenters' Techniques, Dyers' Hands, and the Electric Planetarium," *Isis* 88 (1997): esp. 469–471.
13. Franklin, "Proposal for Promoting Useful Knowledge," 381; Carl and

Jessica Bridenbaugh, *Rebels and Gentlemen: Philadelphia in the Age of Franklin* (New York, 1942), 322–328; Brooke Hindle, "The Quaker Background and Science in Colonial Philadelphia," in Brooke Hindle, ed., *Early American Science* (New York, 1976), 73–80; D. W. Meinig, *The Shaping of America*, vol. 1: *Atlantic America, 1492–1800* (New Haven, 1986), 140–144; Theodore Thayer, "Town into City, 1746–1765," in Russell F. Weigley, ed., *Philadelphia: A Three Hundred Year History* (New York, 1982), 68–108. On American consumption of British material culture and its consequences, see Richard L. Bushman, *The Refinement of America: Persons, Houses, Cities* (New York, 1992); and T. H. Breen, *The Marketplace of Revolution: How Consumer Politics Shaped American Independence* (New York, 2004); on the commercialization of leisure in the eighteenth century, see chapters 6 and 7 by John H. Plumb in Neil McKendrick, John H. Plumb, and John Brewer, eds., *The Birth of a Consumer Society* (London, 1982). One of the most readable single-volume biographies of Franklin is still Carl Van Doren, *Benjamin Franklin* (New York, 1938); on his Philadelphia years, see chapters 3–7. On Franklin's life as a product of Atlantic travel and Atlantic knowledge, see Joyce E. Chaplin, *The First Scientific American: Benjamin Franklin and the Pursuit of Genius* (New York, 2006). For an idea of the range of recent academic interpretations of Franklin's biography, compare Gordon S. Wood, *The Americanization of Benjamin Franklin* (New York, 2004) with David Waldstreicher, *Runaway America: Benjamin Franklin, Slavery, and the American Revolution* (New York, 2004). There is, of course, a perpetual stream of popular biographies, too.
14. I. Bernard Cohen, *Benjamin Franklin's Science* (Cambridge, Mass., 1990), 47; N. H. de V. Heathcote, "Franklin's Introduction to Electricity," in Hindle, *Early American Science*, 43–49. "A Greater Number of Gentlemen [have] subscribed to Dr. Spencer's first Course of Experimental Philosophy, than can be conveniently accommodated": *PG*, Apr. 26, 1744.
15. Albrecht von Haller, "An Historical Account of the Wonderful Discoveries, Made in Germany, &c. Concerning Electricity," first published in the Dutch-edited *Bibliothèque Raisonnée des Ouvrages des Savans de l'Europe*, and reprinted in both *Gentleman's Magazine* 15 (1745): 193–197, and *American Magazine and Historical Chronicle* 2 (1745): 530–537; John L. Heilbron, "Franklin, Haller, and Franklinist History," *Isis* 68 (1977): 539–549. On Collinson, see *"Forget Not Mee and My Garden . . .": Se-*

lected Letters of Peter Collinson, FRS, 1725–1768, ed. Alan Armstrong (Philadelphia, 2002).

16. Franklin described the Wistarburg glassworks for Thomas Darling, a tutor at Yale College, who wished to set up his own factory in Connecticut in early 1747: see BFP, 3:108–115; PG, Aug. 10, 1769; see also Arlene Palmer, "Benjamin Franklin and the Wistarburg Glassworks," Antiques (Jan. 1974): 207–210; and Harrold E. Gillingham, "Pottery, China, and Glass Making in Philadelphia," Pennsylvania Magazine of History and Biography 54 (1930): 126–127.

17. Franklin to Collinson, July 11, 1747, and Mar. 28, 1747, EO, 11, 2; Benjamin Franklin, The Autobiography of Benjamin Franklin: A Genetic Text, ed. J. A. Leo Lemay and P. M. Zall (Knoxville, Tenn., 1981), 75; Franklin to Collinson, 1748, EO (dated Apr. 1749 in James Bowdoin Papers, Massachusetts Historical Society, Boston), 37. On gentlemanly cultivation in British America, see Bushman, Refinement of America, part 1; on genteel social pleasure, see David S. Shields, Civil Tongues and Polite Letters in British America (Chapel Hill, 1997).

18. Franklin to Collinson, July 11, 1747, EO, 7; "Opinions and Conjectures, Concerning the Properties and Effects of the Electrical Matter, Arising from Experiments and Observations Made at Philadelphia, 1749," EO, 55; Isaac Newton, Opticks; or, A Treatise of the Reflections, Refractions, Inflections and Colours of Light, 4th ed. (1730; London, 1952), queries 8 and 31 on pp. 341 and 376, respectively. Newton also discussed electricity in unpublished drafts of the General Scholium of the revised Principia (1713). See Heilbron, Electricity in the Seventeenth and Eighteenth Centuries, 239–241, and Ernan McMullin, Newton on Matter and Activity (Notre Dame, Ind., 1978), 94–101.

19. Franklin, "Opinions and Conjectures," EO, 55; Franklin to Collinson, July 11, 1747, EO, 8–9.

20. Franklin to Collinson, July 11, 1747, and July 27, 1750, EO, 9, 11, 92; Willem D. Hackmann, Electricity from Glass: The History of the Frictional Electrical Machine, 1600–1850 (Alphen aan den Rijn, Neth., 1978), 113. The electrical apparatus donated by Penn may well have been made by the leading London instrument-maker George Adams. The gift included a copy of Adams's Micrographia Illustrata; or, The Knowledge of the Microscope Explained (London, 1746) and "a large Pair of Globes; a large reflecting Telescope; a large double Microscope; a large Camera Obscura": BFP, 3:3.

21. Franklin to Collinson, Sept. 1, 1747, *EO*, 16–18; "Opinions and Conjectures," *EO*, 54.
22. On the relationship between gentlemanliness and epistemological credit, see Steven Shapin, *A Social History of Truth: Civility and Science in Seventeenth-Century England* (Chicago, 1994); on seventeenth-century arguments about the greater fitness of English rather than native bodies to inhabit North America, see Chaplin, *Subject Matter,* esp. chaps. 4–5; on Creole intellectual inferiority, see Bauer, *Cultural Geography of Colonial American Literatures*. Although the humidity of some parts of British America could physically inhibit the performance of experiments, especially in the West Indies, there was no explicit discourse concerning the effect of climate on either the bodies and minds of settlers or their instruments, which denigrated the ability of New World inhabitants to perform electrical experiments.
23. Franklin to Collinson, Mar. 28, 1747, *EO*, 2; Isaac Newton, *Principia*, in *Newton: A Norton Critical Edition*, ed. I. Bernard Cohen and Richard S. Westfall (New York, 1995), 116; *EO*, 1st and 2d eds. (London, 1751 and 1754), title page.
24. Franklin, *Autobiography*, 154.
25. *EO*, 1st, 4th, and 5th eds. (London, 1751, 1769, 1774), title pages. For other early British publications on electricity by Franklin, see Franklin to Collinson, Oct. 19, 1752, *Gentleman's Magazine* 22 (Dec. 1752): 560–561; "A Letter from Mr. Franklin to Mr. Peter Collinson, FRS, Concerning the Effects of Lightning," *Phil. Trans.* 47 (1751–1752): 289–291; "A Letter of Benjamin Franklin, Esq; to Mr. Peter Collinson, FRS, Concerning an Electrical Kite," *Phil. Trans.* 47 (1751–1752): 565–567; and William Watson, "An Account of Mr. Benjamin Franklin's Treatise, Lately Published," *Phil. Trans.* 47 (1751–1752): 202–211. The *Experiments and Observations on Electricity* ran to five editions (1751, 1754, 1760, 1769, 1774), the last two of which were dramatically expanded. Several translations were also made.
26. Heinz Otto Sibum, "The Bookkeeper of Nature: Benjamin Franklin's Electrical Research and the Development of Experimental Natural Philosophy in the Eighteenth Century," in J. A. Leo Lemay, ed., *Reappraising Benjamin Franklin: A Bicentennial Perspective* (Newark, Del., 1993), 221–242.
27. Edward Cave, preface to Franklin, *EO*, 1st ed., 1751; Riskin, *Science in the Age of Sensibility*, 69–103; see also Sutton, *Science for a Polite Society*, chap. 8.

28. Benjamin Franklin, *A Dissertation on Liberty and Necessity, Pleasure and Pain* (London, 1725) and *A Modest Enquiry into the Nature and Necessity of a Paper-Currency* (Philadelphia, 1729), BFP, 1: esp. 67 and 142–145; Franklin, *Autobiography*, 80; Sibum, "Bookkeeper of Nature," 221–242. See also Anke te Heesen, "Accounting for the Natural World: Double-Entry Bookkeeping in the Field," in Schiebinger and Swan, *Colonial Botany*, 237–251; John L. Heilbron, "Franklin as an Enlightened Natural Philosopher," in Lemay, *Reappraising Benjamin Franklin*, 196–220; and Douglas Anderson, *The Radical Enlightenments of Benjamin Franklin* (Baltimore, 1997), 120–155. On French Physiocratic views that wealth should circulate like Franklinist electric fire rather than mechanically, see Riskin, *Science in the Age of Sensibility*, esp. 112 and 135–137; for Franklin's discussions of the circulation of money and the Gulf Stream, see Chaplin, *The First Scientific American*, esp. chap. 4.
29. Franklin, "Opinions and Conjectures," EO, 62; John Winthrop, *A Lecture on Earthquakes* (Boston, 1755), 35, for example, refers to Franklin's "laws of electricity." On laws of nature, see Daston and Park, *Wonders and the Order of Nature*, 110, 120, 324, 350–352; on doubts about the lawfulness of American nature, see Joyce E. Chaplin, "Mark Catesby: A Skeptical Newtonian in America," in Meyers and Pritchard, *Empire's Nature*, 37–39.
30. Franklin to Collinson, Sept. 1, 1747, and July 29, 1750: EO, 14, 56. For contrasting views of Franklin's religion, from its deistic to more mystical elements, see Alfred O. Aldridge, *Benjamin Franklin and Nature's God* (Durham, N.C., 1967) and Jean-Paul de Lagrave, *La Vision Cosmique de Benjamin Franklin* (Paris, 2003).
31. Franklin to Ebenezer Kinnersley, Feb. 20, 1762; Franklin to John Franklin, Dec. 17, 1750; Franklin to James Bowdoin, Dec. 21, 1751: EO, 420, 82–83, 179. See also Benjamin Franklin, "Loose Thoughts on a Universal Fluid" (1784) in Benjamin Franklin, *Writings*, ed. J. A. Leo Lemay (New York, 1987), 988–990. I have not, incidentally, uncovered any evidence directly linking Franklin's later career in Freemasonry and his views on electricity.
32. Franklin, *Autobiography*, 119; Joseph Priestley, *A Familiar Introduction to the Study of Electricity* (1768; Millwood, N.Y., 1978), advertisement, viii; a notice in the *Pennsylvania Packet* on Sept. 29, 1778, advertised the sale of "a very compleat portable Electric Apparatus" and "an Electric battery" with a copy of Priestley's *History and Present State of Electricity* (1767) as

"a book especially necessary for young Electricians." On experimental philosophy using machines, see Simon Schaffer, "Machine Philosophy: Demonstration Devices in Georgian Mechanics," *Osiris* 9 (1994): 157–182.

33. Baldwin advertisement in *Boston Gazette,* Jan. 1, 1770; Porter in *PG,* July 13, 1774; see also an advertisement by the apothecary Robert Bass, *PG,* June 15, 1769; Atlanticus, "Description of a New Electrical Machine," *Pennsylvania Magazine* (Jan. 1775): 31–32; Stiles, June 8, 1784, *The Literary Diary of Ezra Stiles,* 3 vols., ed. Franklin Bowditch Dexter (New York, 1901), 3:125. On the commerce in eighteenth-century philosophical instruments, see James E. Bennett, "Shopping for Instruments in Paris and London," in Pamela H. Smith and Paula Findlen, eds., *Merchants and Marvels: Commerce, Science, and Art in Early Modern Europe* (New York, 2002), 370–395.

34. Franklin to Collinson, Feb. 4, 1751, *BFP,* 4:113; M'Cabe in *Maryland Journal,* Apr. 1774; Claggett in *Boston Evening Post,* Aug. 24–31, 1747; Hiller in *Boston Gazette,* supplement, Mar. 1–8, 1756. Franklin occasionally requested selected items from Collinson: a letter from Feb. 1751 asks for "a large Glass Globe for an Electrical Machine" and "a large Glass Cylinder for Ditto." While in London, he later obtained such pieces from leading metropolitan instrument makers like George Adams: on Apr. 8, 1758, he "bought of George Adams sundry electric implements" (Franklin's fellow electrician Ebenezer Kinnersley also acquired equipment from Adams). See George Simpson Eddy, "Account Book of Benjamin Franklin, Kept by Him during His First Mission to England as Provincial Agent, 1757–1762," *Pennsylvania Magazine of History and Biography* 55 (Apr. 1931): 110. On artisanal techniques in electrical performance, see Schaffer, "Experimenters' Techniques, Dyers' Hands, and the Electric Planetarium," 456–483; on the distinctive social and political position of artisans in colonial and revolutionary Philadelphia, see Gary B. Nash, *The Urban Crucible: Social Change, Political Consciousness and the Origins of the American Revolution* (Cambridge, Mass., 1979).

2. Lightning Rods and the Direction of Nature

1. Constantin F. Volney, *A View of the Soil and Climate of the United States of America,* trans. Charles Brockden Brown (Philadelphia, 1804), 198. On Volney's generally unflattering view of American nature, see Antonello Gerbi, *The Dispute of the New World: The History of a Polemic, 1750–*

1900, trans. Jeremy Moyle (Pittsburgh, 1973), 339–341. On South American responses to European disparagement of American nature and civilization, see Jorge Cañizares-Esguerra, *How to Write the History of the New World: Historiographies, Epistemologies, and Identities in the Eighteenth-Century Atlantic World* (Stanford, 2001).
2. Volney, *View of the Soil and Climate of the United States*, 199; on American lightning, see also *Moreau de St. Méry's American Journey, 1793–1798*, ed. and trans. Kenneth Roberts and Anna M. Roberts (New York, 1947), 321–322; and Franklin to Horace Bénédict de Saussure, Oct. 8, 1772, *BFP*, 19:325.
3. I. Bernard Cohen, *Benjamin Franklin's Science* (Cambridge, Mass., 1990), 6, 158; Cohen, "Prejudice against the Introduction of Lightning Rods," originally published in *Journal of the Franklin Institute* 253 (1952): 393–440, republished as chap. 8 of *Benjamin Franklin's Science*. Bruno Latour cites conflicting interpretations of thunder and lightning as a classic line of division between modern and premodern mentalities in *Nous N'Avons Jamais Été Modernes: Essai d'Anthropologie Symétrique* (Paris, 1991), 57–58, but the heroic perspective persists in some quarters: see, for example, Stefano Casati, "Storie di Folgori: Il Dibattito Italiano sui Conduttori Elettrici nel Settecento," *Nuncius* 13 (1998): 494–512. Recent accounts of early lightning rod controversies have, by contrast, focused primarily on social interests and rhetorical strategies: see Trent A. Mitchell, "The Politics of Experiment in the Eighteenth Century: The Pursuit of Audience and the Manipulation of Consensus in the Debate over Lightning Rods," *Eighteenth-Century Studies* 31 (1998): 307–331; and Jessica Riskin, *Science in the Age of Sensibility: The Sentimental Empiricists of the French Enlightenment* (Chicago, 2002), chap. 5. On the sound of thunder in colonial America, see Richard Cullen Rath, *How Early America Sounded* (Ithaca, 2003); see also Michael Brian Schiffer, *Draw the Lightning Down: Benjamin Franklin and Electrical Technology in the Age of Enlightenment* (Berkeley, 2003), chap. 9.
4. Throughout, I use the term "laboratory" conceptually rather than literally in any modern sense of a fixed space designed specifically and exclusively for institutionally sponsored experimental science. Franklin's "laboratory" (literally, a place where work is performed) was a private yet makeshift early modern space for experiment that he improvised both in the Library Company of Philadelphia, originally situated in the Philadelphia Statehouse, and his own home. On the ambiguity between

privacy and publicity in early modern laboratories, see Steven Shapin and Simon Schaffer, *Leviathan and the Air-Pump: Hobbes, Boyle, and the Experimental Life* (Princeton, 1985), 39, 57, 334–336, 339–341. For an influential account of science as an attempt to describe the world through a stable set of "metrological" techniques, languages, and instruments, see Bruno Latour, *Science in Action: How to Follow Scientists and Engineers through Society* (Cambridge, Mass., 1987), esp. 247–251.

5. *EO*, 44–48; John L. Heilbron, "Benjamin Franklin as an Enlightened Natural Philosopher," in J. A. Leo Lemay, ed., *Reappraising Benjamin Franklin* (Newark, Del., 1993), 196–220; Simon Schaffer, "Fish and Ships: Models in the Age of Reason," in Nick Hopwood and Soraya de Chadarevian, eds., *Models: The Third Dimension in Science* (Stanford, 2004), 71–105. On the connection between Atlantic storms and Franklin's lightning experiments, see Joyce E. Chaplin, *The First Scientific American: Benjamin Franklin and the Pursuit of Genius* (New York, 2006), chap. 5.

6. Franklin credited his co-experimenter Thomas Hopkinson with discovering the "power of points" in drawing and communicating electricity. Franklin, "Opinions and Conjectures, Concerning the Properties and Effects of the Electrical Matter, Arising from Experiments and Observations, Made at Philadelphia, 1749," *EO*, 66; Franklin to Peter Collinson, Sept. 1, 1747, *EO*, 17.

7. Joseph Priestley, *The History and Present State of Electricity, with Original Experiments*, 2 vols., 3d ed. (London, 1775), reprinted as *The Sources of Science*, no. 18 (New York, 1966), 1:216; Franklin to Collinson, Oct. 19, 1752, *EO*, 118.

8. Priestley, *History and Present State of Electricity*, 1:217; Franklin to Collinson, July 11, 1747, *EO*, 9. On Christian asceticism in English science, see Steven Shapin, *A Social History of Truth: Civility and Science in Seventeenth-Century England* (Chicago, 1994). Whether Franklin actually performed the kite experiment—see Tom Tucker, *Bolt of Fate: Benjamin Franklin and His Electric Kite Hoax* (New York, 2003)—is ultimately a less important question than how his experimental accounts persuaded skeptical readers to take their claims seriously.

9. Abbé Mazéas to Stephen Hales, May 30, 1752, *EO*, 106–107. Mazéas's letter to Hales does not specify whether sparks were drawn manually or via a mediating apparatus in the Marly experiment. For the competing claims to priority mounted by the French experimenter Jacques de Romas, see Cohen, *Benjamin Franklin's Science*, 100–107. Cohen argues

that Romas's work was independent of Franklin's, but did not predate it (77–92).
10. Earl of Macclesfield, Copley award speech, Nov. 30, 1753, *BFP*, 5:130–131; Benjamin Franklin, *The Autobiography of Benjamin Franklin: A Genetic Text*, ed. J. A. Leo Lemay and P. M. Zall (Knoxville, Tenn., 1981), 154. On philosophical modesty, see Shapin and Schaffer, *Leviathan and the Air-Pump*, 65–69.
11. Franklin to Collinson, Sept. 1753, *EO*, 120; John Winthrop, July 11, 1768, reprinted in *Virginia Gazette* (Rind), Aug. 11, 1768; Winthrop to Franklin, Mar. 4, 1773, *BFP*, 20:90–95; Franklin "white stream" passage quoted in Cohen, *Benjamin Franklin's Science*, 90; Ezra Stiles, "Electrical Bells," June 29, 1753, Ezra Stiles Papers, microfilm edition, miscellaneous papers, reel 14, Yale University Library, New Haven; William Watson, "An Answer to Dr. Lining's Query Relating to the Death of Professor Richmann," *Phil. Trans.* 48 (1754): 769–770; John L. Heilbron, *Electricity in the Seventeenth and Eighteenth Centuries: A Study of Early Modern Physics*, rev. ed. (Mineola, N.Y., 1999), 312–316; *PG*, Mar. 26, 1754.
12. Ezra Stiles, June 10, 1789, *The Literary Diary of Ezra Stiles*, ed. Franklin Bowditch Dexter, 3 vols. (New York, 1901), 3:356–357; Loammi Baldwin, "An Account of a Very Curious Appearance of the Electrical Fluid, Produced by Raising an Electrical Kite in the Time of a Thunder-Shower," *Memoirs of the American Academy of Arts and Sciences* 1 (1785): 259–260.
13. Franklin, "Of Lightning, and the Method (Now Used in America) of Securing Buildings and Persons from Its Mischievous Effects" (Sept. 1767), *EO*, 502. Some began recommending modified hairpins as portable lightning rods for ladies, whose bodies were considered peculiarly susceptible to electricity: "Hair-pins and other metalics may, if placed in proper directions, preserve the lives of people, by conveying the lightning from their bodies and limbs." See Samuel Stearns, *The American Oracle, Comprehending an Account of Recent Discoveries in the Arts and Sciences* (New York, 1791), 139. Others thought them hazardous: see Anonymous, "On the Great Danger of Ladies Wearing Wires in Their Caps, and Pins in Their Hair," *Pennsylvania Magazine* (July 1776): 318–319. See also Sergio Perosa, "Franklin to Frankenstein: A Note on Lightning and Novels," in Sergio Rossi, ed., *Science and Imagination in Eighteenth-Century British Culture* (Milan, 1987), 321–322.
14. On natural theology in British America, see Nina Ruth Reid-Maroney, *Philadelphia's Enlightenment, 1740–1800: Kingdom of Christ, Empire of*

Reason (Westport, Conn., 2001), chap. 3; on providential wonders in New England during the era of Cotton and Increase Mather, see David D. Hall, *Worlds of Wonder, Days of Judgment: Popular Religious Belief in Early New England* (New York, 1989); and Michael P. Winship, *Seers of God: Puritan Providentialism in the Restoration and Early Enlightenment* (Baltimore, 1996). By 1741, the Library Company of Philadelphia owned several core works of English natural theology: William Wollaston's *Religion of Nature Delineated* (1726), John Ray's *Wisdom of God* (1728), William Derham's *Astro-Theology* (1731) and *Physico-Theology* (1732), and the Abbé Noel-Antoine Pluche's *Spectacle de la Nature; or, Nature Displayed* (trans. 1733). See *A Catalogue of Books Belonging to the Library Company of Philadelphia* (Philadelphia, 1956).

15. Increase Mather, *Essay for the Recording of Illustrious Providences* (Boston, 1684), 109–110, 99, 82; Cotton Mather, "Of the Thunder and Lightning," in Mather, *The Christian Philosopher* (1721; Urbana, Ill., 1994), 71–73; Gilbert Tennent, *All Things Come Alike to All: A Sermon . . . Occasioned by a Person's Being Struck by the Lightning of Thunder* (Philadelphia, 1745), 34, 37, 13, 19. On Increase Mather's moral interpretation of lightning in his sermons, see Winship, *Seers of God*, 67; on early American sound and religion, see Leigh Eric Schmidt, *Hearing Things: Religion, Illusion, and the American Enlightenment* (Cambridge, Mass., 2000).

16. Mather Byles, *Divine Power and Anger Displayed in Earthquakes* (Boston, 1755), 7, 5, 15, 13; Thomas Prince, *Earthquakes the Works of God and Tokens of His Just Displeasure* (Boston, 1755), 9, 8, 14, 18. Colonial New England also possessed a rich literature on the moral purpose of earthquakes: see, for example, Stephen Mix, *Extraordinary Displays of the Divine Majesty and Power* (New London, Conn., 1728). On earthquakes in British America (including the Prince-Winthrop quarrel), see Michael N. Shute, "Earthquakes and the Early American Imagination: Decline and Renewal in Eighteenth-Century Puritan Culture," Ph.D. diss., University of California, Berkeley, 1977, esp. 116–236; and Theodore Hornberger, "The Science of Thomas Prince," *New England Quarterly* 9 (1936): 26–42.

17. Prince, *Earthquakes the Works of God*, 20–22. For the influence of Franklin's speculations on William Stukeley and his *Philosophy of Earthquakes* (London, 1750), see Simon Schaffer, "Natural Philosophy and Public Spectacle in the Eighteenth Century," *History of Science* 21 (Mar. 1983): 18–19.

18. Adams marginalia dated Dec. 1758, *The Diary and Autobiography of John*

Adams, ed. Lyman H. Butterfield et al., 2 vols. (Cambridge, Mass., 1961), 1:61–62; John Winthrop, "Extract of a Letter from J. W. Esq; Professor of Natural Philosophy at Cambridge, in New England, Jan. 6, 1768," *EO*, 503; see also Franklin's reply, *EO*, 503–509; Junto Minute Book, Jan. 18, 1760, American Philosophical Society Archives, Philadelphia; and Schiffer, *Draw the Lightning Down*, 189–190.

19. John Winthrop, *A Letter to the Publishers of the Boston Gazette, &c. Containing an Answer to the Rev. Mr. Prince's Letter* (Boston, 1756), 6; on electricity and volcanoes, see Sir William Hamilton, "An Account of the Late Eruption of Mount Vesuvius, in a Letter from the Right Honourable Sir William Hamilton," *Phil. Trans.* 85 (1795): 80–81, and "Extract from a Letter from Naples," *PG*, Apr. 16, 1788. See also John Winthrop, *A Lecture on Earthquakes* (Boston, 1755), 35.

20. Winthrop, *Letter*, 2, 4, and *Lecture on Earthquakes*, 36–37; Louis Graham, "The Scientific Piety of John Winthrop of Harvard," *New England Quarterly* 46 (1973): 112–118.

21. Ebenezer Gay, *The Sovereignty of God, in Determining Man's Days, or the Time and Manner of His Death* (Hartford, 1767), 20–23; see also Nathan Fiske, *The Sovereignty of God in Determining the Boundaries of Human Life* (Worcester, Mass., 1784); Johannes Martinet, *The Catechism of Nature* (Boston, 1790), 27–28; Charles Reiche, *Fifteen Discourses on the Marvellous Works in Nature, Delivered by a Father to His Children* (Philadelphia, 1791), 22, 24, 25.

22. *South Carolina Gazette*, July 31, 1755; Cohen, *Benjamin Franklin's Science*, 145; Winthrop to Franklin, Oct. 26, 1770, *BFP*, 17:263; Jane Mecom to Franklin, Aug. 25, 1786, Franklin-Bache Collection, American Philosophical Society; James Bowdoin to Franklin, *BFP*, 4:271; Charles Woodmason, "To Benjamin Frankin Esq; of Philadelphia, on His Experiments and Discoveries in Electricity," Sept. 20, 1753, *BFP*, 5:61; Ezra Stiles, Aug. 18, 1785, *Literary Diary*, 3:176–177 (a lightning strike on Stiles's own house was reported by the *PG* on June 24, 1789); Winthrop, *Letter*, 7.

23. Deborah J. Warner, "Lightning Rods and Thunder Houses," *Rittenhouse* 11 (1997): 124–127; I. Bernard Cohen, *Some Early Tools of American Science* (Cambridge, Mass., 1950), 161–162; *Analysis of Certain Parts of a Compendious View of Natural Philosophy, for the Use of Both Sexes* (Boston, 1796), 35.

24. *Poor Richard's Almanack*, 1753, *BFP*, 4:308; *EO*, 510–516. "As the season approaches wherein such accidents may be frequent, and it is a subject

of no small moment, Mr. [William] Johnson has been requested to exhibit his course of Experiments in Electricity, with explanatory lectures on the nature and properties of Lightning," *South Carolina Gazette,* Mar. 21–28, 1768; *Evening Fire-Side; or, Literary Miscellany* 1 (July 13, 1805); letter about conductors in South Carolina written at St. John's, Antigua, reprinted in *PG,* Sept. 13, 1753; remarks by "Philanthropos" reprinted in the *New-Haven Gazette, and the Connecticut Magazine,* July 10, 1788; *Early Proceedings of the American Philosophical Society* (Philadelphia, 1884): 203 (Feb. 3, 1792). Both popular and college lecturers disseminated practical knowledge of lightning and lightning rods orally and in print: see, for example, Thaddeus Dod, "Notes Taken from Lectures on Electricity, Delivered at Nassau Hall by . . . Houston, A.M. and Math. and Phil. Nat. Prof." (1772–1773), manuscript lecture notes, box 54, Seeley Mudd Library, Princeton University; Henry Moyes, *Heads of a Course of Lectures on the Natural History of the Celestial Bodies* (Boston, 1784), 5–6; *The Newtonian System of Philosophy; Explained by Familiar Objects, in an Entertaining Manner, for the Use of Young Ladies and Gentlemen, by Tom Telescope,* ed. Robert Patterson, 2d ed. (Philadelphia, 1808), 66–68, 71–72.

25. Deborah J. Warner, "Edward Nairne: Scientist and Instrument-Maker," *Rittenhouse* 12 (1998): 77–78; Bern Dibner, *Benjamin Franklin, Electrician* (Norwalk, Conn., 1976), 40; Georg Forster, *Observations [Made during a] Voyage Round the World* (London, 1777), 119; James E. McClellan III, *Colonialism and Science: Saint Domingue in the Old Regime* (Baltimore, 1992), 172–173; *In Miserable Slavery: Thomas Thistlewood in Jamaica, 1750–1786,* ed. Douglas Hall (London, 1989), 160, 225; Richard L. Bushman, *The Refinement of America: Persons, Houses, Cities* (New York, 1992), 178–179. On Jefferson's lightning rod(s) at Monticello, see James A. Bear Jr., ed., "Memoirs of a Monticello Slave as Dictated to Charles Campbell by Isaac," *Jefferson at Monticello* (Charlottesville, 1967), 4. The Academy building of the College of Philadelphia and the Pennsylvania Statehouse appear to have been the first American public buildings fitted with protective conductors: see Cohen, *Benjamin Franklin's Science,* 87. On lightning rods in eighteenth-century New Spain, see Bruno de Vecchi Appendini and Carmen Espinosa de los Monteros de de Vecchi, "La Difusión de los Adelantos de la Electricidad en la Nueva España," *Quipu* 13 (Sept.–Dec. 2000): 359–377. See also Schiffer, *Draw the Lightning Down,* 203.

26. Robert Blair St. George, *Conversing by Signs: Poetics of Implication in Co-*

lonial New England Culture (Chapel Hill, 1998), 115–208; Bushman, *Refinement of America*, 100–117, 132–133. While passing through Norfolk, Virginia, in 1794, the Saint-Domingue exile Moreau de St. Méry noted that several large two- and three-story houses possessed conductors: see *Moreau de St. Méry's American Journey*, 47. On Mar. 17, 1779, the *Pennsylvania Gazette* contained an advertisement for the sale of a two-and-a-half-story brick house, with brick kitchen, on Arch Street in Philadelphia, whose roof was "well secured with iron bars, and an electrical conductor." A lightning strike on the three-story, two-chimney house of a Mr. Blanchard in Philadelphia is mentioned in David Rittenhouse and John Jones, "Accounts of Several Houses in Philadelphia, Struck with Lightning, on June 7, 1789," *Transactions of the American Philosophical Society* 3 (1793): 122.

27. Franklin to Collinson, Sept. 1753, *EO*, 124; John Lathrop, "The Fatal Effects of Lightning; in a Letter to the Reverend Joseph Willard" (1798), *Memoirs of the American Academy of Arts and Sciences* 2 (1804): 89. Earlier, in 1753, Franklin had made a similar "Request for Information on Lightning" in *PG* and other newspapers: see *BFP*, 4:510. As a student at the College of New Jersey, Lathrop had studied the Franklinist system in relation to atmospheric electricity: he describes the action of thunderclouds in Franklinist terms in his commonplace book for 1761. See Lathrop, Sermons, 1758–1816, Miscellaneous Papers, box 11, 82–84, Massachusetts Historical Society, Boston. He also gave lectures on electricity and natural philosophy: see his *Synopsis of a Course of Lectures, on the Following Branches of Natural Philosophy, viz. Matter, and Its Properties, Mechanics, Electricity, Hydrostatics, Pneumatics, and Astronomy* (Boston, 1811), 7–10. Modern observations of lightning using high-speed photography indicate that lightning strikes are in fact anticipated by a two-pronged prestroke, with electricity issuing from earth and sky simultaneously.

28. Rittenhouse and Jones, "Accounts of Several Houses in Philadelphia, Struck with Lightning," 119–120. Compare with the diagram accompanying Arthur Lee's account of a lightning strike on the house of William Shippen, "An Account of the Effects of Lightning on Two Houses in the City of Philadelphia" (1781), *Memoirs of the American Academy of Arts and Sciences* 1 (1785): facing 256. On the materialistic implications of Franklin's account of lightning, see Riskin, *Science in the Age of Sensibility*, 84–86.

29. Rittenhouse and Jones, "Accounts of Several Houses in Philadelphia, Struck with Lightning," 120–121; Lathrop, "Fatal Effects," 86; and John Lathrop, "The Effects of Lightning on Several Persons in the House of Samuel Carey," *Memoirs of the American Academy of Arts and Sciences* 3 (1809): 82.
30. Lathrop, "Fatal Effects," 86–87, 93–95; Lee, "An Account of the Effects of Lightning on Two Houses," 250. Bell wires also played a prominent role in John Lathrop, "Effects of Lightning on the House of Capt. Daniel Merry, and Several Other Houses in the Vicinity," *Memoirs of the American Academy of Arts and Sciences* 3 (1809): 87; see also Jeremy Belknap, Meteorological Journal (1798), Jeremy Belknap Papers, Massachusetts Historical Society.
31. Franklin to Marsilio Landriani, Oct. 14, 1787, Osterreichische National-Bibliothek, Vienna; John Lathrop, "An Account of the Effects of Lightning on the House of Jonathan Mason, Esq., in Boston," *Memoirs of the American Academy of Arts and Sciences* 3 (1809): 95; David Rittenhouse and Francis Hopkinson, "An Account of the Effects of a Stroke of Lightning on a House Furnished with Two Conductors," *Transactions of the American Philosophical Society* 3 (1793): 123–124. Franklin recommended extending the rod six to eight feet above the tops of buildings, and sinking its bottom end three or four feet into moist earth by the side of the structure, to which it should be stapled: see *Poor Richard's Almanack* (1753), *BFP,* 4:408–409.
32. Rittenhouse and Hopkinson, "Effects of a Stroke of Lightning on a House," 124; Rittenhouse and Jones, "Accounts of Several Houses in Philadelphia, Struck with Lightning," 125; see also Franklin, "Mr. William Maine's Account of the Effects of Lightning on his Rod, dated at *Indian Land,* in South Carolina, Aug. 28, 1760," *EO,* 427–429; and Franklin's reply, *EO,* 429–435. John Winthrop's letter describing the failure of a lightning rod on Harvard Hall is printed in the *Virginia Gazette* (Rind), Aug. 11, 1768; Loammi Baldwin, "Observations on Electricity, and an Improved Mode of Constructing Lightning Rods," *Memoirs of the American Academy of Arts and Sciences* 2 (1804): 298; Robert Patterson, "An Improvement on Metalic Conductors or Lightening-Rods, in a Letter to Dr. David Rittenhouse," *Transactions of the American Philosophical Society* 3 (1793): 322–323; and Jacob Green and Erskine Hazard, *An Epitome of Electricity and Galvanism, by Two Gentlemen of Philadelphia* (Philadelphia, 1809), 48–53.

33. American electricians did sometimes disagree about technical specifications, such as when Aaron Putnam of Charlestown questioned Loammi Baldwin's design for a conductor with multiple points, which Putnam thought might send lightning in a number of different directions when electrified, or even risk "emptying clouds" if made too tall. See Loammi Baldwin, "Observations on Electricity," and Aaron Putnam, "Remarks on the Hon. Loammi Baldwin's Proposed Improvement in Lightning Rods," *Memoirs of the American Academy of Arts and Sciences* 2 (1804): 96–99, 101–104. Commercial competition among practitioners offering to mount conductors appears to have begun in the nineteenth century. The first American patent for a lightning rod was granted to Elisha Callender of Boston on Oct. 3, 1808. See Paul Fleury Mottelay, *Bibliographical History of Electricity and Magnetism* (London, 1922), 400. "The Lightning-Rod Man" (1854) by Herman Melville, which portrays a lightning rod salesman in the manner of a religious huckster, is thus a story of the nineteenth rather than the eighteenth century.
34. *PG*, June 25, 1788. The author of this report insisted that this disaster should not be taken as evidence against the utility of conductors, "so necessary an appendage to every house in southerly latitudes." In this particular instance, he insisted, lightning had struck the house at the end opposite to that on which the conductor was mounted. In 1810, Williamson was still engaged in defending conductors against those who questioned "whether such rods can be deemed a sure defence against the violent explosions of the electric fluid," citing the by-then standard defenses—shoddy workmanship, imperfect conduction, and improper grounding: Hugh Williamson, "Remarks upon the Incorrect Manner in Which Iron Rods Are Sometimes Set up for Defending Houses from Lightning," *American Medical and Philosophical Register* (July 1810): 7–8.
35. Latour, *Science in Action*, chap. 3, esp. 103–106.
36. Landon Carter, Apr. 17, 1773, *The Diary of Colonel Landon Carter of Sabine Hall, 1752–1778*, 2 vols., ed. Jack P. Greene (Richmond, 1987), 2:749–750. Compare with the account by William Maine (1760) in Franklin, *EO*, 427–429. In his "Poetical Epistle" to Franklin, Woodmason imagined a scene where "Keen light'nings dart, and threat'ning clouds appear: / Now fly the negroes from the impending storm!" *BFP*, 5:60. On Carter more generally, see Rhys Isaac, *Landon Carter's Uneasy Kingdom: Revolution and Rebellion on a Virginia Plantation* (New York, 2004).
37. Carter, *Diary*, 2:750–751.

38. Ibid., 2:789, 748, 751, and also the entry for Oct. 26, 1774, 2:888–889. Carter published instructions for the treatment of lightning victims in the *Virginia Gazette* (Rind), Apr. 14, 1774. Nearly two generations later, the evangelical John Randolph of Roanoke cast his belief in Franklinist electricity as part of a youthful seduction into deism by his stepfather St. George Tucker. In a letter recalling a lightning strike on the family house when he was a child, Randolph wrote to Tucker: "You taught me . . . to ascribe our salvation to principles of electricity. Franklin's name was honoured—not God's and we became little . . . philosophers." John Randolph to St. George Tucker, Feb. 28, 1817, Bryan Family Papers, University of Virginia, Richmond.

3. Wonderful Recreations

1. Joseph Hiller, "Two Lectures on Electricity" (1757), Octavo Volumes "H," American Antiquarian Society, Worcester, Mass., 18; Franklin to Peter Collinson, July 11, 1747, *EO,* 11; Kinnersley, advertisement in the *New-York Gazette, Revived in the Weekly Post-Boy,* June 1, 1752. This advertisement appeared many times in colonial newspapers with minimal variation in the quarter century before the American Revolution, most often in the *PG.* Representations of Mahomet's tomb were common in British magnetic demonstrations: see Patricia Fara, *Sympathetic Attractions: Magnetic Practices, Beliefs, and Symbolism in Eighteenth-Century England* (Princeton, 1996), 31–33.
2. See Ebenezer Kinnersley, *New-York Gazette,* June 1, 1752, and *A Course of Experiments, in That Curious and Entertaining Branch of Natural Philosophy, Called Electricity; Accompanied with Explanatory Lectures: In Which Electricity and Lightning, Will Be Proved to Be the Same Thing* (Philadelphia, 1764), 6.
3. Simon Schaffer, "Natural Philosophy and Public Spectacle in the Eighteenth Century," *History of Science* 21 (Mar. 1983): 1–43; the quadrille comment by Albecht von Haller is quoted in John L. Heilbron, *Electricity in the Seventeenth and Eighteenth Centuries: A Study of Early Modern Physics,* rev. ed. (Mineola, N.Y., 1999), 261; Joseph Priestley, *Experiments and Observations on Different Kinds of Air,* 3 vols., 2d ed. (1775–1777), 1:xiii–xiv. The foundational work on the eighteenth-century public sphere is Jürgen Habermas, *The Structural Transformation of the Public Sphere: An Inquiry into a Category of Bourgeois Society,* trans. Thomas

Burger (Cambridge, Mass., 1991). On "wonders tamed," see Lorraine Daston and Katharine Park, *Wonders and the Order of Nature, 1150–1750* (New York, 1998), 357.

4. See chapters 6 and 7 by Plumb in Neil McKendrick, John H. Plumb, and John Brewer, eds., *The Birth of a Consumer Society* (London, 1982); Giambattista della Porta, *Natural Magick by John Baptista Porta, A Neapolitane* (London, 1658), reprinted in *The Collector's Series in Science*, ed. Derek J. Price (New York, 1957). Paula Findlen acknowledges enlightened electrical demonstrations as descendants of this Renaissance tradition: see her "Jokes of Nature and Jokes of Knowledge: The Playfulness of Scientific Discourse in Early Modern Europe," *Renaissance Quarterly* 43 (1990): 320. See also Barbara Maria Stafford, *Artful Science: Enlightenment Entertainment and the Eclipse of Visual Education* (Cambridge, Mass., 1994), esp. 47–58; Gilles Chabaud, "Entre Sciences et Sociabilités: Les Expériences de l'Illusion Artificielle en France à la Fin du XVIIIe Siècle," *Bulletin de la Société d'Histoire Moderne et Contemporaine* 44 (1997): supplement, 3–4, 36–49.

5. William Hooper, *Rational Recreations, in Which the Principles of Numbers and Natural Philosophy Are Clearly and Copiously Elucidated, by a Series of Easy, Entertaining, Interesting Experiments*, 4 vols., 2d ed. (London, 1782–1783), 1:ii; Baldassare Castiglione, *The Book of the Courtier*, trans. George Bull (1528; Harmondsworth, Eng., 1967), 48. Key works in the genre include Jacques Ozanam, *Récréations Mathématiques et Physiques* (Paris, 1692); and Edmé Guyot, *Nouvelles Récréations* (Paris, 1769–1770). The 1708 English translation of Ozanam is listed in the 1741 *Catalogue of Books Belonging to the Library Company of Philadelphia* (Philadelphia, 1956); see also William Eamon, *Science and the Secrets of Nature: Books of Secrets in Medieval and Early Modern Culture* (Princeton, 1994), esp. chap. 6.

6. Larry Stewart, *The Rise of Public Science: Rhetoric, Technology and Natural Philosophy in Newtonian Britain, 1660–1750* (Cambridge, 1992), esp. chap. 7.

7. For Greenwood and optical performances in Philadelphia, see I. Bernard Cohen, *Benjamin Franklin's Science* (Cambridge, Mass., 1990), 57–59; manuscript note written on the reverse of the title page of Henry Moyes, *Heads of a Course of Lectures on the Natural History of the Celestial Bodies* (Boston, 1784), Library Company of Philadelphia copy; for notes on a lecture Moyes gave in Boston on Oct. 21, 1785, see Loammi

Baldwin's notebook in his Miscellaneous Papers, box 1783–1787, Massachusetts Historical Society, Boston; Franklin to John Lining, Mar. 18, 1755, *EO*, 328–329; *South Carolina Gazette*, Oct. 24–31, 1748, quoted in Demetrius Dvoichenko-Markov, "A Rumanian Priest in Colonial America," *American Slavic and East European Review* 15 (Oct. 1955): 384. On visual deception as entertainment in eighteenth-century America, see Wendy Bellion, "Pleasing Deceptions," *Common-Place* 3 (2002), www.common-place.org.

8. Benjamin Franklin, *The Autobiography of Benjamin Franklin: A Genetic Text*, ed. J. A. Leo Lemay and P. M. Zall (Knoxville, Tenn., 1981), 153; Kinnersley, "A Letter from Ebenezer Kinnersley to His Friend in the Country," *Postscript to the Pennsylvania Gazette*, July 15, 1740; J. A. Leo Lemay, *Ebenezer Kinnersley: Franklin's Friend* (Philadelphia, 1964), 21, 66, 81; Ebenezer Kinnersley to Mrs. Kinnersley, Mar. 13, 1773, photostat, Society Collection, Historical Society of Pennsylvania, Philadelphia; Lemay, *Ebenezer Kinnersley*, 60–61; Carl and Jessica Bridenbaugh, *Rebels and Gentlemen: Philadelphia in the Age of Franklin* (New York, 1942), 22; advertisement by James Pearson, *PG*, Apr. 18, 1765; College of Philadelphia notice, *PG*, Dec. 17, 1767.

9. Ian K. Steele, *The English Atlantic, 1675–1740: An Exploration of Culture and Community* (New York, 1986); Susan O'Brien, "A Transatlantic Community of Saints: The Great Awakening and the First Evangelical Network, 1735–1755," *American Historical Review* 91 (Oct. 1986): 811–832, and Frank Lambert, *Inventing the "Great Awakening"* (Princeton, 1999).

10. On provincial enlightenment in British America, see Ned C. Landsman, *From Colonials to Provincials: American Thought and Culture, 1680–1760* (New York, 1997), esp. 57–63; on improvement and British colonialism, see Joyce E. Chaplin, *Subject Matter: Technology, Science, and the Body on the Anglo-American Frontier, 1500–1676* (Cambridge, Mass., 2001), chap. 6, and Richard H. Drayton, *Nature's Government: Science, Imperial Britain, and the "Improvement" of the World* (New Haven, 2000), esp. chap 4; on practical improvement in Britain and British America, see Stewart, *Rise of Public Science*, esp. part 3, and Bridenbaugh and Bridenbaugh, *Rebels and Gentlemen*, respectively.

11. Kinnersley in *New York Gazette*, June 1, 1752.

12. Isaac Greenwood, *An Experimental Course of Mechanical Philosophy* (Boston, 1726), 2; Ebenezer Kinnersley, "A Course of Experiments on the Newly Discovered Electrical Fire, by Mr E. Kinnersly of Philadelphia,

1752," Ms. am 089, Historical Society of Pennsylvania. Nollet's 1746 display is described in Heilbron, *Electricity in the Seventeenth and Eighteenth Centuries*, 318.
13. Hiller, "Two Lectures on Electricity," 24, 17, 20, 44. I have found no evidence of a personal connection between Hiller and Kinnersley. I quote from Hiller's second lecture where Kinnersley's is not extant. Like all independent demonstrators, Hiller advertised in local newspapers, for example: "*That the* Electrical Experiments *with Methodicall Lectures, exhibited last Winter near the* Blue-Ball, *and now exhibited in* Orange-Street, *a little below* Concert-Hall *in the House where the* Wax-Work *is shewn: Price,* One Pistarene. *By* Joseph Hiller, *Jeweller*": *Boston Gazette* (supplement), March 1–8, 1756. Kinnersley inspired several other imitators, whose published outlines suggest that they reproduced his displays virtually verbatim. Compare Kinnersley, *A Course of Experiments* with William Johnson, *a Course of Experiments, in That Curious and Entertaining Branch of Natural Philosophy, Call'd Electricity* (New York, 1765) with David Mason's broadside, *A Course of Experiments in That Instructive and Entertaining Branch of Natural Philosophy, Called Electricity* (Boston, 1765). For Mason, see also the 1765 *Boston Gazette* advertisement reproduced in William Northrop Morse, "Lectures on Electricity in Colonial Times," *New England Quarterly* 7 (1934): 371–372. A list of American electrical demonstrators is contained in Raymond P. Stearns, *Science in the British Colonies of America* (Urbana, Ill., 1970), 510–511.
14. Hiller, "Two Lectures on Electricity," 47, 46; Dudley Herschbach, *Benjamin Franklin's Scientific Amusements* (Cambridge, Mass., video, 1995).
15. James Bowdoin to Franklin, Dec. 21, 1751, *EO*, 174; anonymous lightning rod letter from St. John's, Antigua, dated July 17, 1753, printed in *PG*, Sept. 13, 1753. On Bowdoin and electricity, see Gordon E. Kershaw, *James Bowdoin II: Patriot and Man of the Enlightenment* (Lanham, Md., 1991), 121; Brandon Brame Fortune and Deborah J. Warner, *Franklin and His Friends: Portraying the Man of Science in Eighteenth-Century America* (Washington, D.C., 1999), 29–31. See also *Journal and Letters of Philip Vickers Fithian, 1773–1774: A Plantation Tutor of the Old Dominion*, ed. Hunter Dickinson Farish (Charlottesville, 1978), 98, 247, n. 137.
16. *New-York Gazette*, June 1, 1752.
17. *Address to Those Who Unnecessarily Frequent the Tavern* (Boston, 1726),

quoted in Bruce C. Daniels, *Puritans at Play: Leisure and Recreation in Colonial New England* (New York, 1995), 19; Benjamin Colman, *The Government of Mirth* (Boston, 1707), quoted in Daniels, *Puritans at Play*, 17; Penn's Law quoted in David S. Shields, *Civil Tongues and Polite Letters in British America* (Chapel Hill, 1997), 63–64; Alexander Hamilton, *Gentleman's Progress: The Itinerarium of Dr. Alexander Hamilton, 1744*, ed. Carl Bridenbaugh (Chapel Hill, 1948), 22–23; Virginia planter quoted in Rhys Isaac, "Evangelical Revolt: The Nature of the Baptists' Challenge to the Traditional Order in Virginia, 1765 to 1775," *WMQ* 31 (July 1974): 358; Continental Congress order quoted in Rhys Isaac, *The Transformation of Virginia, 1740–1790* (Chapel Hill, 1982), 247.

18. On electricity and earthquakes, see Schaffer, "Natural Philosophy and Public Spectacle," 15–21; Samuel Williams, "Observations and Conjectures on the Earthquakes of New-England," *Memoirs of the American Academy of Arts and Sciences* 1 (1785): 310, 308.

19. Wesley quoted in Schaffer, "Natural Philosophy and Public Spectacle," 6; Hiller, "Two Lectures on Electricity," 43 (emphasis added, and compare with Franklin, *EO*, 14).

20. *PG*, Dec. 29, 1773; "An Account of the *Gymnotus Electricus*, or Electrical Eel, in a Letter from Alexander Garden, MD, FRS, to John Ellis, Esq., FRS," Aug. 14, 1774, *Phil. Trans.* 65 (1775): 102 (emphasis added); Anonymous, "Of Superstitious Fears, and Their Causes Natural and Accidental," *American Magazine* (May 1744): 372–373; Joseph Pope, "A Letter from Mr. Pope to Edward Blount, *Esq.*," *American Magazine and Monthly Chronicle* (May 1745), 211.

21. A copy of *The History and Present State of Electricity* was acquired by the Library Company of Philadelphia sometime between 1764 and 1770; see *The Charter, Laws, and Catalogue of Books of the Library Company of Philadelphia* (Philadelphia, 1770). An advertisement in the *Pennsylvania Packet*, Sept. 29, 1778, offered "a very compleat portable Electric Apparatus, an Electric battery, and the history and present state of Electricity, with original experiments, by Joseph Priestly, LL.D. F.R.S. (a book especially necessary for young Electricians, and the most esteemed of any treatise upon Electricity.)"

22. Joseph Priestley, *The History and Present State of Electricity, with Original Experiments*, 2 vols., 3d ed. (London, 1775), reprinted as *The Sources of Science*, no. 18 (New York, 1966), 1:xii–xiii, 1:xxiii–xxiv, and 2:295–307; *The New and Complete American Encyclopaedia; or, Universal Dictionary*

of Arts and Sciences, 7 vols. (New York, 1807), 3:275. On Priestley in America, see Jenny Graham, *Revolutionary in Exile: The Emigration of Joseph Priestley to America, 1794–1804* (Philadelphia, 1995).

23. Thomas Prince, *Earthquakes the Works of God and Tokens of His Just Displeasure* (Boston, 1755), appendix "Concerning the Operation of God in *Earthquakes* by Means of the *Electrical Substance*," 20–23; Proverbs 22:3, quoted by Kinnersley in *New-York Gazette*, June 1, 1752; Hiller, "Two Lectures on Electricity," 63; *PG*, Dec. 22, 1763.

24. Ebenezer Kinnersley, *A Letter from Ebenezer Kinnersley to His Friend in the Country*, postscript to the *Pennsylvania Gazette*, July 15, 1740, quoted in Lemay, *Ebenezer Kinnersley*, 20–21; and Lemay, *Ebenezer Kinnersley*, 34–35, 39.

25. Kinnersley, in *New-York Gazette*, June 1, 1752; Kinnersley, "Course of Experiments," lecture one (the same wording may be found in Hiller, "Two Lectures on Electricity," 31–32); Lemay, *Ebenezer Kinnersley*, 75; Williams, "Observations and Conjectures," 310–311; Henry Muhlenberg quoted in Sara S. Gronim, "At the Sign of Newton's Head: Astrology and Cosmology in British Colonial New York," *Explorations in Early American Culture/Pennsylvania History* 66 (1999): 68. On religion and science in Kinnersley, see Nina-Ruth Reid-Maroney, *Philadelphia's Enlightenment, 1740–1800: Kingdom of Christ, Empire of Reason* (Westport, Conn., 2001), 52–60.

26. *American Magazine* (May 1744): 416–417.

27. David Hume, "Of the Rise and Progress of the Arts and Sciences" (1742), in *Political Writings*, ed. Knud Haakonssen (Cambridge, 1994), 71–75; James Forrester, *The Polite Philosopher* (Philadelphia, 1781), 48–50; Charles Allen, *The Polite Lady; or, A Course of Female Education: In a Series of Letters, from a Mother to a Daughter* (Philadelphia, 1798), 87, 121. Forrester's book was a ninth edition; a 1758 London edition had been reprinted in New York that same year; a later edition was also reprinted in Boston in 1787.

28. On polite science, see Geoffrey V. Sutton, *Science for a Polite Society: Gender, Culture and the Demonstration of Enlightenment* (Boulder, Colo., 1995), 143–187, and Alice N. Walters, "Conversation Pieces: Science and Politeness in Eighteenth-Century England," *History of Science* 35 (1997): 121–154. On natural history and Wakefield (reprinted in America in 1799), see Ann B. Shteir, *Cultivating Women, Cultivating Science: Flora's Daughters and Botany in England, 1760 to 1860* (Baltimore, 1996),

82–83; *Analysis of Certain Parts of a Compendious View of Natural Philosophy, for the Use of Students of Both Sexes* (Boston, 1796).

29. Benjamin West, *New-England Almanack; or, Lady's and Gentleman's Diary, for the Year of Our Lord Christ, 1769* (Boston, 1769). A similar entry from Drinker's diary for 1773 reveals that these experiments were performed by "Js. B"—James Bringhurst, a carpenter—and notes the presence of five of the Drinker family, including Elizabeth's husband, Henry, a sometime botanist: *The Diary of Elizabeth Drinker*, vol. 1, ed. Elaine Forman Crane (Boston, 1991), 43, 47, 182.

30. Hannah Webster Foster, *The Coquette; or, The History of Eliza Wharton: A Novel; Founded on Fact* (Harmondsworth, Eng., 1996), 196; *Extracts from the Journal of Mrs. Ann Ashby Manigault, 1754–1781*, ed. Mabel J. Webber (Charleston, 1920), entry for May 10, 1765; see also the entry for Mar. 8, 1770, in the diary of Mary Vial Holyoke—*The Holyoke Diaries, 1709–1865*, ed. George Francis Dow (Salem, Mass., 1911). Henry Laurens, sometime vice president of the Charleston Library Society, held a meeting to determine whether the society would allow Johnson to use their rooms: see *The Papers of Henry Laurens*, 10 vols., ed. Philip M. Hamer (Columbia, S.C., 1968–1985), 5:223. Silvio A. Bedini identifies the Charleston Library Society as the first museum in North America; it acquired an electrical machine in 1774 (this perished in a fire in 1778); see Bedini, *Thinkers and Tinkers: Early American Men of Science* (New York, 1975), 182–183. See also *Guide to the Philadelphia Museum* (Philadelphia, 1805), 2 (the museum opened in 1785); and Robert M. McClung and Gale S. McClung, "Tammany's Remarkable Gardiner Baker: New York's First Museum Proprietor, Menagerie Keeper, and Promoter Extraordinary," *New York Historical Society Quarterly* 42 (Apr. 1958): 142–169.

31. *The Newtonian System of Philosophy; Explained by Familiar Objects, in an Entertaining Manner, for the Use of Young Ladies and Gentlemen, by Tom Telescope*, ed. Robert Patterson, 2d ed. (Philadelphia, 1808), 1–3, 64–72; see James A. Secord, "Newton in the Nursery: Tom Telescope and the Philosophy of Tops and Balls, 1761–1838," *History of Science* 23 (1985): 127–151.

32. Samuel Miller, *A Brief Retrospect of the Eighteenth Century*, 2 vols. (New York, 1803), 2:278–285; Kinnersley in the *New-York Gazette*, June 1, 1752; Isaac Greenwood III, *A Brilliant Electrical Exhibition* (Providence, 1793), and *Isaac Greenwood, Dentist, No. 49 Marlborough-Street, Boston* (Boston, 1788), both broadsides; John Winthrop, *A Letter to the Pub-*

lishers of the Boston Gazette, &c. Containing an Answer to the Rev. Mr. Prince's Letter (Boston, 1756), 7; *Analysis of Certain Parts of a Compendious View of Natural Philosophy,* 6.

33. Hiller, "Two Lectures on Electricity," 32, 46; Kinnersley in the *New-York Gazette,* June 1, 1752.
34. Erasmus Darwin, *The Botanic Garden, a Poem, in Two Parts,* 2d ed. (New York, 1807), 25–26. This passage describing the Venus was excerpted in the *New York Magazine and Literary Repository* 4 (Aug. 1793): 506–508; on Darwin in America, see Fredrika J. Teute, "The Loves of the Plants; or, the Cross-Fertilisation of Science and Desire at the End of the Eighteenth Century," *Huntington Library Quarterly* 63 (2000): 319–345.
35. Albrecht von Haller, "An Historical Account of the Wonderful Discoveries, Made in Germany, &c. Concerning Electricity," *American Magazine and Historical Chronicle* 2 (Boston, 1745): 533; John Cleland, *Fanny Hill; or, Memoirs of a Woman of Pleasure* (1749; Harmondsworth, Eng., 1985), 211, and, for example, *An Elegy on the Lamented Death of the Electrical Eel* (1778) by "Lucretia Lovejoy," quoted in Peter Wagner, *Eros Revived: Erotica of the Enlightenment in England and America* (London, 1988), 196–197; Joseph Aignan Sigaud de la Fond, *Précis Historique et Expérimental des Phénomènes Electriques* (Paris, 1781), cited in Heilbron, *Electricity in the Seventeenth and Eighteenth Centuries,* 320. On eroticism and materialism in the Enlightenment, see Margaret C. Jacob, "The Materialist World of Pornography," in Lynn Hunt, ed., *The Invention of Pornography: Obscenity and the Origins of Modernity, 1500–1800* (New York, 1993), 157–202.
36. Charles Rabiqueau, *Le Spectacle du Feu Elémentaire ou Cours d'Electricité Expérimentale* (1753), quoted in Gaston Bachelard, *The Psychoanalysis of Fire,* trans. Alan Ross (1938; Boston, 1964), 26; John Shebbeare, *The Practice of Physic,* 2 vols. (London, 1755), 1:165–167 and 2:350; James Walker, *An Inquiry into the Causes of Sterility in Both Sexes; with Its Methods of Cure* (Philadelphia, 1797), 7–8, 15–22; T. Gale, *Electricity, or Ethereal Fire, Considered* (Troy, N.Y., 1802), 141, 182–184. Electricity may also have been used to induce abortions: see Michael Brian Schiffer, *Draw the Lightning Down: Benjamin Franklin and Electrical Technology in the Age of Enlightenment* (Berkeley, 2003), 155–156.
37. Paula Findlen, "Science as a Career in Enlightenment Italy: The Strategies of Laura Bassi," *Isis* 84 (1993): 440–469; Lorraine Daston, "The Naturalised Female Intellect," *Science in Context* 5 (1992): 229–230;

Londa Schiebinger, *Nature's Body: Gender in the Making of Modern Science* (Boston, 1993), chap. 1; Linda Kerber, *Women of the Republic: Intellect and Ideology in Revolutionary America* (Chapel Hill, 1980), 282; Ludmilla Jordanova, *Nature Displayed: Gender, Science and Medicine, 1760–1820* (New York, 1999), esp. chap. 2; Diary of Alexander Anderson Jr., 1793–1799, Ms. 119, Columbiana Collection, Columbia University Rare Books and Special Collections; Thomas Green Fessenden, *Terrible Tractoration!!: A Poetical Petition against Galvanising Trumpery, and the Perkinistic Institution in Four Cantos*, 3d ed. (Philadelphia, 1806), 265. Fessenden became involved in the marketing of medical electricity at the turn of the nineteenth century (see chap. 7).

38. In insisting on the persistence of wonder and play as legitimate nonvulgar modes of scientific interaction with nature in the Enlightenment, my interpretation diverges from Daston and Park, *Wonders and the Order of Nature*, 316–363, and Paula Findlen, "Between Carnival and Lent: The Scientific Revolution at the Margins of Culture," *Configurations* 6 (1998): 243–267.

39. Johann Huizinga, *Homo Ludens: A Study of the Play-Element in Culture* (Boston, 1955), 186; Stafford, *Artful Science*, esp. chap. 3; Schiffer, *Draw the Lightning Down*, 44. The classic analysis of early modern politeness is Norbert Elias, *The Civilizing Process: Sociogenetic and Psychogenetic Investigations*, rev. ed., trans. Edmund Jephcott (Oxford, 2000).

40. Franklin to Collinson, n.d., 1748, *EO*, 37–38; *PG*, Mar. 19, 1761.

41. Morse, "Lectures on Electricity in Colonial Times," 364–365; Claggett, in *Boston Gazette*, Sept. 1, 1747; Richard Brickell, in *New York Gazette and Weekly Post-Boy*, May 2, 1748; Allen, *Polite Lady*, 117; on machines and monsters, see Simon Schaffer, "Enlightened Automata" and Michael Hagner, "Enlightened Monsters," in William Clark, Jan Golinski, and Simon Schaffer, eds., *The Sciences in Enlightened Europe* (Chicago, 1999), 126–165 and 175–217; for curiosities exhibited in eighteenth-century New York (including electrical demonstrations), see Rita Gottesman, ed., *The Arts and Crafts in New York*, 3 vols. (New York, 1938–1949); Claggett in the *Independent Advertiser* (Boston), Feb. 8, 1748; Hamilton, *Itinerarium*, 11; Diary of Alexander Anderson Jr., Feb. 27, 1795. The estate of William Faris, an Annapolis tavern keeper, contained "one electrifying machine and apparatus compleat" (1805): see Kym S. Rice, *Early American Taverns: For the Entertainment of Friends and Strangers* (Chicago, 1983), 117.

42. Jonathan Dewald has argued that in early modern France, "the court

was not a scene of heightened self-control only, but also one of continuous ruptures of self-control": see Dewald, *Aristocratic Experience and the Origins of Modern Culture: France, 1570–1715* (Berkeley, 1993), 9. Cultural historians have articulated the logic underlying laughter in several different social contexts. Most notably, Mikhail Bakhtin has explored the subversive and carnivalesque laughter of Rabelaisian comedy, while Robert Darnton has examined humor produced by the social antagonism between artisans and bourgeois masters in ancien régime France. See Mikhail Bakhtin, *Rabelais and His World*, trans. Hélène Iswolsky (Cambridge, Mass., 1968) and Robert Darnton, "Workers' Revolt: The Great Cat Massacre," *The Great Cat Massacre and Other Episodes in French Cultural History* (New York, 1984), 75–107. Peter Burke has suggested that early modern European elites, especially after the Counter-Reformation, increasingly preferred verbal wordplay to the full-bodied expressiveness of the carnival: see Burke, *Varieties of Cultural History* (Ithaca, 1997), 77–93. On sensory disorientation ("ilinx") and play, see Roger Caillois, *Man, Play and Games*, trans. Meyer Barash (New York, 1961), 23–24; on play as an expression of cosmopolitan confidence, see Kay Dian Kriz, "'Stare-cases': Engendering the Public's Two Bodies at the Royal Academy of Arts," in David Solkin, ed., *Art on the Line* (New Haven, 2001), 55–63.

43. Priestley, *History and Present State of Electricity*, 1:xxv, 2:150–153; M. M. McDowell, "A Cursory View of Cheating at Whist in the Eighteenth Century," *Harvard Library Bulletin* 22 (Apr. 1974): 162–175; on forks see Elias, *Civilizing Process*, 107–109; William Watson, *Sequel to the Experiments and Observations Tending to Illustrate the Nature and Properties of Electricity* (London, 1746), 22; Kinnersley in the *New-York Gazette*, June 1, 1752.

44. My emphasis departs somewhat from Simon Schaffer's characterization of polite knowledge as the alienation of experience from the body, which he describes as "the cartesianism of the genteel": see Schaffer, "Self Evidence," in James Chandler, Arnold I. Davidson, and Harry Harootunian, eds., *Questions of Evidence: Proof, Practice, and Persuasion across the Disciplines* (Chicago, 1994), 68.

45. Priestley, *History and Present State of Electricity*, 2:151; Franklin to Cadwallader Colden, n.d., 1751, *EO*, 93; *PG*, Dec. 22, 1763.

46. Per Kalm, *Peter Kalm's North American Journey: Its Ideological Background and Results*, ed. Martti Kerkkonen (Helsinki, 1959), 192; Charles

Caldwell in Johann Friedrich Blumenbach, *Elements of Physiology; Translated from the Original Latin, and Interspersed with Occasional Notes by Charles Caldwell, to Which Is Subjoined, by the Translator an Appendix, Exhibiting a Brief and Compendious View of the Existing Discoveries Relative to the Subject of Animal Electricity*, 2 vols. (Philadelphia, 1795), 1:245.

47. Greenwood, *A Brilliant Electrical Exhibition*; Priestley, *History and Present State of Electricity*, 1:xxi; William Smith, speech in Charleston, South Carolina, Jan. 1, 1772, *PG*, Mar. 12, 1772. See also Henry F. May, *The Enlightenment in America* (New York, 1976), 80–86, and on the British-American empire for liberty, Landsman, *From Colonials to Provincials*, 149–180.

48. *PG*, Nov. 11, 1772; André Morellet, "Anecdotes sur Francklin," *Gazette Nationale, ou le Moniteur Universel*, July 15, 1779, quoted in P. M. Zall, ed., *Benjamin Franklin Laughing: Anecdotes from Original Sources by and about Benjamin Franklin* (Berkeley, 1980), 101; Edward Bancroft, *The History of Charles Wentworth*, 3 vols. (London, 1770), 2:261–262; "Letter from a Gentleman at Tillicheery in the East Indies," *Virginia Gazette* (Richmond), Dec. 9, 1780. The eighteenth-century Jamaica planter Thomas Thistlewood owned an electrical machine that he may well have demonstrated to his slaves: see Trevor Burnard, *Mastery, Tyranny, and Desire: Thomas Thistlewood and His Slaves in the Anglo-American World* (Chapel Hill, 2004), 119. Thomas Jefferson invited Native Americans to view his curiosities at Monticello in the hope of inspiring "astonishment or surprise" in them: see Anthony F. C. Wallace, *Jefferson and the Indians: The Tragic Fate of the First Americans* (Cambridge, Mass., 1999), 128.

49. Kinnersley broadside advertisement, St. John's (Antigua), Apr. 25, 1753, reproduced in Douglas C. McMurtrie, *Early Printing on the Island of Antigua* (Evanston, Ill., 1943); Deborah J. Warner, "Lightning Rods and Thunder Houses," *Rittenhouse* 11 (1997): 125; *PG*, Dec. 12, 1754 (for the "model negroe" display in Philadelphia); *House of Commons Sessional Papers of the Eighteenth Century*, 140 vols., ed. Sheila Lambert (Wilmington, Del., 1975), 2:219; Vincent Brown, "Spiritual Terror and Sacred Authority in Jamaican Slave Society," *Slavery and Abolition* 24 (Apr. 2003): 24–53; Peter Linebaugh and Marcus Rediker, *The Many-Headed Hydra: Sailors, Slaves, Commoners and the Hidden History of the Revolutionary Atlantic* (Boston, 2000), 221–224. "Volt(a)," an exhibition

at the Musée des Arts et Métiers, Paris, in 2000–2001, exhibited an eighteenth-century Italian model of an African boy, whose head was designed to separate from his body on the application of a strong electric charge: see *La Revue des Arts et Métiers* 31 (Dec. 2000).
50. Priestley, *History and Present State of Electricity*, 2:134.

4. Electrical Politics and Political Electricity

1. Franklin to Peter Collinson, n.d., 1748, *EO*, 29–30.
2. Jan Golinski, *Science as Public Culture: Chemistry and Enlightenment in Britain, 1760–1820* (Cambridge, 1992), 176–179; Jessica Riskin, *Science in the Age of Sensibility: The Sentimental Empiricists of the French Enlightenment* (Chicago, 2002), 273–277; Max Horkheimer and Theodor Adorno, *Dialectic of Enlightenment*, trans. John Cumming (New York, 1972), 3–42; on the "Newtonian constitution," see I. Bernard Cohen, *Science and the Founding Fathers: Science in the Political Thought of Thomas Jefferson, Benjamin Franklin, John Adams and James Madison* (New York, 1995), 237–280. In this book, Cohen in fact rejected his own thesis that science shaped the political thought of the Founders. See also Brooke Hindle, *The Pursuit of Science in Revolutionary America, 1735–1789* (Chapel Hill, 1956), esp. chap. 12; Garry Wills, *Inventing America: Jefferson's Declaration of Independence* (New York, 1978); and the response in Ronald Hamowy, "Jefferson and the Scottish Enlightenment: A Critique of Garry Wills' *Inventing America: Jefferson's Declaration of Independence*," *WMQ* 36 (Oct. 1979): 503–523; and Joyce E. Chaplin, "Nature and Nation: Natural History in Context," in Sue Ann Prince, ed., *Stuffing Birds, Pressing Plants, Shaping Knowledge: Natural History in North America, 1730–1860* (Philadelphia, 2003), 75–95.
3. For a discussion of magnetism as a "powerful language," see Patricia Fara, *Sympathetic Attractions: Magnetic Practices, Beliefs, and Symbolism in Eighteenth-Century England* (Princeton, 1996), 171–207.
4. Anonymous ex-slave quoted in John H. Wigger, "Taking Heaven by Storm: Enthusiasm and Early American Methodism, 1770–1820," *Journal of the Early Republic* 14 (Summer 1994): 167; Caleb Rich, "A Narrative of the Elder Caleb Rich," *Candid Examiner* 2 (Apr. 30–June 18, 1827); and John Bishop in *Testimonies on the Life, Character, Revelations and Doctrines of Our Every Blessed Mother Ann Lee* (Hancock, Mass., 1816), quoted in Susan Juster, "Mystical Pregnancy and Holy Bleeding: Vision-

ary Experience in Early Modern Britain and America," *WMQ* 57 (Apr. 2000): 267–268; Samuel Goodrich, *Recollections of a Lifetime; or, Men and Things I Have Seen*, 2 vols. (New York, 1856), quoted in Wigger, "Taking Heaven by Storm," 192. The links between electricity and Protestant theology in early modern Germany are examined in Ernst Benz, *The Theology of Electricity: On the Encounter and Explanation of Theology and Science in the Seventeenth and Eighteenth Centuries*, trans. Wolfgang Taraba (Allison Park, Penn., 1989).

5. It is also fair to assume that the figure of Franklin, now a revolutionary statesman as well as electrician, helped to inspire these rhetorical links between electricity and politics.

6. There is evidence of at least one electrical demonstrator who performed during the Revolutionary period: John McPherson, a Scottish privateer based in Philadelphia who advertised "a course of lectures on Astronomy, and every other branch of Natural Philosophy, particularly Mechanicks and Electricity" in the *Pennsylvania Packet*, Dec. 25, 1781. McPherson claimed to want to avoid partisanship, though he was apparently a Patriot: "I'm high church, not low church, no tory, a whig . . . / No party I serve, in no quarrels I join; Nor damn the opinion that differs from mine: / No corruption I screen, no treason I sing; / I'm a friend to this country, but wish for no King."

7. On British sensibility, see Christopher Lawrence, "The Nervous System and Society in the Scottish Enlightenment," in Barry Barnes and Steven Shapin, eds., *Natural Order: Historical Studies in Scientific Culture* (Beverly Hills, 1979), 19–40, and G. J. Barker-Benfield, *The Culture of Sensibility: Sex and Society in Eighteenth-Century Britain* (Chicago, 1992); for revolutionary America, see Sarah Knott, "Sensibility and the War for American Independence," *American Historical Review* 109 (Feb. 2004): 19–40, as well as Andrew Burstein, "The Political Character of Sympathy," *Journal of the Early Republic* 21 (Winter 2001): 601–632; on sensibility and enlightened science, see Riskin, *Science in the Age of Sensibility*.

8. Quoted in Peter Oliver, *The Origin and Progress of the American Rebellion* (1781), reprinted as *Peter Oliver's Origin and Progress of the American Rebellion: A Tory View*, ed. Douglass Adair and John A. Schutz (Stanford, 1961), 55. Non-Americans were not long in appropriating this usage. Reflecting on Admiral Rodney's victory at the Saints (1782), Sir Nathaniel Wraxall recalled that this was "an event which electrified the whole population of Great Britain." See Wraxall, *The Historical and*

Posthumous Memoirs of Sir Nathaniel Wraxall, 5 vols. (New York, 1884), 2:314, quoted in Stephen Conway, "'A Joy Unknown for Years Past': The American War, Britishness and the Celebration of Rodney's Victory at the Saints," *History* 86 (Apr. 2001): 199.

9. William Henry Drayton, notes for a speech before Congress on a wartime resolution banning correspondence between Americans and Britons, June 17 [?], 1778, *Letters of Delegates to Congress*, 26 vols., ed. Paul H. Smith (Washington, D.C., 1976–), 10:115; John Adams to James Warren, Apr. 22, 1776, *Letters of Delegates*, 3:569; Thomas Jefferson, "Autobiography," in *Thomas Jefferson: Writings*, ed. Merrell D. Peterson (New York, 1984), 9; the Sheffield pamphlet is *Observations on the Commerce of the American States with Europe and the West Indies* (London, 1783); John Francis Mercer to Henry Tazewell, Sept., 13, 1783, *Letters of Delegates*, 20:671.

10. *PG*, Dec. 15, 1778, and May 26, 1779. Before the eighteenth century, electricians routinely referred to charges as electrical "virtues": see John L. Heilbron, *Electricity in the Seventeenth and Eighteenth Centuries: A Study of Early Modern Physics*, rev. ed. (Mineola, N.Y., 1999), 215–218; on metaphorical invocations of magnetic virtue, see Fara, *Sympathetic Attractions*, 176–177.

11. Charles Royster, *A Revolutionary People at War: The Continental Army and American Character, 1775–1783* (Chapel Hill, 1979), 25–53; Wheeler Case, *Poems, Occasioned by Several Circumstances and Occurrences, in the Present Grand Contest of America for Liberty* (New Haven, 1778), quoted in Royster, *A Revolutionary People at War*, 30; *PG*, Feb. 13, 1782; William Duer to the New York Convention, Apr. 17, 1777, *Letters of Delegates*, 6: 602. On male sympathy in the early republic, see Caleb Crain, *American Sympathy: Men, Friendship, and Literature in the New Nation* (New Haven, 2001).

12. *PG*, May 18, 1791; Erasmus Darwin, *The Botanic Garden*, 2 vols. (New York, 1798), 1:56; Timothy Dwight, *Greenfield Hill* (New York, 1794), 15; David Humphreys, "A Poem on the Future Glory of the United States of America," in *The Miscellaneous Works of David Humphreys* (New York, 1804), 56. On the relationship between factional stage management and the image of spontaneous universalism in the political festival in revolutionary France, see Mona Ozouf, *Festivals and the French Revolution*, trans. Alan Sheridan (Cambridge, Mass., 1988), 19–28; for American festivals, see David Waldstreicher, *In the Midst of Perpetual Fetes: The Making of American Nationalism, 1776–1820* (Chapel Hill, 1997).

13. Humphreys, "A Poem on the Love of Country, in Celebration of the Twenty-Third Anniversary of the Independence of the United States of America" (1799), *Miscellaneous Works*, 124.
14. Robert Merry, "Ode for the Fourteenth of July, 1791," *PG*, Sept. 21, 1791; see also *New York Magazine and Literary Repository* 4 (May 1793), 315; St. John Honeywood, "Poetical Address to the Citizen Adet," *Poems by St. John Honeywood* (New York, 1801), 72; Rufus King to Timothy Pickering, Dec. 31, 1799, extract printed in *PG*, Apr. 23, 1800; John Adams (1790), quoted in Cohen, *Science and the Founding Fathers*, 211–212.
15. Otto Mayr, *Authority, Liberty and Automatic Machinery in Early Modern Europe* (Baltimore, 1986).
16. On the elevation of powerful feeling above rational persuasion in the revolutionary United States, see Jay Fliegelman, *Declaring Independence: Jefferson, Natural Language, and the Culture of Performance* (Stanford, 1993), esp. 36.
17. John Adams, "A Dissertation on the Canon and Feudal Law" (1765), quoted in David S. Lovejoy, *Religious Enthusiasm in the New World: Heresy to Revolution* (Cambridge, Mass., 1985), 228; on the positive reevaluation of enthusiasm in the eighteenth century, see Michael Heyd, *"Be Sober and Reasonable": The Critique of Enthusiasm in the Seventeenth and Early Eighteenth Centuries* (Leyden, 1995), 221–227, and n. 41; on electricity and romanticism in nineteenth-century American literature, see Paul Gilmore, "Romantic Electricity, or the Materiality of Aesthetics," *American Literature* 76 (Sept. 2004): 467–494.
18. On the relationship between secular and sacred idioms in the American Founding, see Robert A. Ferguson, *The American Enlightenment, 1750–1820* (Cambridge, Mass., 1997), 44–79.
19. On changing definitions of Newton's genius, from the methodical to the Romantic, see Richard Yeo, "Genius, Method, and Morality: Images of Newton in Britain, 1760–1860," *Science in Context* 2 (1988): 257–284, and Patricia Fara, *Newton: The Making of Genius* (London, 2002); see also Simon Schaffer, "Genius in Romantic Natural Philosophy," in Andrew Cunningham and Nick Jardine, eds., *Romanticism and the Sciences* (Cambridge, 1990), 82–98. For an extended appraisal of Franklin's genius, see Joyce E. Chaplin, *The First Scientific American: Benjamin Franklin and the Pursuit of Genius* (New York, 2006).
20. *The Nightingale*, June 11, 1796; review of *A View of the History of Great Britain during the Administration of Lord North* in *English Review* 1 (Jan.

1783): 28–29, reprinted in the *Boston Magazine* (May 1784): 294; on Rittenhouse's genius, see Thomas Jefferson, *Notes on the State of Virginia* (1785), ed. William Peden (Chapel Hill, 1954), 64. Carolus Linnaeus regarded Bartram as the finest natural botanist of his day: see Thomas P. Slaughter, *The Natures of John and William Bartram* (New York, 1996), 51. Yeo has convincingly shown how eighteenth-century accounts of Newton's genius stressing his obedience to true philosophical method were challenged in the nineteenth century by redefinitions of his genius as a transcendence of method: see "Genius, Method, and Morality," 257–284.

21. On philosophical modesty, see Steven Shapin and Simon Schaffer, *Leviathan and the Air-Pump: Hobbes, Boyle, and the Experimental Life* (Princeton, 1985), 65–69, and on Franklin in particular, Riskin, *Science in the Age of Sensibility*, 91.

22. *BFP*, 7:74; on Franklin versus Newton, see J. G. Crowther, *Famous American Men of Science* (Freeport, N.Y., 1965), 135–155; Philip Freneau and Hugh Henry Brackenridge, *A Poem, on the Rising Glory of America* (Philadelphia, 1772), 18; William Smith, *Eulogium on Benjamin Franklin* (Philadelphia, 1792), 8, 28, 23; Francis Jeffrey, review of *The Complete Works in Philosophy, Politics, and Morals, of the late Dr. Benjamin Franklin* (London, 1806), in *Edinburgh Review, or Critical Journal* 8 (July 1806): 327–344 (reprinted in New York in 1815).

23. Smith, *Eulogium*, 25, 22. My discussion of Franklin's experimental body has been informed by the issues raised in Christopher Lawrence and Steven Shapin, eds., *Science Incarnate: Historical Embodiments of Natural Knowledge* (Chicago, 1998); on Robert Boyle as an ascetic Christian natural philosopher, see Shapin, *A Social History of Truth: Civility and Science in Seventeenth-Century England* (Chicago, 1994); on the enlightened philosopher as a bon vivant rather than an ascetic, see Shapin, "The Philosopher and the Chicken: On the Dietetics of Disembodied Knowledge," in Lawrence and Shapin, *Science Incarnate*, 43, and Roy Porter, *Flesh in the Age of Reason* (New York, 2003), 203–204. See also Betsy Erkkila, "Franklin and the Revolutionary Body," *English Literary History* 67 (2000): 717–741, and Deborah J. Warner, "Portrait Prints of Men of Science in Eighteenth-Century America," *Imprint* 25 (Spring 2000): 26–33; on French views of genius and virtue in Franklin, see Riskin, *Science in the Age of Sensibility*, 135–136.

24. *Virginia Gazette* (Purdie), Dec. 12, 1777, quoted in Richard A. Overfield,

"Science in the *Virginia Gazette, 1736–1780,*" *Emporia State Research Studies* 16 (Mar. 1968): 44; *Pennsylvania Packet and General Advertiser,* July 4, 1775; on Franklin's political affiliations in prerevolutionary London, see Verner Crane, "The Club of Honest Whigs: Friends of Science and Liberty," *WMQ* 23 (Apr. 1966): 210–233.

25. Earl of Macclesfield, Nov. 30, 1753, *BFP,* 5:130–131; Giambattista Toderini to Franklin, Aug. 15, 1772, *BFP,* 19:242 (Toderini was a committed Franklinist and author of *Filosofia Frankliniana delle Punte Preservatrici dal Fulmine* [Modena, 1771], a copy of which he sent to Franklin); *PG,* Aug. 22, 1771; May 17, 1780; *Royal Pennsylvania Gazette,* Apr. 17, 1778.

26. See Wayne Craven, "The American and British Portraits of Benjamin Franklin," in J. A. Leo Lemay, ed., *Reappraising Benjamin Franklin: A Bicentennial Perspective* (Newark, Del., 1993), 247–271, and Brandon Brame Fortune and Deborah J. Warner, *Franklin and His Friends: Portraying the Man of Science in Eighteenth-Century America* (Washington, D.C., 1999). Surprisingly little is known about the composition of West's painting, which dates to the period between 1805 and 1815, or its author's motives: see Charles Coleman Sellers, *Benjamin Franklin in Portraiture* (New Haven, 1962), 401–402. On Franklin's Americanization, see Gordon S. Wood, *The Americanization of Benjamin Franklin* (New York, 2004).

27. Smith, *Eulogium,* 24; "The Prospect of America," printed in the *American Museum* 1 (May 1787), 479; Joel Barlow, *The Prospect of Peace* (New Haven, 1778), 7; *Time-Piece; and Literary Companion,* Aug. 16, 1798; *Nightingale,* June 11, 1796; Darwin, *Botanic Garden,* 1:23–24, 56.

28. Edward Larkin, "Seeing through Language: Narrative, Portraiture, and Character in Peter Oliver's *The Origin and Progress of the American Rebellion,*" *Early American Literature* 36 (2001): 427, 449; on conspiracy, see Bernard Bailyn, *The Ideological Origins of the American Revolution,* rev. ed. (Cambridge, Mass., 1992), 144–159; Ira D. Gruber, "The American Revolution as a Conspiracy: The British View," *WMQ* 26 (July 1969): 360–372; Gordon S. Wood, "Conspiracy and the Paranoid Style: Causality and Deceit in the Eighteenth Century," *WMQ* 39 (July 1982): 401–441; Ed White, "The Value of Conspiracy Theory," *American Literary History* 14 (Spring 2002): 1–31.

29. Anonymous, *A Letter to Benjamin Franklin, LL.D., Fellow of the Royal Society, in Which His Pretensions to the Title of Natural Philosopher Are Considered* (London, 1777), reproduced in I. Bernard Cohen, *Benjamin Franklin's Experiments: A New Edition of Franklin's Experiments and Ob-*

servations on Electricity (Cambridge, Mass., 1941), 424, 428–431. There were, of course, many "Newtons" invoked in the eighteenth century—the Newtons of the *Principia* and the *Opticks*, for example, were mathematical and experimental in their respective emphases. Franklin has long been identified as an exponent of "speculative Newtonian experimental science": see I. Bernard Cohen, *Franklin and Newton: An Inquiry into Speculative Newtonian Experimental Science and Franklin's Work in Electricity as an Example Thereof* (Philadelphia, 1956). But on the plasticity of this label, see Simon Schaffer, "Newtonianism," in R. C. Olby et al., eds., *Companion to the History of Modern Science* (New York, 1990), 610–626; on polemics of systematic spirit versus spirit of system, as they pertained to Franklin and Nollet, see Geoffrey V. Sutton, *Science for a Polite Society: Gender, Culture and the Demonstration of Enlightenment* (Boulder, Colo., 1995), 287–336, and Riskin, *Science in the Age of Sensibility*, 69–103; on the role of imagination in system-building, see Lorraine Daston, "Fear and Loathing of the Imagination in Science," *Daedalus* 127 (Winter 1998): 73–95.

30. *Letter to Benjamin Franklin*, 432–433, 427; William Jones, *Six Letters on Electricity* (London, 1800), 5–7. For definitions of natural and experimental philosophy in the eighteenth century, see Simon Schaffer, "Natural Philosophy," and John L. Heilbron, "Experimental Natural Philosophy," in Roy Porter and G. S. Rousseau, eds., *The Ferment of Knowledge: Studies in the Historiography of Eighteenth-Century Science* (Cambridge, 1980), 55–91 and 357–388; and Thomas L. Hankins, *Science and the Enlightenment* (Cambridge, 1985), 46–50. Recent scholarship has emphasized the indebtedness of early modern experiment to artisanal dexterity. On manual dexterity in electrical performance, see in particular Simon Schaffer, "Experimenters' Techniques, Dyers' Hands, and the Electric Planetarium," *Isis* 88 (1997): 456–483.

31. Jonathan Boucher, *A View of the Causes and Consequences of the American Revolution* (New York, 1967), 438, 448; the invocation of the label *"fur"* is derived from a passage in Plautus's *Aulularia, BFP*, 21:49 and n. 4, and as discussed in Michael Warner, *The Letters of the Republic: Publication and the Public Sphere in Eighteenth-Century America* (Cambridge, Mass., 1990), 95; Keith Arbour, "One Last Word: Benjamin Franklin and the Duplessis Portrait of 1778," *Pennsylvania Magazine of History and Biography* 118 (1994): 183–208; Adams quoted, n.d., in Simon Schama, *Citizens: A Chronicle of the French Revolution* (New York, 1989), 46.

32. *PG*, May 4, 1774; J. A. Leo Lemay identifies Kinnersley as the orchestrator of this display in *Ebenezer Kinnersley, Franklin's Friend* (Philadelphia, 1964), 111; on revolutionary effigies, see Paul A. Gilje, *The Road to Mobocracy: Popular Disorder in New York City, 1763–1834* (Chapel Hill, 1987), esp. 25–29, 39–41. Jacob Hiltzheimer, a German immigrant and Philadelphia livestock merchant, noted in his diary that the effigies were "set in flames by *electric* fire and consumed to ashes about six o'clock in the evening." "Extracts from the Diary of Jacob Hiltzheimer, 1768–1798," *Pennsylvania Magazine of History and Biography* 16 (1892): 97.
33. Oliver, *Origin and Progress of the American Rebellion*, 79; Jonathan Odell, verse from the *Gentleman's Magazine* 47 (1777), quoted in Oliver, *Origin and Progress of the American Rebellion*, 79–80; *BFP*, 21:60, 65–66. Odell was also quoted by Boucher, who noted that these lines "were inscribed on a chamber-stove, which was made in the form of an urn, invented by the Doctor; and so contrived, that the flame, instead of ascending, descended": see Boucher, *View of the Causes and Consequences of the American Revolution*, 448–449.
34. Wilson quoted in Cohen, *Franklin and Newton*, 417–418; poem reprinted in *Boston Evening Post*, July 20, 1782; Trent A. Mitchell, "The Politics of Experiment in the Eighteenth Century: The Pursuit of Audience and the Manipulation of Consensus in the Debate over Lightning Rods," *Eighteenth-Century Studies* 31 (1998): 307–331; Simon Schaffer, "Fish and Ships: Models in the Age of Reason," in Nick Hopwood and Soraya de Chadarevian, eds., *Models: The Third Dimension in Science* (Stanford, 2004), 71–105.
35. *BFP*, 21:56, 54, 47–50; Oliver, *Origin and Progress of the American Rebellion*, 78; Johann Caspar Lavater, *Essays on Physiognomy, Designed to Promote the Knowledge and Love of Mankind*, 2 vols., trans. Henry Hunter (London, 1789), 2:318; *BFP*, 21:28. The Philadelphia crowd that burned Wedderburn and Hutchinson with electric fire responded in kind: "On the Breast of 'Wedderburne' the following Label was fixed, Such horrid Monsters are a Disgrace to human Nature. On Governor Hutchinson's Breast was fixed the following Label, GOVERNOR HUTCHINSON, whom we now consign to the Gallows and Flames, as the only proper Reward for DOUBLE DEALING AND TREACHERY to his native Country": *PG*, May 4, 1774.
36. Bancroft quoted in *Memoirs of the Life and Writings of Benjamin Franklin*, ed. William Temple Franklin, 2 vols., 2d ed. (London, 1818), 1:358n;

Public Advertiser, Feb. 2, 1774; *BFP,* 21:40; Franklin to Thomas Cushing, Feb. 15, 1774, *BFP,* 21:93; *PG,* May 18, 1774.

37. Valentine Rathbun, *An Account of the Matter, Form, and Manner of a New and Strange Religion* (Providence, 1781), 10 (emphasis added).
38. There are now excellent literary and cultural histories of the Founding, such as Fliegelman, *Declaring Independence;* Ferguson, *American Enlightenment;* and Warner, *Letters of the Republic,* to name just three, but they barely touch on Loyalist writing, which continues to lack a good recent treatment.
39. My account challenges the claim that Burke's treatment of secular enthusiasm was wholly novel—a claim made by J. G. A. Pocock in "Enthusiasm: The Antiself of Enlightenment," *Huntington Library Quarterly* 60 (1998): 25–27.
40. Josiah Tucker, *A Series of Answers to Certain Popular Objections, Against Separating from the Rebellious Colonies, and Discarding Them Entirely* (Gloucester, 1776), ix; Oliver, *Origin and Progress of the American Rebellion,* 52–53; *The Delusive and Dangerous Principles of the Minority, Exposed and Refuted* (London, 1778), xxviii, iv, 52–53; Oliver, *Origin and Progress of the American Rebellion,* 65, 145–146; *Virginia Gazette* (Purdie), Dec. 12, 1777, quoted in Overfield, "Science in the Virginia Gazette," 44; *BFP,* 21:56.
41. "Political Electricity" is described in the *Catalogue of Prints and Drawings in the British Museum,* Division 1: *Political and Personal Satires,* vol. 4 (London, 1883), 649–660; see also James Delbourgo, "Political Electricity: The Occult Mechanism of Revolution," *Common-Place* 5 (2004), www.common-place.org.

5. How to Handle an Electric Eel

1. *South Carolina Gazette,* June 20, 1774.
2. Garden to Ellis, Aug. 14, 1774, published as "An Account of the Gymnotus Electricus, or Electrical Eel, in a Letter from Alexander Garden," *Phil. Trans.* 65 (1775): 102. The eel was named the *Gymnotus* by Linnaeus in the thirteenth edition of the *Systema Naturae* (Vienna, 1767), 1:427–428, citing a letter by Governor Laurens Storm van 's Gravesande from Dutch Guiana in 1754. Since the eighteenth century, the eel has been reclassified as *Electrophorus electricus.* Strictly speaking, the electric "eel" is a fish because it has fins.

3. Vere T. Daly, *A Short History of the Guyanese People* (London, 1975), 19–68; on Dutch alliances with the Caribs, see Neil L. Whitehead, *Lords of the Tiger Spirit: A History of the Caribs in Colonial Venezuela and Guyana, 1498–1820* (Dordrecht, 1988), chap. 7.
4. Edward Bancroft, An *Essay on the Natural History of Guiana, in South America* (London, 1769), 353–355; Daly, *Short History of the Guyanese People,* 72–74. John Gabriel Stedman's *Narrative of a Five Years' Expedition against the Revolted Negroes of Surinam* (1790), ed. Richard and Sally Price (Baltimore, 1988), provides a notably vivid late-eighteenth-century account of Guiana's plantation regime, though it is anticipated in some ways by Bancroft's lesser-known published accounts in 1769–1770.
5. Bancroft, *Natural History,* 355–357; Daly, *Short History of the Guyanese People,* 70–71.
6. Bancroft to Thomas Williams, Dec. 21, 1763, quoted in Godfrey Tryggve Anderson and Dennis Kent Anderson, "Edward Bancroft, M. D., F.R.S., Aberrant 'Practitioner of Physic,'" *Medical History* 17 (1973): 358; Edward Bancroft, *The History of Charles Wentworth,* 3 vols. (London, 1770), on which see James Delbourgo, "Natural History as a Good (Colonial) Novel: Science and Enlightenment in the Revolutionary Atlantic," forthcoming; Bancroft to Williams, Dec. 4, 1763, quoted in Julian Boyd, "Silas Deane: Death by a Kindly Teacher of Treason?" *WMQ* 16 (1959): 177; on Caribbean slavery and medicine, see Richard B. Sheridan, *Doctors and Slaves: A Medical and Demographic History of Slavery in the British West Indies, 1680–1834* (Cambridge, 1985), 42–48, 292–295. For evidence of Bancroft's relations with Paul Wentworth and the British government, see Bancroft to Lord North, July 22, 1783, British Library, Add. Mss. 61863, f. 118. Settling permanently in Britain, Bancroft also published on the science of colors, secured a royal patent for South American dyes, and speculated (like his diplomatic colleagues) on American western lands at the time of the Revolution. His son, Edward Nathaniel Bancroft, became a physician in Jamaica during the early nineteenth century.
7. Boyd, "Silas Deane," 165–187, 319–342, 515–550. Bancroft wrote at length on native poisons in the *Natural History* (270–300). See also Godfrey Tryggve Anderson and Dennis Kent Anderson, "The Death of Silas Deane: Another Opinion," *New England Quarterly* 57 (1984): 98–105. On early modern natural history, see Nick Jardine, James Secord, and

Emma Spary, eds., *Cultures of Natural History* (Cambridge, 1996); Lisbet Koerner, *Linnaeus: Nature and Nation* (Cambridge, Mass., 1999); Barbara J. Shapiro, *A Culture of Fact: England, 1550–1720* (Ithaca, 2000), 63–85; on American natural history, see Susan Scott Parrish, *American Curiosity: Cultures of Natural History in the Colonial British Atlantic World* (Chapel Hill, 2006).

8. Bancroft, *Natural History,* 170–171, 181, 250–251, 342. Janet Browne has offered a taxonomy of traveling naturalists for the nineteenth century, identifying three main types: freelance entrepreneurs, naval and military personnel, and paid collectors. Bancroft would fit best in the first category, although there is no evidence he planned to support himself primarily through the sale of specimens (part of Browne's definition of the category): see Browne, "Biogeography and Empire," in Jardine, Secord, and Spary, *Cultures of Natural History,* 306–307. Richard Drayton emphasizes that Banks inaugurated a new age of professionalism in natural history, creating careers for young men abroad that had not existed before (Banks himself exemplified the usefulness of exotic travel for metropolitan careers): see Drayton, *Nature's Government: Science, Imperial Britain, and the "Improvement" of the World* (New Haven, 2000), 127–128. On Banksian science and empire generally, see Drayton, *Nature's Government,* 85–128; John Gascoigne, *Science in the Service of Empire: Joseph Banks, the British State and the Uses of Science in the Age of Revolution* (Cambridge, 1998); and David Philip Miller and Peter Hanns Reill, eds., *Visions of Empire: Voyages, Botany, and Representations of Nature* (Cambridge, 1996). On Linnaean agents and travel writing, see Koerner, *Linnaeus,* and Mary Louise Pratt, *Imperial Eyes: Travel Writing and Transculturation* (New York, 1992).

9. Bancroft, *Natural History,* iii, 108–109, 34–35; on the use of eyewitness accounts in early modern travel writing, see Shapiro, *Culture of Fact,* 63–85.

10. Bancroft, *Natural History,* 57–58, 247, 189–190; on danger in American natural history, see Kay Dian Kriz, "Curiosities, Commodities, and Transplanted Bodies in Hans Sloane's 'Natural History of Jamaica,'" *WMQ* 57 (2000): 35–78, and Joyce E. Chaplin, "Mark Catesby, a Skeptical Newtonian in America," in Amy R. W. Meyers and Margaret Beck Pritchard, eds., *Empire's Nature: Mark Catesby's New World Vision* (Chapel Hill, 1998), 34–90.

11. Bancroft, *Natural History,* 203, 146, 223–224.

12. Ibid., 118, 232, 337–344, 401, 308, 337–338. Compare Bancroft with Cotton Mather's Puritan emphasis on the rational orderliness, intelligibility, and utility of nature in *The Christian Philosopher: A Collection of the Best Discoveries in Nature, with Religious Improvements* (1721; Gainesville, Fla., 1968), 122–141. On emergent perceptions of the fragility of colonial ecosystems in the eighteenth century, see Richard Grove, *Green Imperialism: Colonial Expansion, Tropical Island Edens, and the Origins of Environmentalism, 1600–1860* (Cambridge, 1995).
13. Brian P. Copenhaver, "A Tale of Two Fishes: Magical Objects in Natural History from Antiquity through the Scientific Revolution," *Journal of the History of Ideas* 52 (1991): 375–376; the Gesner quotation (1558) and Charleton quotation (1654) are also from Copenhaver's article, pages 385 and 391, respectively. The Lorenzini quotation (1705) can be found in W. Cameron Walker, "Animal Electricity before Galvani," *Annals of Science* 2 (1937): 87–88; see also Copenhaver, "A Tale of Two Fishes," 394–396.
14. Peter Kellaway, "The Part Played by Electric Fish in the Early History of Bioelectricity and Electrotherapy," *Bulletin of the History of Medicine* 20 (June 1946): 112–115, 124–125, 128–131; David C. Schechter, "Origins of Electrotherapy: Part One," *New York State Journal of Medicine* 71 (May 1971): 999.
15. Jean de Léry, *History of a Voyage to the Land of Brazil*, trans. Janet Whatley (Berkeley, 1990), 96; on Richer, see Nicholas Dew, "The Eel and the Pendulum: Natural History and Natural Philosophy in the French American Expeditions of the Late Seventeenth Century" (unpublished paper); Charles-Marie de la Condamine, *Relation Abregée d'un Voyage fait dans l'Intérieur de l'Amérique* (1745), quoted in Bancroft, *Natural History*, 193 (my translation). John Greenwood, nephew of the Harvard experimenter and electrician Isaac Greenwood, appears to have been the first North American to notice the eel's electrical properties. Traveling in Guiana in 1757, he recorded how an allegedly eleven-foot eel "shocked" a seaborne companion: John Greenwood, Memorandum Book, Dec. 1752–Apr. 1758, 168, New York Historical Society Manuscripts Collection. Greenwood is best known as an early American genre painter for works like *Sea Captains Carousing in Surinam* (1758).
16. See Michel Adanson, *A Voyage to Senegal, the Isle of Goree, and the River Gambia* (London, 1759), 244. Alexander von Humboldt, who experimented with American eels in 1799, believed Adanson "possess[ed] the honor of having been the first to recognize in 1751 the analogy of the ef-

fects of these electric fish with that of a Leyden jar, an honor which has been falsely attributed to 's Gravesande and Walsh." Alexander Von Humboldt and Aimé Bonpland, "Observations sur l'Anguille Electrique (*Gymnotus Electricus,* Lin.) du Nouveau Continent," *Recueil d'Observations de Zoologie et d'Anatomie Comparée* (Paris, 1811), 1:59. If Adanson was the first to believe that such fish generated electricity based on direct experience and observation, he was not the first to form such a hypothesis. That distinction appears to belong to the British natural philosopher Robert Turner. In *Electricology; or, A Discourse upon Electricity* (London, 1746), Turner discharged shocks from an "electrified flounder" connected to an artificial generator (he could not procure a torpedo) to argue for the natural electricity of the torpedo. See Philip C. Ritterbush, *Overtures to Biology: The Speculations of Eighteenth-Century Naturalists* (New Haven, 1964), 35–36.

17. Van 's Gravesande to Allamand, Nov. 22, 1754, quoted in Ritterbush, *Overtures to Biology,* 36; Walker, "Animal Electricity before Galvani," 89; Frans Van der Lott, "Kort Bericht van den Conger-Aal, afte Drilvisch," *Verh. Holl. Maatsch. Weten. Haarlem* 6 (1762), part 2:87–93, quoted in Kellaway, "Part Played by Electric Fish," 135–136. For van Musschenbroek's view of the eel, see Joseph Priestley, *The History and Present State of Electricity,* 2 vols., 3d ed. (London, 1775), 1:496–497; on Dutch electricity, see Lissa Roberts, "Science Becomes Electric: Dutch Interaction with the Electrical Machine during the Eighteenth Century," *Isis* 90 (1999): 680–714.
18. Bancroft, *Natural History,* 226–227, 191–192, 194–196. Van der Lott's electrical experiments with the eel in 1761 had also challenged Réaumur's mechanical account of the torpedo: see Ritterbush, *Overtures to Biology,* 37.
19. Bancroft, *Natural History,* 196–197.
20. Ibid., 197; John Hunter, "An Account of the Gymnotus Electricus," *Phil. Trans.* 65 (1775): 395–407; see also Steve Cross, "John Hunter, the Animal Economy and Late Eighteenth Century Physiological Discourse," *Studies in the History of Biology* 5 (1981): 1–110.
21. Bancroft, *Natural History,* 198, 200.
22. William Bryant, "Account of an Electrical Eel, or the *Torpedo* of Surinam," *Transactions of the American Philosophical Society* 2 (1786): 167 (the date of this paper is noted in the *Early Proceedings of the American Philosophical Society* [Philadelphia, 1884], 76); Van der Lott had carried

out similar insulation tests: see Ritterbush, *Overtures to Biology,* 37; Stedman, June 29, 1773, *Narrative of a Five Years Expedition,* 115.

23. My account of the very visible role of natives and Africans in Bancroft's text departs from Pratt, *Imperial Eyes,* which argues for the careful exclusion of such actors from enlightened natural histories. On the role of non-European expertise in European natural history practice, see Londa Schiebinger, *Plants and Empire: Colonial Bioprospecting in the Atlantic World* (Cambridge, Mass., 2004). On slavery in Guiana, see Gert Oostindie and Alex van Stipriaan, "Slavery and Slave Cultures in a Hydraulic Society: Suriname," in Stephan Palmié, ed., *Slave Cultures and the Cultures of Slavery* (Knoxville, Tenn., 1995), 78–99; for an oral history of Afro-American Maroon memory about "First-Time" (the eighteenth century), see Richard Price, *First-Time: The Historical Vision of an Afro-American People* (Baltimore, 1983).

24. Bancroft, *Natural History,* 309–315, 131, 214–215; *Charles Wentworth,* 2:254–262. In 1800, in Calabozo (in modern-day Venezuela), von Humboldt encountered "in the middle of the desert of the *Llanos* an individual [named Carlos el Pozo] who possessed a beautiful electrical machine which he had made himself" (my translation): see von Humboldt and Bonpland, "Observations sur l'Anguille Electrique," 64. While the two-headed serpent described by Guiana's natives was a chimerical product of the indigenous imagination, an image of the "monstrous" Champlain serpent verified by a British gentleman provided the sole illustration—the frontispiece—of Bancroft's *Natural History.*

25. Bancroft, *Natural History,* 136–137, 221, 258–259, 263–264, 375–377.

26. Ibid., 373–374, 269–270, 107. "This gymnotus, I suppose, is a different fish from the *Anguille tremblante, the trembling eel,* which is also a native of Surinam, and lives in marshy places, from whence it cannot be drawn, except when it is intoxicated": Priestley, *History and Present State of Electricity,* 1:498. Compare this method with the dramatic capture of eels described by von Humboldt, known as "*embarbascar con caballos*" ("poisoning with horses"), in which natives led a team of horses into an eel-infested body of water to induce the fish to expend all their electricity in a volley of defensive discharges, only to leave themselves exhausted and subject to easy capture. See von Humboldt and Bonpland, "Observations sur l'Anguille Electrique," 55. On native consumption of *Gymnoti,* see Humboldt and Bonpland, "Observations sur l'Anguille Electrique," 53, and Stedman, *Narrative of a Five Years Expedition,* 115.

27. Bancroft, *Natural History*, 367, 220, 183; Bryant, "Account of an Electrical Eel," 168; Philippe Fermin, *Déscription Générale, Historique, Géographique et Physique de la Colonie de Surinam*, 2 vols. (Amsterdam, 1769), 2:261 (my translation).
28. Henry Collins Flagg, letter from South Carolina, Oct. 8, 1782, published as "Observations on the Numb Fish, or Torporific Eel," *Transactions of the American Philosophical Society* 2 (1786): 170–173. Flagg cited Behn's account in his article. See Aphra Behn, *Oroonoko, and Other Writings* (Oxford, 1994), 51; on Behn in Guiana, see Mary Baine Campbell, *Wonder and Science: Imagining Worlds in Early Modern Europe* (Ithaca, 1999), 257–283; on African insensibility, see Joyce E. Chaplin, *Subject Matter: Technology, the Body, and Science on the Anglo-American Frontier, 1500–1676* (Cambridge, Mass., 2001), 160, and on Native American delicacy, throughout Chaplin's book; on sensibility's role in enlightened hierarchies, see Christopher Lawrence, "The Nervous System and Society in the Scottish Enlightenment," in Barry Barnes and Steven Shapin, eds., *Natural Order: Historical Studies in Scientific Culture* (Beverly Hills, 1979), 19–40.
29. Bancroft, *Natural History*, 199–200; van 's Gravesande to Allamand, Nov. 22, 1754, published in Allamand, "Kort verhaal van de Uitwerkzelen, Welke een Americanise vis Veroorsaakt op de Geenen die hem Aanraaken," *Verh. Holl. Maatsch. Weten. Haarlem* 2 (1756), quoted in Kellaway, "Part Played by Electric Fish," 135; Van der Lott, letter from Essequibo, June 7, 1761, published in "Kort Bericht van den Conger-Aal," and quoted in Kellaway, "Part Played by Electric Fish," 135 (emphasis added). Kellaway claims that the naturalist Fahlberg conducted medical experiments in Sweden with a live eel successfully brought back from Guiana (136).
30. *PG*, Dec. 29, 1773.
31. Bancroft, *Natural History*, 180–181, 229, 174, 300, 220; Louise E. Robbins, *Elephant Slaves and Pampered Parrots: Exotic Animals in Eighteenth-Century Paris* (Baltimore, 2002), 9–36.
32. Bancroft, *Natural History*, 380; Garden, "Account of the Gymnotus Electricus," 109; entry for July 30, 1773, *Early Proceedings of the American Philosophical Society*, 8.
33. Flagg, "Observations on the Numb Fish," 173, unnumbered note; Fermin, *Déscription Générale*, 262 (my translation); Hugh Williamson to John Walsh, Sept. 3, 1773, "Experiments and Observations on the

Gymnotus Electricus, or Electrical Eel," *Phil. Trans.* 65 (1775): 96–97; Bryant, "Account of an Electrical Eel," 168; Williamson, "Experiments and Observations on the *Gymnotus Electricus,*" 99. Van der Lott had in fact noted similar inconsistencies: see Ritterbush, *Overtures to Biology,* 37, and "Experiments on the *Gymnotus Electricus,* or the Electric Eel, made at Philadelphia," *Philadelphia Medical and Physical Journal* 1 (1805): 97, 159–160. The article dates these experiments to August 1770, but they are dated July 1773 in the *Early Proceedings of the American Philosophical Society,* 82. The American Philosophical Society group also included Isaac Bartram, Levi Hollingsworth, and Owen Biddle.

34. Williamson, "Experiments and Observations on the *Gymnotus Electricus,*" 95; Bancroft, *Natural History,* 200–201; Williamson, "Experiments and Observations on the *Gymnotus Electricus,*" 94–95; Bryant, "Account of an Electrical Eel," 168; Flagg, "Observations on the Numb Fish," 172.

35. Garden, "Account of the Gymnotus Electricus," 103, 108; Garden to Ellis, Mar. 12, 1775, and Daniel Solander to John Ellis, Nov. 7, 1774, in *A Selection of the Correspondence of Linnaeus, and Other Naturalists from the Original Manuscripts,* ed. Sir James Edward Smith, 2 vols. (London, 1821), 1:604–605 and 2:21. According to von Humboldt, only one other eel reached Europe alive in the century (it was described by the Swede Fahlberg). Both von Humboldt and Williamson refer to a live specimen brought over in winter 1774 by Williamson, but it is unclear whether this was a different eel from Baker's: see von Humboldt and Bonpland, "Observations sur l'Anguille Electrique," 51, unnumbered note.

36. Von Humboldt and Bonpland, "Observations sur l'Anguille Electrique," 52, 51 (my translation). Von Humboldt insisted that the bodily sensations produced by electrical machines and electrical fish were in fact different (84). "Lucretia Lovejoy, Sister of Mr. Adam Strong," *An Elegy on the Lamented Death of the Electrical Eel* (1778), quoted in Peter Wagner, *Eros Revived: Erotica of the Enlightenment in England and America* (London, 1988), 196–197; see also Adam Strong, *The Electric Eel; or, Gymnotus Electricus,* 3d ed. (London, 1777); Marquis de Sade, *Philosophy in the Boudoir* (1795), trans. "Meredith X," rev. ed. (London, 1995), 183.

37. Bancroft also excerpted his own account in the *Monthly Review* 40 (Feb. 1769): 206–207.

38. John Hunter, "Anatomical Observations on the Torpedo," *Phil. Trans.* 63 (1773–1774): 481–489; Benjamin Franklin, "Instructions for Testing the

Torpedo Fish," Aug. 12, 1772, *BFP*, 19:233–235; Walsh to Franklin, July 12 and Aug. 27, 1772, in Walsh, "Of the Electric Property of the Torpedo," *Phil. Trans.* 63 (1773): esp. 462 and 464; Walker, "Animal Electricity before Galvani," states "it is almost certain that [Bancroft's] more than any other source secured the interest of John Walsh" in the electricity of fish (90–92). Marco Piccolino, *The Taming of the Ray: Electric Fish Research in the Enlightenment from John Walsh to Alessandro Volta* (Florence, 2003), mentions that Walsh asked Franklin for Bancroft's London address in order to discuss the electric eel with him: 14, n. 22.

39. Henry Cavendish, "An Account of Some Attempts to Imitate the Effects of the Torpedo by Electricity," *Phil. Trans.* 66 (1776): 196–225; Alessandro Volta, "On the Electricity Excited by the Mere Contact of Conducting Substances of Different Kinds," *Phil. Trans.* 90 (1800): 403–431; on analogical models' mediation between nature and art, see Walker, "Animal Electricity before Galvani," 95–100, and Simon Schaffer, "Fish and Ships: Models in the Age of Reason," in Nick Hopwood and Soraya de Chadarevian, eds., *Models: The Third Dimension in Science* (Stanford, 2004), 71–105. In demonstrating a perpetual flow of "weak electricity" between these two metals, Volta transformed electrical science, reorienting understandings of electricity away from the electrostatic model and toward that of a constantly circulating current: see Giuliano Pancaldi, *Volta: Science and Culture in the Age of Enlightenment* (Princeton, 2003), esp. chap. 6. See also John L. Heilbron, *Electricity in the Seventeenth and Eighteenth Centuries: A Study of Early Modern Physics*, rev. ed. (New York, 1999), xxv, inspired by Giuliano Pancaldi, "Electricity and Life: Volta's Path to the Battery," *Historical Studies in the Physical and Biological Sciences* 21 (1990): 123–160.

6. Electrical Humanitarianism

1. The term "humanitarian" (1819) originally denoted those Protestants who believed in the humanity of Christ, but not his divinity. Shortly thereafter (1831), the term assumed its modern connotation of dedication to the welfare of human beings: see *The Oxford English Dictionary*, rev. ed. (New York, 1989).
2. T. Gale, *Electricity, or Ethereal Fire, Considered* (Troy, N.Y., 1802), 129–130.
3. Ibid., 5, 68–69; Peter Benes, "Itinerant Physicians, Healers, and Surgeon-

Dentists in New England and New York, 1720–1825," *Dublin Seminar for New England Folklife Annual Proceedings* (July 1990): 96 and appendix. The only article I have found on Gale is Charles C. Dennie, "T. Gale, The Man of Mystery," *New York State Journal of Medicine* 54 (Feb. 1954): 400–405. Gale also published *Brief Instructions for Administering Medical Electricity* (Hartford, Conn., 1805), a brief digest recapitulating his advice of 1802.

4. Gale, *Electricity, or Ethereal Fire, Considered*, 3.
5. Ibid., 95, 59. According to the works he cited, Gale's awareness of European electricity did not extend past the 1780s, with his greatest debts being to Franklin's midcentury writings, and works by the London-based Neapolitan natural philosopher Tiberius Cavallo, "the latest and best" writer on medical electricity (whose attempt to situate medical electricity in a philosophical framework Gale may well have been consciously imitating). Cavallo became a leading publicist for Galvanism in London, urging its potential as a philosophical foundation for medical electricity, and discussing animal electricity in the fourth edition of his *Complete Treatise* in 1795. But Gale evidently worked from Cavallo's earlier *Essay on the Theory and Practice of Medical Electricity* (1780–1781), before Galvanism had made animal electricity central to research in the field. See Cavallo, *A Complete Treatise on Electricity, in Theory and Practice*, 3 vols., 4th ed. (London, 1795), and the *Essay on the Theory and Practice of Medical Electricity*. On Cavallo's career, see Paola Bertucci, "Medical and Animal Electricity in the Work of Tiberius Cavallo, 1780–1795," in Marco Bresadola and Giuliano Pancaldi, eds., *Luigi Galvani International Workshop Proceedings* (Bologna, 1999), 147–166.
6. Franklin to James Logan, Dec. 16, 1749, *BFP*, 3:433; Jonathan Belcher to Franklin, Jan. 20, 1752, *BFP*, 4:255; "A Relation of a Cure Performed by Electricity, from Dr. Cadwallader Evans, Student in Physic at Philadelphia, Communicated October 21, 1754," *Medical Observations and Inquiries* 1 (1763): 83; Franklin to Sir John Pringle, Dec. 21, 1757, *EO*, 367–369; Franklin to La Sablière de la Condamine, Mar. 19, 1784, Library of the Bakken Museum of Electricity and Life, Minneapolis. See also Whitfield Bell Jr., "Benjamin Franklin and the Practice of Medicine," in his *The Colonial Physician and Other Essays* (New York, 1975), 118–130. On Nollet's tour, see Simon Schaffer, "Self Evidence," in James Chandler, Arnold I. Davidson, and Harry Harootunian, eds., *Questions of Evidence: Proof, Practice, and Persuasion across the Disciplines* (Chicago,

1991), 73–78, and Paola Bertucci, "The Electrical Body of Knowledge: Medical Electricity and Experimental Philosophy in the Mid-Eighteenth Century," in Paola Bertucci and Giuliano Pancaldi, eds., *Electric Bodies: Episodes in the History of Medical Electricity* (Bologna, 2001), 43–68. For imagination and experimental therapy, see Chapter 7.

7. One advertisement for a Kinnersley demonstration noted, "Electricity is now become a considerable Article in the Materia Medica, Directions will be given for the proper Application thereof." This demonstration took place at the College of Philadelphia in 1770. The advertisement continued, quoting Joseph Priestley: "And as 'Antonius de Haem, one of the most eminent Physicians of the present Age, after six Years uninterrupted Use of it, reckons it among the most valuable Assistances of Medical Art; and expressly says, that though it has often been applied in vain, it has often afforded Relief, where another Application would have been effectual'": *PG*, Dec. 20, 1770.

8. Dellap in *PG*, Oct. 20, 1784; *Isaac Greenwood, Dentist, No. 49 Marlborough-Street, Boston*, broadside dated 1788, *Evans Early American Imprints*, 1st ser., item no. 45488, Readex Digital Archives, www.readex.com; King quoted in Benes, "Itinerant Physicians," 100; Niderburg in *New York Evening Post*, Apr. 4, 1803 (for more such itinerants, see Benes, "Itinerant Physicians," appendix); Richard Willmott Hall, *An Inaugural Essay on the Use of Electricity in Medicine* (Philadelphia, 1806), 9. Like Gale, Hall did not mention Galvani, an even stranger omission for an academically trained physician. American physicians were clearly experimenting with electricity before the Revolution: witness the brief descriptions of doctors Greenhill of Virginia, Pillins of South Carolina, and Jerningham of Maryland later. But there is no evidence of their being professionally devoted to electrotherapy in particular.

9. Gale, *Electricity, or Ethereal Fire, Considered*, 13–14, 6, 84, 73, 77, 131, 78, 90, 140.

10. Greenhill quoted in Wyndham B. Blanton, *Medicine in Virginia in the Eighteenth Century* (Richmond, 1941), 6–7; Gale, *Electricity, or Ethereal Fire, Considered*, 107–109, 150–151, 162–163. For a general treatment of British-American medicine, see John Duffy, *From Humours to Medical Science: A History of American Medicine*, 2d ed. (Urbana, Ill., 1993).

11. Gale, *Electricity, or Ethereal Fire, Considered*, 144, 192–193, 97–99. For an overview of European medical electricity in this era, see Margaret

Rowbottom and Charles Susskind, *Electricity and Medicine: A History of Their Interaction* (San Francisco, 1984), 15–54. Gale wrote of Cavallo, "I have not that author's writing by me at this time, or I would copy their manner of treatment," so strongly did he endorse his techniques: see Gale, *Electricity, or Ethereal Fire, Considered,* 89, 190, 203.

12. Gale, *Electricity, or Ethereal Fire, Considered,* 99, 102–103. On sympathetic healing in the Revolutionary era, see Sarah Knott, "A Cultural History of Sensibility in the Era of the American Revolution," D.Phil. diss., Oxford University, 1999, 83–84.

13. Gale, *Electricity, or Ethereal Fire, Considered,* 140–149, 122–123, 125–128, 181; Edward Cutbush, *An Inaugural Dissertation on Insanity* (Philadelphia, 1794), 33 (compare Cutbush with John Vaughan's account of animal electricity in Chapter 7); Thomas Law, *Ballston Springs* (New York, 1806), 24; Bertucci, "Electrical Body of Knowledge," 49. Law's poem continues: "The shocks which came from Emma's eye, / Afflicted instantly my breast; / My wounds all remedies defy, / So deep, so firmly they're imprest. / Yet would I pleas'd resign my breath, / A pang from Emma to restrain, / And my last pray'r should be in death, / May Emma never know a pain."

14. There was, however, a review in the *Medical Repository,* where the editors of the early republic's leading medical journal denounced Gale as a "Hutchinsonian" (see later) and "one of those zealots in physics who ascribe almost every phenomenon in nature to the operation of electricity": *Medical Repository* 10 (1807): 277. On the contours of early republican science, see John C. Greene, *Science in the Age of Jefferson* (Ames, Iowa, 1984), and Joyce E. Chaplin, "Nature and Nation: Natural History in Context," in Sue Anne Prince, ed., *Stuffing Birds, Pressing Plants, Shaping Knowledge: Natural History in North America, 1730–1860* (Philadelphia, 2003), 75–95.

15. Gale, *Electricity, or Ethereal Fire, Considered,* 13–15, 25, 30.

16. Ibid., 65–66.

17. Ibid., 32, 34, 23, 9–10, 65.

18. Ibid., 34, 65–69.

19. On Behmenism (the Christian movement based on the work of Jakob Böhme) in Georgian Britain, and electricity as a spiritualized active power, see Simon Schaffer, "The Consuming Flame: Electrical Showmen and Tory Mystics in the World of Goods," in John Brewer and Roy Porter, eds., *Consumption and the World of Goods* (New York, 1993), 489–

526; on Hutchinson, see C. B. Wilde, "Hutchinsonianism, Natural Philosophy and Religious Controversy in Eighteenth-Century Britain," *History of Science* 18 (1980): 1–24; George Berkeley, *Siris: A Chain of Philosophical Reflexions and Inquiries Concerning the Virtues of Tar Water* (London, 1744); on the relation between ether, the sun, and primitive Christianity, see Betty Jo Teeter Dobbs and Margaret C. Jacob, *Newton and the Culture of Newtonianism* (Atlantic Highlands, N.J., 1995), 36, 47–48; on hermeticism in Europe, see Margaret C. Jacob, *The Radical Enlightenment: Pantheists, Freemasons and Republicans* (London, 1981), and in America, John L. Brooke, *The Refiner's Fire: The Making of Mormon Cosmology, 1644–1844* (Cambridge, 1994); see also Paul E. Johnson and Sean Wilentz, *The Kingdom of Matthias: A Story of Sex and Salvation in Nineteenth-Century America* (New York, 1994).

20. On religion and provincial science in Europe, see the essays by Lissa Roberts and Lisbet Koerner in William Clark, Jan Golinski, and Simon Schaffer, eds., *The Sciences in Enlightened Europe* (Chicago, 1999), 350–422; on Protestantism and the American Enlightenment, see Henry F. May, *The Enlightenment in America* (New York, 1976); on religious views of electricity in nineteenth-century America, see the accounts of the professional chemist Robert Hare and the Mormon cosmologist Orson Pratt in Craig J. Hazen, *The Village Enlightenment in America: Popular Religion and Science in the Nineteenth Century* (Urbana, Ill., 2000), chaps. 1–2, and Brooke, *Refiner's Fire*, 273, 285, 287; for the German context, see Ernst Benz, *The Theology of Electricity: On the Encounter and Explanation of Theology and Science in the Seventeenth and Eighteenth Centuries*, trans. Wolfgang Taraba (Allison Park, Pa., 1989); and as a general introduction, Dennis Stillings, "Electricity," in Gary B. Ferngren et al., eds., *The History of Science and Religion in the Western Tradition: An Encyclopedia* (New York, 2000), 384–386.

21. Samuel Johnson, "Memoirs of the Life of the Rev. Dr. Johnson, and Several Things Relating to the State Both of Religion and Learning in His Times" (1768–1770), in *Samuel Johnson, President of King's College, His Career and Writings*, 4 vols., ed. Herbert and Carol Schneider (New York, 1929), 1:45. On materialism, see Cadwallader Colden to Johnson, May 18, 1747, in *Samuel Johnson*, 2:297, and Herbert Leventhal, *In the Shadow of the Enlightenment: Occultism and Renaissance Science in Eighteenth-Century America* (New York, 1976), 179–188; Roy N. Lokken, "Cadwallader Colden's Attempt to Advance Natural Philosophy beyond

the Eighteenth-Century Mechanistic Paradigm," *Proceedings of the American Philosophical Society* 122 (Dec. 1978): 365–376; Joseph J. Ellis III, *The New England Mind in Transition; Samuel Johnson of Connecticut, 1696–1772* (New Haven, 1973). On competing interpretations of Newton (among which materialism figured prominently), see Margaret C. Jacob, *The Newtonians and the English Revolution, 1689–1720* (Ithaca, 1976).

22. Gale, *Electricity, or Ethereal Fire, Considered*, 216.
23. Michel Foucault, *Discipline and Punish: The Birth of the Prison*, trans. Alan Sheridan (New York, 1979); on humanitarian sensibility and antislavery, see Thomas Bender, ed., *The Antislavery Debate: Capitalism and Abolitionism as a Problem in Historical Interpretation* (Berkeley, 1992). A useful if celebratory overview of humanitarianism across the British Atlantic is Michael Kraus, *The Atlantic Civilisation: Eighteenth-Century Origins* (Ithaca, 1949), 123–158; on abolitionism and pornography, see Karen Halttunen, "Humanitarianism and the Pornography of Pain in Anglo-American Culture," *American Historical Review* 100 (1995): 303–334; on humanitarian narratives, see Thomas F. Laqueur, "Bodies, Details, and the Humanitarian Narrative," in Lynn Hunt, ed., *The New Cultural History* (Berkeley, 1989), 176–204.
24. Joseph Hiller, "Two Lectures on Electricity" (1757), Octavo volumes "H," American Antiquarian Society, 20; untitled Humane Society of Massachusetts broadside, July 21, 1789, *Evans Early American Imprints*, 1st ser., item no. 45499, Readex Digital Archives, www.readex.com; Thomas Thacher, *A Discourse Delivered at Boston, before the Humane Society of the Commonwealth of Massachusetts* (Boston, 1800), 14; Thomas Barnard, *A Discourse Delivered before the Humane Society of the Commonwealth of Massachusetts* (Boston, 1794), 15, 9–10, 18; John Lathrop, *A Discourse before the Humane Society in Boston* (Boston, 1787), 34. On the new rhetorical emphasis on life, and the transition from natural history to biology at the turn of the nineteenth century, see Michel Foucault, *The Order of Things* (New York, 1970), chap. 5.
25. Thacher, *Discourse*, 20, 15–16; Humane Society of Massachusetts broadside, July 21, 1789; Thacher, *Discourse*, 9; Barnard, *Discourse*, 12–14. On ideological constructions of commerce in this period, see Joyce O. Appleby, *Capitalism and a New Social Order: The Republican Vision of the 1790s* (New York, 1984); on party politics, see Richard Hofstadter, *The Idea of a Party System: The Rise of a Legitimate Opposition in the United States, 1780–1840* (Berkeley, 1969).

26. Barnard, *Discourse*, 20; Richard Willmott Hall cited Darwin's *Zoonomia; or, The Laws of Organic Life* (New York, 1796) as the first text to include electricity in its materia medica: Hall, *Inaugural Essay on the Use of Electricity*, 7; John Fleet, *A Discourse Relative to the Subject of Animation* (Boston, 1797), 13, 7, 15; for Kite, see Rowbottom and Susskind, *Electricity and Medicine*, 23. John Bartlett also recommended electrical reanimation in *A Discourse on the Subject of Animation* (Boston, 1792), 13.

27. David Hosack, *An Inquiry into the Causes of Suspended Animation from Drowning; with the Means of Restoring Life* (New York, 1792), 24; Jedediah Morse, *A Sermon Preached before the Humane Society of the Commonwealth of Massachusetts* (Boston, 1801), 50; Eliphalet Porter, *A Discourse Delivered before the Humane Society of the Commonwealth of Massachusetts* (Boston, 1802), 44; Bartlett, *Discourse on the Subject of Animation*, epigraph; *The Institution of the Humane Society of the Commonwealth of Massachusetts* (Boston, 1788), 5; Lathrop, *Discourse before the Humane Society*, 20, 34; *Philadelphia Medical Museum* 2 (1806): 113. See also the broadsides *Directions for Recovering Persons, Who Are Supposed To Be Dead . . . Published by Order of the Humane Society of Philadelphia* (Philadelphia, 1789), and *The Institution of the Merrimack Humane Society, with the Rules for Regulating Said Society, and the Methods of Treatment to Be Used with Persons Apparently Dead* (Newburyport, Conn., 1803), 9.

28. Although Gale did not practice such techniques himself, he encouraged the use of shocks to "move the blood by springing the vessels," noting that "persons that have appeared to die suddenly, without disease, that have swooned or fainted, apparently into death, or have been hanged, or in a fit of apoplexy, may many times be restored to life": Gale, *Electricity, or Ethereal Fire, Considered*, 234–237, 268–273, 9, 69.

29. On Benjamin Rush, see Michael Meranze, *Laboratories of Virtue: Punishment, Revolution, and Authority in Philadelphia, 1760–1835* (Chapel Hill, 1996); and Jacquelyn C. Miller, "The Body Politic and the Body Somatic: Benjamin Rush's Fear of Social Disorder and His Treatment of Yellow Fever," in Janet Moore Lindmann and Michelle Lise Tarter, eds., *A Centre of Wonders: The Body in Early America* (Ithaca, 2001), 61–74. On early American medicine, see Richard H. Shryock, *Medicine and Society in Early America, 1660–1860* (New York, 1960); and Helen Brock, "North America: A Western Outpost of European Medical Culture," in Andrew Cunningham and Roger French, eds., *The Medical Enlighten-*

ment of the Eighteenth Century (Cambridge, 1990), 194–216. For the British scene, see Roy Porter, *Health for Sale: Quackery in England, 1660–1850* (New York, 1989), and *Doctor of Society: Thomas Beddoes and the Sick Trade in Late-Enlightenment England* (New York, 1992), esp. 22–36.

30. Gale, *Electricity, or Ethereal Fire, Considered*, 11, 227–228, 105, 6. Wesley's *Primitive Physic* was published in numerous editions in Britain and the colonies. See Samuel J. Rogal, "Electricity: John Wesley's 'Curious and Important Subject,'" *Eighteenth-Century Life* 13 (1989): 79–90.

31. Gale, *Electricity, or Ethereal Fire, Considered*, 243–246, 249–257, 69.

32. Moses Willard, "Treatise on Medical Electricity," and replies, Benjamin Rush Papers, box 8, Historical Society of Pennsylvania, Philadelphia; James Cunningham to Rush, Mar. 7, 1789, and reply, Mar. 16, 1789, SP 23 Rush MS 33, Historical Society of Pennsylvania. For auction notices pertaining to the estates of Dr. William Pillins in South Carolina and Dr. Henry Jerningham of Maryland, see *South Carolina Gazette*, Nov. 10, 1767, and *Maryland Gazette* (Annapolis), Sept. 9, 1773, respectively. See also Thomas Dancer, *The Medical Assistant; or, Jamaica Practice of Physic: Designed Chiefly for the Use of Families and Plantations* (Kingston, 1801), 186; and *The New and Complete American Encyclopaedia, or, Universal Dictionary of Arts and Sciences*, 7 vols. (New York, 1807), 3:292–295.

33. Gale, *Electricity, or Ethereal Fire, Considered*, 143, 60. This was only slightly more than Kinnersley had asked at midcentury.

34. Donald Fleming, *Science and Technology in Providence, 1760–1914: An Essay on the History of Brown University in the Metropolitan Community* (Providence, 1952), 53–54; Isaac Greenwood III, *Sublime Entertainment . . . a Brilliant Electrical Exhibition* (Providence, 1793), broadside; entries for May 26, 1796, to Feb. 7, 1798, diary of Alexander Anderson Jr., 1793–1799, Ms. 119, Columbiana Collection, Columbia University Rare Books and Special Collections. W. and S. Jones, the London instrument makers whose products were imported by Americans, sold machines (with conductors and jars) starting at the price of 2 pounds, 12 shillings, 6 pence. They also listed machines specifically for medical use, with cylinders of 7–10 inches in diameter, for between 6 pounds, 6 shillings and almost 13 pounds. Their most expensive machine, with a 17.5-inch cylinder, cost 100 guineas: see W. and S. Jones, *A Catalogue of Optical, Mathematical, and Philosophical Instruments* (London, 1795), 7, 16. The key works on early modern replication and public science are Steven Shapin and Si-

mon Schaffer, *Leviathan and the Air-Pump: Hobbes, Boyle, and the Experimental Life* (Princeton, 1985); Jan Golinski, *Science as Public Culture: Chemistry and Enlightenment in Britain, 1760–1820* (Cambridge, 1992); and Larry Stewart, *The Rise of Public Science: Rhetoric, Technology and Natural Philosophy in Newtonian Britain, 1660–1750* (Cambridge, 1992).
35. Gale, *Electricity, or Ethereal Fire, Considered*, 144 (emphasis added), 93–94; Simon Schaffer, "Machine Philosophy: Demonstration Devices in Georgian Mechanics," *Osiris* 9 (1994): 159.
36. Gale, *Electricity, or Ethereal Fire, Considered*, 278, 81, 284–285; Ruth Bloch, *Visionary Republic: Millennial Themes in American Thought, 1756–1800* (Cambridge, 1985); Drew R. McCoy, *The Elusive Republic: Political Economy in Jeffersonian America* (Chapel Hill, 1980).
37. Gale, *Electricity, or Ethereal Fire, Considered*, 284–285.
38. Ibid., 255, 279. Richard Hall spoke of the way in which the shock could be "regulated with great accuracy by the electrometer": see Hall, *Inaugural Essay on the Use of Electricity*, 11.
39. Henry Miles, "A Letter from the Reverend Henry Miles . . . Containing Observations of Luminous Emanations from Human Bodies, and from Brutes; with Some Remarks on Electricity," *Phil. Trans.* 43 (1745): 441–447; Charles Vancouver, *A General Compendium; or, Abstract of Chemical, Experimental and Natural Philosophy*, 4 vols. (Philadelphia, 1785), 1:25–26. Zangari is also discussed in David Brewster's classic of enlightened debunking, *Letters on Natural Magic* (London, 1832), chap. 12.
40. David Ross, "History of a Case of Palsy, Cured by Lightning," *Philadelphia Medical Museum* 1 (1805): 420–422; Vancouver, *General Compendium*, 1:27; Anonymous, "An Account of Some Uncommon Burnings, Lucid Appearances, and Fiery Emanations," *New England Magazine of Knowledge and Pleasure* (May 1758): 45–48.
41. Anonymous, "Letter Respecting an Italian Priest, Killed by an Electric Commotion, the Cause of Which Resided in His Own Body," *American Museum, or Universal Magazine* (Apr. 1792): 146–149; Noah Webster, *A Brief History of Epidemic and Pestilential Diseases* (Hartford, Conn., 1799), 316; see also John Vaughan, *The Valedictory Lecture Delivered before the Philosophical Society of Delaware* (Wilmington, Del., 1800), 19–20. Ebenezer Hazard, secretary of the North American Insurance Company, blamed the yellow fever on the proliferation of conductors in the city: "Perhaps the large number of lightning rods fixed to houses in the city (one of Dr. Franklin's lesser legacies), 'by imperceptibly drawing off

the electric fluid from the clouds, and thereby preventing the thunder,' contributed to the increase of the disease." Quoted in John Harvey Powell, *Bring Out Your Dead: The Great Plague of Yellow Fever in Philadelphia in 1793* (Philadelphia, 1949), 49. On the connection between atmospheric and bodily electricity in *jettatura* (folk belief in the "evil eye") in eighteenth-century Italy, see Ernesto de Martino, *Sud e Magia* (Milan, 1959), 152.

42. Harry R. Warfel, *Charles Brockden Brown: American Gothic Novelist* (Gainesville, Fla., 1949), 204, 206–208, 216–217; Peter Kafer, *Charles Brockden Brown's Revolution and the Birth of American Gothic* (Philadelphia, 2004); for recent reapproaches to Brown, see Bryan Waterman, "The Bavarian Illuminati, the Early American Novel, and Histories of the Public Sphere," *WMQ* 62 (Jan. 2005): 9–30, and more generally, Philip Barnard, Mark L. Kamrath, and Stephen Shapiro, eds., *Revising Charles Brockden Brown: Culture, Politics, and Sexuality in the Early Republic* (Knoxville, Tenn., 2004).

43. Charles Brockden Brown, *Wieland; or, The Transformation: An American Tale* (Kent, Ohio, 1977), 11–12, 17–19; Warfel, *Charles Brockden Brown*, 34, 38, 44. Kafer, *Brockden Brown's Revolution*, 113–124, is especially good on the biographical context for these scenes in *Wieland;* on the Kelpius connection, see Jon Butler, *Awash in a Sea of Faith: Christianizing the American People* (Cambridge, Mass., 1990), 76–77, 225–226. The spontaneous combustion footnote reads: "A case, in its symptoms exactly parallel to this, is published in one of the Journals of Florence. See, likewise, similar cases reported by Messrs. Merille and Muraire, in the 'Journal de Medicine [sic],' for February and May, 1783. The researches of Maffei and Fontana have thrown some light upon this subject": Brown, *Wieland*, 19.

44. Alexander Von Humboldt and Aimé Bonpland, "Observations sur l'Anguille Electrique (*Gymnotus Electricus*, Lin.) du Nouveau Continent," *Recueil d'Observations de Zoologie et d'Anatomie Comparée* (Paris, 1811), 1:68.

7. Electricity as Common Sense

1. Elisha Perkins, *Evidences of the Efficacy of Doctor Perkins's Patent Metallic Instruments* (New London, Conn., 1797), 10.
2. N.d., quoted in Frederick J. Curley, "Elisha Perkins's Patent Metallic

Tractors," *Synthesis* 2 (1975): 14; Elisha Perkins to unknown, June 22, 1796, in Elisha Perkins Letter and Account Book (1788–1816) MS 18th+, Yale University Medical School Historical Library; John Corry, *Quack Doctors Dissected; or, A New, Cheap, and Improved Edition of Corry's Detector of Quackery* (London, 1810), 25; Walter R. Steiner, "Dr. Elisha Perkins of Plainfield, Conn., and His Metallic Tractors," *Bulletin of the Society of Medical History of Chicago* 3 (Jan. 1923): 92, 95; Marcus A. McCorison, "Thomas Green Fessenden, 1771–1837: Not in *BAL*," *Papers of the Bibliographical Society of America* 89 (Mar. 1995): 9.

3. Steven Shapin and Simon Schaffer, *Leviathan and the Air-Pump: Hobbes, Boyle, and the Experimental Life* (Princeton, 1985), 60–65.

4. Steiner, "Dr. Elisha Perkins," 79–95; Howard W. Haggard, "The First Published Attack on Perkinism: An Anonymous Eighteenth Century Poetical Satire," *Yale Journal of Biology and Medicine* 9 (1936): 137–153; Curley, "Elisha Perkins' Patent Metallic Tractors," 8–21; Eric T. Carlson and Meribeth M. Simpson, "Perkinism vs. Mesmerism," *Journal of the History of Behavioral Sciences* 6 (Jan. 1970): 16–24; Jacques M. Quen, "Elisha Perkins, Physician, Nostrum-Vendor, or Charlatan?" *Bulletin of the History of Medicine* 37 (1963): 159–166; and "Case Studies in Nineteenth Century Scientific Rejection: Mesmerism, Perkinism, and Acupuncture," *Journal of the History of Behavioral Sciences* 11 (Apr. 1975): 149–156; L. S. Jacyna, "Galvanic Influences: Themes in the Early History of British Animal Electricity," in Marco Bresadola and Giuliano Pancaldi, eds., *Luigi Galvani International Workshop Proceedings* (Bologna, 1999), 177–185; and Marco Bresadola, "Early Galvanism as Technique and Medical Practice," in Paola Bertucci and Giuliano Pancaldi, eds., *Electric Bodies: Episodes in the History of Medical Electricity* (Bologna, 2001), 170–172.

5. Peter Dear, "*Totius in Verba*: Rhetoric and Authority in the Royal Society," *Isis* 76 (1985): 145–161; Shapin and Schaffer, *Leviathan and the Air-Pump*; Bruno Latour, *Science in Action: How to Follow Scientists and Engineers through Society* (Cambridge, Mass., 1987); Lorraine Daston, *Classical Probability in the Enlightenment* (Princeton, 1988); Mary Poovey, *A History of the Modern Fact: Problems of Knowledge in the Sciences of Wealth and Society* (Chicago, 1998); Barbara J. Shapiro, *A Culture of Fact: England, 1550–1720* (Ithaca, 2000). Jessica Riskin has argued that Maximilien Robespierre's successful legal defense of lightning rods in France in 1783 turned not on technical questions, but on Robespierre's

dismissal of expert scientific testimony in favor of commonsense evaluations based on untheorized, hence allegedly uncontroversial, facts. See Riskin, *Science in the Age of Sensibility: The Sentimental Empiricists of the French Enlightenment* (Chicago, 2002), 139–187. On self-evidence, see Simon Schaffer, "Self Evidence," in James Chandler, Arnold I. Davidson, and Harry Harootunian, eds., *Questions of Evidence: Proof, Practice, and Persuasion across the Disciplines* (Chicago, 1991), 56–91.

6. Voltaire, *Lettres Philosophiques*, 2 vols. (Paris, 1964), 2:175–177; Meyer Reinhold, "The Quest for 'Useful Knowledge' in Eighteenth-Century America," *Proceedings of the American Philosophical Society* 119 (Apr. 1975): 111–112, 131; Lorraine Daston, "Afterword: The Ethos of Enlightenment," in William Clark, Jan Golinski, and Simon Schaffer, eds., *The Sciences in Enlightened Europe* (Chicago, 1999), 495–504; Benjamin Franklin, "Proposal for Promoting Useful Knowledge among the British Plantations in America" (1743), *BFP*, 2:380–383.

7. David Hume, "Of the Rise and Progress of the Arts and Sciences" (1742), in *Political Essays*, ed. Knud Haakonssen (Cambridge, 1994), 67; Alexis de Tocqueville, *Democracy in America*, trans. George Lawrence, 2 vols. (New York, 1966), 2:459–465. See also Drew McCoy, *The Elusive Republic: Political Economy in Jeffersonian America* (New York, 1980), 76–119; Edwin T. Layton Jr., "Newton Confronts the American Millwrights; or, Action and Reaction Are Not Always Equal," in Elizabeth Garber, ed., *Beyond History of Science: Essays in Honor of Robert E. Schofield*, (London, 1990), 179–193; John L. Larson, *Internal Improvement: National Public Works and the Promise of Popular Government in the Early United States* (Chapel Hill, 2001); Joyce E. Chaplin, "Nature and Nation: Natural History in Context," in Sue Anne Prince, ed., *Stuffing Birds, Pressing Plants, Shaping Knowledge: Natural History in North America, 1730–1860* (Philadelphia, 2003), 75–95.

8. Jefferson to John Adams, Aug. 15, 1820, *The Adams-Jefferson Letters*, vol. 2: *1812–1826*, ed. Lester J. Cappon (Chapel Hill, 1959), 567–569; Dennie quoted in Gilman M. Ostrander, *Republic of Letters: The American Intellectual Community, 1776–1865* (Madison, Wis., 1999), 108–109; Otto and Latrobe quoted in Gordon S. Wood, *The Radicalism of the American Revolution* (New York, 1992), 241, 295; for Mitchill, see Ostrander, *Republic of Letters*, 82–83. On Burke and the "sentimental empiricist" critique of instrumental reason in revolutionary French science and politics, see Jan Golinski, *Science as Public Culture: Chemistry and En-*

lightenment in Britain, 1760–1820 (Cambridge, 1992), 176–179, and especially Riskin, *Science in the Age of Sensibility,* 273–277. See also Steven Shapin, *A Social History of Truth: Science and Civility in Seventeenth-Century England* (Chicago, 1994); on "anti-science," see Roger Cooter and Stephen Pumfrey, "Separate Spheres and Public Spaces: Reflections on the History of Science Popularization and Science in Popular Culture," *History of Science* 32 (1994): 246.

9. Anthony Ashley Cooper, Third Earl of Shaftesbury, *Characteristics of Men, Manners, Opinions, Times* (1711), ed. Lawrence E. Klein (Cambridge, 1999), 29–69; Alexander Hamilton, *The Tuesday Club: A Shorter Edition of the History of the Ancient and Honorable Tuesday Club,* ed. Robert Micklus (Baltimore, 1995), 12–14. See also Klein, *Shaftesbury and the Culture of Politeness: Moral Discourse and Cultural Politics in Early Eighteenth-Century England* (Cambridge, 1994); on pleasure and sociability in British America, see David S. Shields, *Civil Tongues and Polite Letters in British America* (Chapel Hill, 1997); on the influence of Shaftesbury and Hutcheson on Franklin, see Douglas Anderson, *The Radical Enlightenments of Benjamin Franklin* (Baltimore, 1997), esp. introduction and chap. 1; on moral sense philosophy in America, see Ned C. Landsman, *From Colonials to Provincials: American Thought and Culture, 1680–1760* (New York, 1997), 73–77, 137–138.

10. Shaftesbury, *Characteristics,* 37. In formal moral philosophy, this tradition is sometimes referred to as "common-sense ethics," and is related to the commonsense epistemology articulated by Thomas Reid and others.

11. Thomas Paine, *Rights of Man, Common Sense and Other Political Writings* (Oxford, 1995), 19; Landsman, *From Colonials to Provincials,* 78–80; de Tocqueville, *Democracy in America,* 2:459. On the American public sphere, see Michael Warner, *The Letters of the Republic: Publication and the Public Sphere in Eighteenth-Century America* (Cambridge, Mass., 1990).

12. These social, political, and epistemological discourses of common sense were by no means unrelated. Reid, for example, used the reliability of common sensory evidence to justify the public arbitration of knowledge claims: see Benjamin W. Redekop, "Thomas Reid and the Problem of Induction: From Common Experience to Common Sense," *Studies in the History and Philosophy of Science* 33 (2002): 35–57.

13. Riskin, *Science in the Age of Sensibility,* 25–26, 207; Thomas Reid, *Essays*

on the *Intellectual Powers of Man* (Edinburgh, 1785), 201 and as quoted in Redekop, "Thomas Reid," 43; Theodore Dwight Bozeman, *Protestants in an Age of Science: The Baconian Ideal and Antebellum American Religious Thought* (Chapel Hill, 1977), 4–21; Elisabeth Flower and Murray G. Murphey, *A History of Philosophy in America* (New York, 1977), 242–269; Jean V. Matthews, *Toward a New Society: American Thought and Culture, 1800–1830* (New York, 1991), 37; Douglas Sloan, *The Scottish Enlightenment and the American College Ideal* (New York, 1971).

14. Elisha Perkins to Benjamin Perkins, Nov. 22, 1795; Elisha Perkins to unknown, June 22, 1796; Elisha Perkins to Dr. Flint, Nov. 2, 1795, all in Elisha Perkins Letter and Account Book; Bryan A. Laver, "Miracles No Wonder! The Mesmeric Phenomena and Organic Cures of Valentine Greatrakes," *Journal of the History of Medicine and Allied Sciences* 33 (1978): 35–46; *Directions for Performing the Metallic Operation with Perkins's Patent Tractors,* British broadside signed by Benjamin Perkins, dated Aug. 22, 1799, available on microfilm series *The Eighteenth Century* (Woodbridge, Conn.), reel 4923, item 2.

15. On the yellow fever, see J. Worth Estes and Billy Smith, eds., *A Melancholy Scene of Devastation: The Public Response to the 1793 Philadelphia Yellow Fever Epidemic* (Canton, Mass., 1997); on Mesmerism, see Robert Darnton, *Mesmerism and the End of the Enlightenment in France* (Cambridge, Mass., 1968); Schaffer, "Self Evidence," 78–87; Adam Crabtree, *From Mesmer to Freud: Magnetic Sleep and the Roots of Psychological Healing* (New Haven, 1993); Patricia Fara, "An Attractive Therapy: Animal Magnetism in Eighteenth-Century England," *History of Science* 33 (June 1995): 127–177; and Riskin, *Science in the Age of Sensibility,* 189–225. Despite the endorsement of the Marquis de Lafayette, Mesmerism failed to take hold in the United States at the height of its French popularity in the 1780s: it lacked a sufficient number of North American advocates, and was subject to a hostile campaign by Jefferson, the American minister in Paris at the time, who sent anti-Mesmerist literature back across the Atlantic. See Louis Gottschalk, *Lafayette between the American and the French Revolution (1783–1789)* (Chicago, 1950), 77–101; Thomas Jefferson, *The Papers of Thomas Jefferson,* ed. Julian P. Boyd (Princeton, 1950–), 7:514, 517–518, 642. Mesmerism appears to have penetrated more deeply into French colonial Saint Domingue. On "creole Mesmerism," see James E. McClellan, *Colonialism and Science: Saint Domingue in the Old Regime* (Baltimore, 1992), 178. On Mesmerism's transformation and

revival after the 1830s, see Alison Winter, *Mesmerized: Powers of Mind in Victorian Britain* (Chicago, 1998).

16. Benjamin Perkins, *The Influence of Metallic Tractors on the Human Body* (London, 1799), 20; Benjamin Perkins, *The Efficacy of Perkins's Patent Metallic Tractors, in Topical Diseases, on the Human Body and Animals* (London, 1800), lv–lvi; Benjamin Perkins, *Influence of Metallic Tractors*, 35; Tilton quoted in John Vaughan, *Observations on Animal Electricity, in Explanation of the Metallic Operation of Doctor Perkins* (Wilmington, Del., 1797), vi–vii; Benjamin Perkins, *Influence of Metallic Tractors*, xiv; as well as his *Experiments with Metallic Tractors* (London, 1799), 318, and *Efficacy of Perkins's Patent Metallic Tractors*, 45; Benjamin Perkins, *Influence of Metallic Tractors*, 83–84. The austerity of Perkinism can also be contrasted with the theatricality of Dr. James Graham's electrical therapies in 1770s London: see Roy Porter, "Sex and the Singular Man: The Seminal Ideas of James Graham," *Studies on Voltaire and the Eighteenth Century* 228 (1984): 3–24. Graham claimed to have learned his electricity in Philadelphia and to have attended Kinnersley's lectures at the College of Philadelphia: see James Graham, *The General State of Medical and Chirurgical Practice, Exhibited* (London, 1779), 37.

17. Elisha Perkins, *Evidences of the Efficacy of Doctor Perkins's Patent Metallic Instruments*, 3–4; Benjamin Perkins, *Influence of Metallic Tractors on the Human Body*, 78 (emphasis added).

18. Benjamin Perkins, *Influence of Metallic Tractors on the Human Body*, 37–39. At times, Elisha Perkins spoke of tractoration as a form of faith healing: "Many people are and more will be brought to the faith daily." See Elisha Perkins to Benjamin Perkins, Oct. 19, 1798, Elisha Perkins Letter and Account Book.

19. Charles Cunningham Langworthy, *A View of the Perkinean Electricity; or, An Enquiry into the Influence of Metallic Tractors*, 2d ed. (Bath, 1798), 28; Benjamin Perkins, *Efficacy of Perkins's Patent Metallic Tractors*, xliv; Abildgaard quoted in Benjamin Perkins, *Experiments with the Metallic Tractors*, 38–40; Thomas Green Fessenden, *Terrible Tractoration!!: A Poetical Petition against Galvanising Trumpery, and the Perkinistic Institution in Four Cantos*, 2d ed. (New York, 1804), 126. On Abildgaard's electrical reanimation, see Michael Brian Schiffer, *Draw the Lightning Down: Benjamin Franklin and Electrical Technology in the Age of Enlightenment* (Berkeley, 2003), 123.

20. Roderick W. Home, "Force, Electricity, and the Powers of Living Matter in Newton's Mature Philosophy of Nature," in Margaret J. Osler and Paul Lawrence Farber, eds., *Religion, Science and Worldview: Essays in Honor of Richard S. Westfall* (Cambridge, 1985), 95–117; on animal electricity and its relation to the invention of the battery, see W. Cameron Walker, "Animal Electricity before Galvani," *Annals of Science* 2 (1937): 84–113, and Giuliano Pancaldi, "Electricity and Life: Volta's Path to the Battery," *Historical Studies in the Physical and Biological Sciences* 21 (1990): 123–126. Bertholon's *De L'Électricité du Corps Humain* (1780) grew out of the prize essay he wrote in response to the following question posed by the Academy at Lyon in 1777: "What are the diseases that depend on an increase or a decrease in the quantity of the electric fluid in the human body, and how may they be treated?" Quoted in Margaret Rowbottom and Charles Susskind, *Electricity and Medicine: History of Their Interaction* (San Francisco, 1984), 25.
21. For general treatments of Galvani, see Marcello Pera, *The Ambiguous Frog: The Galvani-Volta Controversy on Animal Electricity*, trans. Jonathan Mandelbaum (Princeton, 1992); Bresadola and Pancaldi, *Luigi Galvani International Workshop Proceedings;* and Marco Piccolino and Marco Bresadola, *Rane, Scintille, e Torpedini* (Turin, 2003). See also Rowbottom and Susskind, *Electricity and Medicine*, 31–44; and John L. Heilbron, *Electricity in the Seventeenth and Eighteenth Centuries: A Study of Early Modern Physics*, rev. ed. (Mineola, N.Y., 1999), 491. On Fowler, see Jacyna, "Galvanic Influences," 167–177.
22. Moses Willard, "Treatise on Medical Electricity" (Worcester, Mass., Sept. 1789), Benjamin Rush Papers, box 8, Historical Society of Pennsylvania, Philadelphia; John Redman Coxe, *An Inaugural Essay on Inflammation* (Philadelphia, 1794), 23; Johann Friedrich Blumenbach, *Elements of Physiology; Translated from the Original Latin, and Interspersed with Occasional Notes by Charles Caldwell, to Which Is Subjoined, by the Translator an Appendix, Exhibiting a Brief and Compendious View of the Existing Discoveries Relative to the Subject of Animal Electricity*, 2 vols. (Philadelphia, 1795), 1:217–239; anonymous review of *Experiments and Observations Relative to the Influence Lately Discovered by M. Galvani, and Commonly Called Animal Electricity by Richard Fowler* in the *American Monthly Review, or Literary Journal* (Jan. 1795): 139–142; and Joseph Macrery, *An Inaugural Dissertation on the Principle of Animation* (Wilmington, Del., 1802), 22–23. Caldwell claimed that the only English

accounts of Galvani in North America were Fowler's and the work of Eusebio Valli: see Blumenbach, *Elements of Physiology*, 1:239. The Paris *Report of the Commissioners Appointed by the National Institute to Repeat the Experiments Which Have Been Made on Galvanism* was excerpted from the *Bulletin des Sciences*, and reprinted in the *Medical Repository* 3 (1800): 75–77.

23. On April 2, 1803, the *New York Herald* reported that Aldini had "shewed the eminent and superior powers of Galvanism to be far beyond any other stimulant in nature." A second article in the *Herald* (June 4, 1803) advocated the use of both oxygen and electricity to restore suspended animation. See S[igismund] N. Niderburg, *Improved Galvanismus, and Its Medical Application* (New York, 1803), 9. Niderburg advertised his services in the local press: see *New York Evening Post*, Apr. 4, 1803. He later went to Cuba: see *Cultivo del Galvanismo y Uso de Sus Virtudes por la Medicina* (Havana, 1807). *Morning Chronicle* (New York), July 11, 1803, and *New York Spectator*, July 11, 1804, as quoted in Rita Gottesman, ed., *The Arts and Crafts in New York, 1800–1804: Advertisements and News Items from New York City Newspapers* (New York, 1965), 408–409; Samuel Miller, *A Brief Retrospect of the Eighteenth Century*, 2 vols. (New York, 1803), 1:21–32; Thomas Dobson, *Supplement to the Encyclopaedia*, 2 vols. (Philadelphia, 1803), 2:73–91. On the persistence of galvanic interpretations and uses of the Voltaic pile among physicians, see Christine Blondel, "Animal Electricity in Paris: From Initial Support, to Its Discredit and Eventual Rehabilitation," in Bresadola and Pancaldi, *Luigi Galvani International Workshop Proceedings*, 187–209.

24. Perkins, *Influence of the Metallic Tractors*, 2–3; Lyman F. Kebler, "United States Patents Granted for Medicines during the Pioneer Years of the Patent Office," *Journal of the American Pharmaceutical Association* 24 (June 1935): 486–487.

25. Vaughan, *Observations on Animal Electricity*, iv, 10, 15–16, 20–21, 25, 27–31. See also Vaughan, *The Valedictory Lecture Delivered before the Philosophical Society of Delaware* (Wilmington, Del., 1800), 16–22. Vaughan appears to have been one of the few Americans who published remarks on the Voltaic pile (and its electrochemical applications) around the year 1800: see his *Valedictory Lecture*, 14–15.

26. Perkins, *Influence of the Metallic Tractors*, 27, 78, 80–81, 90–100; Matthew Yatman, *Animal Electricity; or, Observations on the Origin and Identity of the Electric and Galvanic Fluids; with a Review of the Use of the Metallic*

Conductors, 2d ed. (London, 1805), 1, 3, 9, 10, 25; Abildgaard quoted in Perkins, *Experiments with the Metallic Tractors*, 41–50.

27. Langworthy, *A View of the Perkinean Electricity*, 25–26; Perkins, *Experiments with the Metallic Tractors*, 73, 305–307.
28. Perkins, *Experiments with the Metallic Tractors*, 50; *Extract from the Minutes of the Proceedings of a Committee on the Establishment of the Perkinean Institution* (London, n.d.), 5–6.
29. Peter Benes, "Itinerant Physicians, Healers, and Surgeon-Dentists in New England and New York, 1720–1825," *Dublin Seminar for New England Folklife Annual Proceedings* (July 1990): 100; *New York Evening Post*, Apr. 4, 1803; Peter Benes, "Itinerant Entertainers in New England and New York, 1687–1830," *Dublin Seminar for New England Folklife: Annual Proceedings, June 16–17, 1984* (Boston, 1986), 108; *The Public Records of the Colony of Connecticut, from October 1772 to April 1775, Inclusive*, ed. Charles J. Hoadly (Hartford, Conn., 1887), 208; Walter J. Meserve, *An Emerging Entertainment: The Drama of the American People to 1828* (Bloomington, Ill., 1977); Barbara Maria Stafford, *Artful Science: Enlightenment Entertainment and the Eclipse of Visual Education* (Cambridge, Mass., 1994), 16. For an early American antitheatrical primer, see *Extracts from the Writings of Divers Eminent Authors, of Different Religious Denominations; and at Various Periods of Time, Representing the Evils and Pernicious Effects of the Stage Plays, and other Vain Amusements* (Philadelphia, 1789).
30. Langworthy, *A View of the Perkinean Electricity*, 28; Charles Wilkinson, *Elements of Galvanism*, 2 vols. (London, 1804), 2:212; Grete de Francesco, *The Power of the Charlatan*, trans. Miriam Beard (New Haven, 1939), 3–12, 74; Jay Fliegelman, *Declaring Independence: Jefferson, Natural Language, and the Culture of Performance* (Stanford, 1993), 79–94; Warner, *Letters of the Republic*, 32–72. On the use of machines to speak authoritatively for nature, see the analysis of Shapin and Schaffer's account of Boyle in *Leviathan and the Air-Pump* by Bruno Latour in *We Have Never Been Modern*, trans. Catherine Porter (Cambridge, Mass., 1993), 22–24.
31. Gale advocated using tractors in one instance but remained skeptical, suggesting that "perhaps a pair of steel-pointed scissors would answer the same purpose." He concluded that they worked through mechanical frictional stimulus or the imagination: see T. Gale, *Electricity, or Ethereal Fire, Considered* (Troy, N.Y., 1802), 9, 118–119. Some electricians patented their designs for machines: see Paola Bertucci, "A Philosophical Busi-

ness: Edward Nairne and the Patent Electrical Machine (1782)," *History of Technology* 23 (2001): 41–58. Industrial entrepreneurs were better able to exploit legal patents than gentlemanly natural philosophers because the credibility of philosophers was often explicitly tied to an image of social disinterestedness. See, for example, the comparison between James Watt and Joseph Black in Golinski, *Science as Public Culture*, 43–44. See also Kebler, "United States Patents Granted for Medicines during the Pioneer Years of the Patent Office," 486–487; Christine MacLeod, *Inventing the Industrial Revolution: The English Patent System, 1660–1800* (Cambridge, 1988); and B. Zorina Khan and Kenneth L Sokoloff, "Patent Institutions, Industrial Organisation, and Early Technological Change: Britain and the United States, 1790–1850," in Maxine Berg and Kristine Bruland, eds., *Technological Revolutions in Europe: Historical Perspectives* (Northampton, Mass., 1998), 292–313.

32. Elisha Perkins, *To All People to Whom These Presents Shall Come, Greeting*, broadside (1796?); Benjamin Perkins, *Was Published, April 1st, 1801, Price 1s. an Entire New Work, Entitled, Cases of Successful Practice with Perkins's Patent Metallic Tractors* (London, 1801). The words "Perkins's Patent Tractors" appear to have been stamped only on tractors sold in Britain, not those in the United States.

33. John Corry, *The Detector of Quackery; or, Analyser of Medical, Philosophical, Political, Dramatic, and Literary Imposture*, 2d ed. (London, 1802), 54; Perkins, *Influence of Metallic Tractors*, viii–x. John Wesley published a book called *The Desideratum; or, Electricity Made Plain and Useful, by a Lover of Mankind, and of Common Sense* (London, 1760).

34. For a list of agents in the United States, see Elisha Perkins, *Certificates of the Efficacy of Doctor Perkins's Patent Metallic Instruments* (Newburyport, Mass., 1796), 24.

35. Perkins, *Influence of the Metallic Tractors*, 69; and Perkins, *Experiments with the Metallic Tractors*, 255–256. Elisha Perkins thought the endorsement of clergymen was crucial in New England: "I have found essential service from the Clergy especially in New England where they are generally pious men and seek the Happiness of their fellow creatures." Elisha Perkins to Benjamin Perkins, Oct. 19, 1798, Elisha Perkins Letter and Account Book. Elisha Perkins quoted Oliver Ellsworth defending him in a letter to John Marshall in 1797: "Strange as his hypothesis may be, experiments give it countenance. In some cases the effects wrought are not easily ascribable to imagination, great and delusive as is its power"

(Ellsworth to Marshall, Mar. 7, 1797, in Benjamin Perkins, *Influence of the Metallic Tractors*, 9). I have not found any external corroboration of these endorsements. The receipt for a purchase of tractors in 1796 in the name of Jeremy Belknap, however, is extant in the collection of broadsides at the Massachusetts Historical Society, Boston; and Rufus King, whom Elisha Perkins asked to write a letter of introduction for his son in London, owned a copy of Benjamin Perkins's *Experiments with the Metallic Tractors*. Its pages, however, were never cut (the copy is now in the library of the New York Historical Society).

36. Benjamin Perkins, *Efficacy of Perkins's Patent Metallic Tractors*, xxix–xxx, lv–lvi; *Transactions of the Perkinean Society* (London, 1804), 2; Elisha Perkins, *Certificates*, 24. On the emergence of institutional settings for medical electricity in late eighteenth-century London, see Rowbottom and Susskind, *Electricity and Medicine*, 27–28.

37. Langworthy, *A View of the Perkinean Electricity*, 11–13; Perkins, *Influence of the Metallic Tractors*, x; Perkins, *Experiments with the Metallic Tractors*, 7 (Tode is quoted on 80–95). See also Chandos Michael Brown, "A Natural History of the Gloucester Sea Serpent: Knowledge, Power, and the Culture of Science in Antebellum America," *American Quarterly* 42 (1990): 424.

38. *Connecticut Courant* quoted in Haggard, "The First Published Attack on Perkinism," 152–153; James Currie, "Remarks on the Effects of Perkins's Metallic Rods in the Cure of Diseases," read Apr. 7, 1797, Manuscript Communications, American Philosophical Society Archives, Philadelphia. On the limited police powers of the predominantly Federalist Connecticut Medical Society, see Chandos Michael Brown, *Benjamin Silliman: A Life in the Young Republic* (Princeton, 1989), 264–265.

39. Lorraine Daston, "Fear and Loathing of the Imagination in Science," *Daedalus* 127 (Winter 1998): 73–95; Lorraine Daston and Katharine Park, *Wonder and the Orders of Nature, 1150–1750* (New York, 1998), 339–341; Riskin, *Science in the Age of Sensibility*, 209–225; Jan Goldstein, "Enthusiasm or Imagination? Eighteenth-Century Smear Words in Comparative National Context," *Huntington Library Quarterly* 60 (1998): 29–49; and Simon Schaffer, "Self Evidence," in James Chandler, Arnold I. Davidson, and Harry Harootunian, eds., *Questions of Evidence: Proof, Practice, and Persuasion across the Disciplines* (Chicago, 1994), 68.

40. John Haygarth, *Of the Imagination as a Cause and as a Cure of Disorders of the Body* (Bath, 1800), 3–5, 15, 23. On Smith, see Jacyna, "Galvanic

Influences," 182–183; and Mary E. Fissell, *Patients, Power, and the Poor in Eighteenth-Century Bristol* (Cambridge, 1991), 55–57. Haygarth was not shy in criticizing American expertise. He characterized the response of the Philadelphia Academy of Medicine to the yellow fever epidemic as "frivolous, inadequate and groundless." This charge produced a nationalist furor among American physicians, leading to the publication of Charles Caldwell's *A Reply to Dr. Haygarth's 'Letter to Dr. Percival on Infectious Fever' . . . Exposing the Medical, Philosophical, and Literary Errors of That Author, and Vindicating the Right to Which the Faculty of the United States Have to Think and Decide for Themselves Respecting the Diseases of Their Own Country, Uninfluenced by the Notions of the Physicians of Europe* (Philadelphia, 1802). On Haygarth, see Francis M. Lobo, "John Haygarth, Smallpox and Religious Dissent in Eighteenth-Century England," in Andrew Cunningham and Roger French, eds., *The Medical Enlightenment of the Eighteenth Century* (Cambridge, 1990), 217–253.

41. Perkins, *Efficacy of Perkins's Patent Metallic Tractors*, 16; *Influence of the Metallic Tractors*, 2; Haygarth, *Of the Imagination*, 15, 17, 25, 28 (emphasis added), and as quoted in Fessenden, *Terrible Tractoration*, 84–86. See also *Rapport des Commissaires chargés par le Roi, de l'Examen du Magnétisme Animal* (Paris, 1784). On witness manipulation in Mesmerism, see Schaffer, "Self Evidence," 89–97.

42. Corry, *Quack Doctors Dissected*, 25; Perkins, *Efficacy of Perkins's Patent Metallic Instruments*, xxxvii–xxxviii; Perkins, *Influence of the Metallic Tractors*, 24–25; Langworthy, *View of the Perkinean Electricity*, 94–95.

43. See Perkins, *Efficacy of Perkins's Patent Metallic Instruments*, xxix–xxx, and *Influence of the Metallic Instruments*, 94–95; Smith quoted in Haygarth, *Of the Imagination*, 5, 13, 18; "Dr. Christopher Caustic," Fessenden's narrator in *Terrible Tractoration*, argued that horses and other animals treated were in fact peculiarly "susceptible of impressions from imagination": Fessenden, *Terrible Tractoration*, 94.

44. Corry, *Detector of Quackery*, title page, 10, 21; Corry, *Quack Doctors*, 25, 27–37; Fessenden, *Terrible Tractoration*, 133.

45. Brown, *Benjamin Silliman*, 151, 220, 241.

46. Lorraine Daston, "Marvelous Facts and Miraculous Evidence in Early Modern Europe," *Critical Inquiry* 18 (1991): 93–124. For recent approaches to the relationship of things and ideas, see Bill Brown, ed., *Things*, special issue of *Critical Inquiry* 28 (2001); and Lorraine Daston,

ed., *Things That Talk: Object Lessons from Art and Science* (New York, 2004).

47. Darnton, *Mesmerism and the End of the Enlightenment*; Miller, *Brief Retrospect of the Eighteenth Century*, 2:410. On early American medicine, see Richard H. Shryock, *Medicine and Society in America, 1660–1860* (New York, 1960); for the British scene, see Roy Porter, *Health for Sale: Quackery in England, 1660–1850* (New York, 1989).

Conclusion

1. On quantification and mechanization in late-eighteenth-century electricity, see John L. Heilbron, *Electricity in the Seventeenth and Eighteenth Centuries: A Study of Early Modern Physics*, rev. ed. (Mineola, N.Y., 1999), 449–500; and Heinz Otto Sibum, "The Bookkeeper of Nature: Benjamin Franklin's Electrical Research and the Development of Experimental Natural Philosophy in the Eighteenth Century," in J. A. Leo Lemay, ed., *Reappraising Benjamin Franklin: A Bicentennial Perspective* (Newark, Del., 1993), 237–239. The classic work on American anti-intellectualism is Richard Hofstadter, *Anti-Intellectualism in American Life* (New York, 1966).
2. Steven Shapin and Simon Schaffer, *Leviathan and the Air-Pump: Hobbes, Boyle, and the Experimental Life* (Princeton, 1985), 332–344.
3. Bruno Latour, *Science in Action: How to Follow Scientists and Engineers through Society* (Cambridge, Mass., 1987), chap. 6; Ralph Bauer, *The Cultural Geography of Colonial American Literatures: Empire, Travel, Modernity* (Cambridge, 2003); James Delbourgo, "Leviathan and the Atlantic," *History of Science* 43 (Mar. 2005): 101–107.

Illustration Sources

1. Marguerite Gérard, after Jean-Honoré Fragonard, *Au Génie de Franklin* (Paris, 1779). Davison Art Center, Wesleyan University.
2. Robert Sayer and John Bennett, *North America and the West Indies with the Opposite Coasts of Europe and Africa* (London, 1775). Map Collection, Rare Books and Special Collections Division, McGill University Libraries, Montreal.
3. William Watson, *Expériences et Observations, pour servir a l'Explication de la Nature et de Propriétés de l'Electricité* (Paris, 1748). Courtesy of the Bakken Library and Museum, Minneapolis.
4. Jean Antoine, Abbé Nollet, *Essai sur l'Electricité des Corps* (Paris, 1746). Courtesy of the Bakken Library and Museum, Minneapolis.
5. Benjamin Franklin, *Experiments and Observations on Electricity* (London, 1774). Courtesy of the Bakken Library and Museum, Minneapolis.
6. Benjamin West, *Benjamin Franklin Drawing Electricity from the Sky*, ca. 1805–1815. Philadelphia Museum of Art: Gift of Mr. and Mrs. Wharton Sinkler, 1956.
7. Edward Fisher, mezzotint (1763), after Mason Chamberlain, *Benjamin Franklin* (1762). Philadelphia Museum of Art: Gift of Mr. and Mrs. Wharton Sinkler, 1956.
8. Dominickus Beck, *Kurzer Entwurf der Lehre von der Electricität* (Salzburg, 1787). Courtesy of the Bakken Library and Museum, Minneapolis.
9. David Rittenhouse and John Jones, "Accounts of Several Houses in Philadelphia, Struck with Lightning," *Transactions of the American Philosophical Society* 3 (1793). Courtesy American Antiquarian Society, Worcester, Mass.

10. Ebenezer Kinnersley, "A Course of Experiments, on the newly-discovered Electrical Fire," broadside (1752). Rosenbach Museum and Library, Philadelphia.
11. Benjamin Martin, *The Young Gentleman and Lady's Philosophy, in a Continued Survey of the Works of Nature and Art* . . . 2 vols. (London, 1781). This item is reproduced by permission of The Huntington Library, San Marino, California.
12. Anonymous, untitled, and undated watercolor on paper dated Auvergne, France, 1788. From the collection of Charles W. Lard.
13. "Political Electricity; or, An Historical and Prophetical Print in the Year 1770," broadside (London, 1770). Courtesy American Antiquarian Society, Worcester, Mass.
14. "Political Electricity," detail. Courtesy American Antiquarian Society, Worcester, Mass.
15. Rigobert Bonne, *Carte de la Terre Ferme, de la Guyane, et du Pays des Amazones* (Bordeaux, 1771). Map Collection, Rare Books and Special Collections Division, McGill University Libraries, Montreal.
16. John Hunter, "An Account of the Gymnotus Electricus," *Philosophical Transactions of the Royal Society* 65 (1775). Courtesy of the Bakken Library and Museum, Minneapolis.
17. Perkins's Patent Metallic Tractors. Courtesy of the New York Academy of Medicine Library.

Acknowledgments

My first thanks go to David Armitage, who has supported this project throughout, and read the manuscript at several stages. His energy, rigor, and generosity have inspired and improved it at every turn. Special thanks also go to Simon Schaffer for providing an exhilarating introduction to the history of science in Cambridge, for repeatedly reading and discussing this material, and for the inspiration of his work. I thank my dissertation readers: Richard Bushman, Joyce Chaplin, Herb Sloan, and Margaret Jacob, who has also been unstinting in her encouragement and generosity. Jorge Cañizares-Esguerra and an anonymous reviewer for Harvard University Press read the revised manuscript in its entirety and made several useful suggestions. In addition to these scholars, colleagues who have given valuable advice on specific chapters include Douglas Anderson, Paola Bertucci, John L. Brooke, Vincent Brown, Nick Dew, Chris Grasso, Fredrik Jonsson, Marjoleine Kars, Sarah Knott, Giuliano Pancaldi, Lissa Roberts, Neil Safier, Tom Schaeper, Andrea Tone, as well as anonymous referees for the Klemperer fellowship at the New York Academy of Medicine and the *William and Mary Quarterly*. Librarians Rob Cox, Norman Fiering, Roy Goodman, Elizabeth Ihrig, Ed Morman, John Pollack, and Robert Scott gave generously of their time and expertise, as did their staffs. For their support over the years, I am grateful to Howard Temperley and Roger Thompson, my mentors at the University of East Anglia; Milton Cantor and Arthur Kinney of the University of Massachusetts Amherst; and at Columbia, Alan Brinkley, Eric Foner, and Eben Moglen. Victoria de Grazia graciously provided a place to stay and write in New York during the summer of 2003, and Dieter Kopp did the same on numerous occasions in Rome. And from start to finish it has been a pleasure to work with my editor at Harvard University Press, Kathleen McDermott; her assistant, Kathi Drummy; and my copyeditor, Julie Carlson. Thanks also go to my indispensable research assistant, Kristen Keerma, who helped proofread the entire manuscript.

I happily acknowledge support from the following institutions during research and writing: the Graduate School of Arts and Sciences, Columbia University; the McNeil Center for Early American Studies, University of Pennsylvania, directed by Dan Richter, where I was a fellow during 2001–2002; the American Philosophical Society; the Library Company of Philadelphia; the John Carter Brown Library; the Massachusetts Historical Society; the Gilder Lehrman Institute of American History; and the American Historical Association (the Kraus Grant in Atlantic History). Earlier versions of the material in this book were presented at the Omohundro Institute for Early American History and Culture annual conference; Brunel University; "Electric Bodies: Episodes in the History of Medical Electricity," a workshop organized by the International Centre for the History of Science (CIS) at the University of Bologna; "The Circulation of Ideas," an International Seminar on the History of the Atlantic World, Harvard University, directed by Bernard Bailyn; "Bacon to Bartram: Early American Inquiries into the Natural World," a conference organized by the OIEAHC at the American Museum of Natural History; the John Carter Brown Library; the Seminar in Social Studies of Medicine, McGill University; "Taming the Electrical Fire," a conference on the history of the lightning rod organized by Peter Heering, Oliver Hochadel, and David Rhees at the Bakken Museum of Electricity and Life, Minneapolis; the McNeil Center for Early American Studies, University of Pennsylvania; the History of Science Society annual conference; history of science seminars at the Universities of Cambridge, Toronto, Bologna, and l'Université de Québec à Montréal; the seminar in Early American History at Columbia University; the quadrennial meeting of the British, American, and Canadian Societies for the History of Science; a seminar on early modern scientific biography at the Folger Shakespeare Library, Washington, D.C., directed by Steven Shapin; and the William Andrews Clark Memorial Library, UCLA. I am grateful to the audiences on all these occasions. I also wish to thank CIS at the University of Bologna and the *William and Mary Quarterly* for allowing me to reprint material that now appears in revised form as Chapters 6 and 7.

My most personal thanks go to my friends in New York and Cambridge over the years, and to my new friends and colleagues in Montreal, for their companionship and intellectual fellowship. I thank my family, especially my mother, Rosella Maria Delbourgo *née* Properzi; my brother, Richard; and my sister, Liz. And finally, I thank Laura Kopp, who gave some of the very best readings of my work, and with whom I have enjoyed wonderful conversations about electricity, and many other things. I dedicate this book also to her, with love, admiration, and gratitude.

Index

Page numbers in italics refer to figures.

Abildgaard, Peter Christian, 255, 260, 261, 262
Adams, George, 293n20, 296n34
Adams, John, 68, 70, 96, 135, 139, 142, 154, 219, 231, 281
Adanson, Michel, 178–179, 328n16
Adorno, Theodor, 130–131
African-Americans. *See* Slavery/Slaves
Aldini, Giovanni, 257, 349n23
Algarotti, Francesco, 111
Allamand, Jean, 179, 188
Allen, Charles, 110, 121
American Academy of Arts and Sciences, 74, 77
American Philosophical Society, 22, 28, 74, 77, 81, 82, 145, 182, 192, 244, 245, 270
American Revolution, 3, 11, 59, 102–103, 131–164, 171, 206, 238, 247, 248, 270, 275, 278, 280, 281, 283, 326n6. *See also* Politics and electricity
Analogy, 42–43, 53–59, 72–74, 77, 85, 100, 179–182, 190, 192, 197–199, 214, 216, 217, 328–329n16
Anderson, Alexander, Jr., 118, 121, 229
Animal electricity, 12, 118, 167, 204, 224, 255–262, 334n5. *See also* Galvanism

Animal magnetism. *See* Mesmer, Franz Anton/Mesmerism
Anti-intellectualism, 243, 246, 280, 283
Aristotle/Aristotelianism, 18, 41, 42
Artemidorus, 177
Atlantic Ocean, circulation of knowledge around, 7, 11, 12, 16–18, *19*, 22, 27–28, 40, 43, 58, 93, 96, 104, 111, 178–179, 182, 190, 195–196, 227, 238, 243, 244, 268, 270, 276–277, 279–282
Atmospheric electricity, 67, 103, 235–238, 303n27, 341n41
Au Génie de Franklin (Fragonard, 1779), 3, 148–149
Autotherapy, 226–231, 266, 280

Bacon, Francis/Baconianism, 3, 16, 18, 243, 244, 249
Baker, Gardiner, 112, 118
Baker, George, 191–195, 332n35
Baldwin, Cyrus, 47
Baldwin, Loammi, 1–4, 60–64, 65, 236, 305n33
Bancroft, Edward, 12, 125, 158, 167–176, 179–186, 188, 191–193, 196–198, 256, 282, 326nn4,6, 327n8, 333n38
Bandi, Cornelia, 234

Banks, Joseph, 17, 75, 172, 174, 327n8
Barker, William, 267
Barlow, Joel, 149
Barnard, Thomas, 221
Bartlett, John, 224
Bartram, John, 21, 29, 144
Bartram, William, 175
Bass, Robert, 296n33
Bassi, Laura, 118
Beccaria, Giambattista, 58
Behn, Aphra, 188
Belcher, Jonathan, 205
Belknap, Jeremy, 352n35
Bellevue Hospital (New York), 227
Bentham, Jeremy, 158
Berkeley, Bishop George, 217, 219
Bertholon, Abbé Pierre, 256, 348n20
Bianchini, Giovanni, 233
Bible, 107, 126, 152, 217, 218, 219
Bishop, John, 133
Bligh, William, 172
Body and electricity, interactions between, 5, 6–9, 11, 14–15, 25–27, 39, 46, 48–49, 55–56, 57, 62–66, 71, 88–89, 98, 119–128, 131–132, 133–134, 140–142, 148–150, 162, 163, 180–182, 186–189, 198, 202–203, 213–216, 223–224, 232–233, 260, 282, 332n36
Böhme, Jakob, 217
Bonnet, Charles, 248
Borelli, Giovanni, 178
Bose, Georg Matthias, 30, 115, 117, 120
Botany. *See* Natural history/botany
Boucher, Jonathan, 153
Bowdoin, James, II, 100, 221
Boyle, Robert, 25, 145, 233
Brackenridge, Hugh Henry, 144
Brain and electricity, 211, 256, 257, 259
Brattle, Thomas, 20
Bray, Thomas, 243–244
Brickell, Richard, 121
Brockenburry, John, 233
Brown, Charles Brockden, 51, 235–238
Brown, John, 207
Browne, Thomas, 24
Bryant, William, 182, 186, 193, 194

Buffon, Georges Louis Leclerc, Comte de, 20, 58
Burke, Edmund, 130–131, 158, 159, 245, 325n39
Burnet, Gilbert, 65, 300n14
Bute, Lord, 162, *163*
Byles, Mather, 66
Byrd, William, II, 20

Cabeo, Niccolò, 25
Caldwell, Charles, 123–124, 257, 259, 348n22, 353n40
Canton, John, 58
Carter, Landon, 75, 84–86
Caruthers, William, 15
Castiglione, Baldassare, 91
Catesby, Mark, 18, 173, 175
Cavallo, Tiberius, 209, 227, 256, 261, 334n5
Cave, Edward, 41, 56
Cavendish, Henry, 197–198, 256
Chamberlain, Mason, 59–60, *61*, 147
Charlatanry, 90, 262–277
Charleston Library Society, 112, 312n30
Charleton, Walter, 178
Children and electricity, 112, *113*
Claggett, William, 48, 120–121
Clarke, Gedney, 170
Clayton, John, 233
Cleland, John, 117
Clive, Robert, 197
Cohen, I. Bernard, 7–8
Colden, Cadwallader, 21, 29, 52, 219
College of Philadelphia, 94, 96, 111, 112, 124, 144, 239, 302n25, 335n7, 347n16
Collinson, Peter, 29–30, 32, 38, 40, 56, 120, 173, 296n34
Colman, Benjamin, 102
Columbus, Christopher, 178
Commerce, 13, 16, 17, 18, 22, 28, 43, 45, 47–48, 91–92, 95, 124, 135, 168, 185, 190–195, 206–207, 212, 222, 235, 277
Common sense, 12, 38, 63, 242, 246–249, 254, 255, 263–264, 266, 269, 270, 272, 274–277, 345n12
Condamine, Charles-Marie de la, 178

Connecticut Medical Society, 264, 269
Conspiracy, 129–130, 159–164
Cook, James, 75, 172
Corry, John, 240, 266, 272, 274–275
Cosmopolitanism, 20–22, 59, 97, 143, 147
Coxe, John Redman, 257
Creole American knowledge, politics of, 20, 21, 38–40, 58–59, 143, 173–176, 180, 204–205, 206, 238, 281–283, 353n40
Cullen, William, 223
Cunningham, James, 228
Currie, James, 269–270
Cutbush, Edward, 211

Dancer, Thomas, 228
Darwin, Erasmus, 116–117, 118, 120, 138, 149, 223
Deane, Silas, 171
Declaration of Independence, 138, 142, 247
Deism, 45, 109, 218
Dellap, Samuel, 206
Dennie, Joseph, 245, 246
Derham, William, 65, 300n14
Desaguliers, Jean T., 23, 26, 92
Descartes, René/Cartesianism, 8, 25, 39, 41, 246
Devotion, John, 260
Dexter, Aaron, 228
Digby, Kenelm, 25
Dioscorides, 177
Dispute of the New World, 20, 50–51, 237–238
Dobson, Thomas, 258
Domjen, Samuel, 93
Doorn, Abraham van, 189
Drayton, William Henry, 135
Drinker, Elizabeth, 111, 312n39
Duer, William, 137
Dufay, Charles, 26, 29
Duplessis, Joseph-Siffred, 153
Dutch West India Company, 168
Dwight, Timothy, 138

Earthquakes, 8, 66–72, 104, 106, 107
Economy of charge. *See* Franklinist electricity
Edwards, Pierpont, 253
Electric battery (Voltaic pile), 12, 167, 198–199, 204, 256, 260, 333n39, 349n25
Electric eel (Gymnotus Electricus), 12, 105, 165–167, 177–197, 204, 238, 256, 325n2, 328nn15,16, 329n18, 330n26, 332n36, 333n38
Electric kiss. *See* Venus electrificata
Electrical demonstrations, 11, 27, 28–29, 48, 72–73, 74, 87–128, 129–130, 154, 165–166, 190, 206, 211–212, 262–263, 302n24, 303n27, 309n13, 312n29, 318n6, 335n7
Electrical machines (electrostatic generators), 7, 14, 34, *35*, 47–48, 106, 121, 124–125, 127, 140–142, 146, 151–152, *155*, 159, *163*, 181, 182, 184, 199, 205, 206, 208–209, 227–230, 293n20, 295n32, 296n34, 312n30, 314n41, 316n48, 329n16, 330n24, 332n36, 340n34
Electricity, early history of (pre-1740s), 24–27
Electrometers, 209, 210, 228, 232, 341n38
Electrostatic generators. *See* Electrical machines
Electrotherapy. *See* Medical electricity
Eliot, Andrew, 109
Elliott, John, 31
Ellis, John, 166
Ellsworth, Oliver, 267, 351n35
Empire/Imperialism, 17, 21, 40, 46, 124, 143, 144, 153, 156, *161*, 162, 172–176, 238, 244, 248, 276, 280, 281
Enlightenment, 3–4, 6–11, 28, 42, 44, 51–52, 75–76, 89–90, 119–120, 130–132, 141–143, 166–167, 183, 202–203, 218, 243, 276–277, 278–283, 314n38
Enthusiasm, 4, 62–63, 68, 94, 107, 131, 136–137, 141–143, 159–164, 205, 206, 236–237, 243, 270–273, 280, 325n39
Entrepreneurialism, 17–18, 47–48, 90, 119, 166, 191, 202, 206, 225, 277, 351n31
Ether, 32, 117, 207, 213–216, 217, 256

Evangelicalism, 96, 102, 142, 159, 204, 213, 218, 235, 306n38
Experiment on a Bird in the Air Pump (Wright of Derby, 1768), 123
Experiment(s), 23–24, 31–42, 46–47, 53–63, 151–153, 173–174, 176–177, 179–183, 186–189, 193, 195, 197–199, 219, 229, 241, 252–253, 256, 261, 278–279, 281, 294n22, 329n18

Fact, matters of, 12, 24, 41, 83, 143, 145, 151, 240, 242–243, 246, 247, 248–249, 252–253, 254, 261, 262, 263, 264, 266, 269, 272, 275–277; facts vs. theories, 249–255, 262
Faraday, Michael, 7–8
Fermin, Philippe, 186–187, 192
Fessenden, Thomas Green, 118, 255, 269, 275
Festival, 137–139
Fisher, Edward, *61*, 147
Flagg, Henry Collins, 187–188, 192
Fleet, John, 223–224
Fontenelle, Bernard de, 111
Foster, Hannah Webster, 111
Fothergill, John, 41, 56, 223
Foucault, Michel, 9, 220, 226
Fowler, Richard, 256, 257
Fragonard, Jean-Honoré, 3, 148–149
Frankenstein (Shelley, 1818), 4, 223, 257
Franklin, Benjamin, 1, 3–4, 5, 6–7, 10–11, 13, 15–16, 22, 27–46, 51, 53–60, *61*, 67, 74, 77, 81, 82, 87, 89, 92, 93, 120, 123, 132, 139, 143–164, 170, 171, 173, 177, 198, 202, 205, 218, 219, 244, 270, 272, 278–279, 281, 282, 283, 293n16, 296n34, 297n4, 298n6, 303n27, 304n31, 333n38
Franklin, William, 267
Franklinist electricity (economy of charge), 31–46, 53–63, 68–72, 85–86, 90, 98–102, 104–105, 112, 143–164, 176–177, 181, 197–198, 203, 207, 211, 214–215, 218–219, 223–224, 225, 232–233, 235, 238, 258, 259–260, 272, 276, 279, 281, 303n27, 306n38, 323n29, 334n5
Freemasonry, 154, *155*, 217

French Revolution, 116, 130, 139, 141, 231, 236, 245, 270, 275
Freneau, Philip, 144

Gale, T., 12, 200–232, 235, 237, 239, 242, 249, 251, 264, 266, 279–280, 281, 334n5, 335n8, 336n14, 339n28, 350n31
Galen, 177, 208
Galvani, Luigi, 204, 207, 256–257, 258, 272, 335n8
Galvanism, 255–262, 272, 276, 334n5, 349n23. *See also* Animal electricity
Garden, Alexander, 21, 165–166, 192, 194
Gassendi, Pierre, 178
Gay, Ebenezer, 71
Genius, Franklin's, 143–147, 154, 156, 244, 245, 246
George III, 21, 156–157, 162
Gerry, Elbridge, 267
Gesner, Konrad, 177
Gilbert, William, 24–25
Goodrich, Samuel, 133
Gorton, Benjamin, 217
Gothic literature, 203, 232, 236–237
Graham, James, 117–118, 263
Gravesande, Laurens Storm van 's, 170, 179, 188, 325n2
Gravesande, Willem Jacob van 's, 179
Gray, Stephen, 26, 29
Great Awakening: First, 94, 96, 107; Second, 133, 217–218
Greatrakes, Valentine, 250
Greenhill, James, 208
Greenwood, Isaac, 23, 68, 92, 97, 328n15
Greenwood, Isaac, III, 114, 124, 206, 229
Guericke, Otto von, 25
Gymnotus Electricus. *See* Electric eel

Hales, Stephen, 58, 298n9
Hall, Richard Willmot, 207, 335n8
Haller, Albrecht von, 30, 117, 120, 248
Hamilton, Alexander, 102, 121, 247
Harvard College, 1, 20, 22, 23, 44, 59, 68, 92, 103, 114
Hauksbee, Francis, 34

Haygarth, John, 270–272, 274, 353n40
Heilbron, John L., 198–199
Herholdt, C. G., 261
Hermetic philosophy, 217
Hiller, Joseph, 48, 99–100, 104–105, 107, 108, 115, 127–128, 215, 220–221, 309n13
Hoffman, Josiah Ogden, 137
Hogarth, William, 26
Hollis, Thomas, 23
Honeywood, St. John, 139
Hooper, William, 91
Hopkinson, Francis, 81, 82
Hopkinson, Thomas, 31, 298n6
Horkheimer, Max, 130–131
Hosack, David, 224
Humane societies, 203, 221–225, 267
Humanitarianism, 200–204, 220–232, 234, 237, 333n1
Humboldt, Alexander von, 195, 238, 257, 328n16, 330nn24,26, 332nn35,36
Hume, David, 110, 245, 248
Humor. *See* Play; Pleasure
Humoralism, 208
Humphreys, David, 138
Hunter, John, 18, 194–195, *196*, 197, 223, 240
Hutcheson, Francis, 247
Hutchinson, John/Hutchinsonianism, 152, 217, 219, 336n14
Hutchinson, Thomas, 154, 157, 324n35

Ibn-Sidah, 177
Imagination, 3–4, 12, 25, 62–63, 68, 85, 151, 176, 184, 205, *241*, 243, 251, 261, 270–273, 330n24, 351n35
Imperialism. *See* Empire/Imperialism
Improvement, 76, 90, 96–97, 101–102, 111–112, 173, 174–176
Industrialization, 244, 245, 264–265, 268, 351n31
Insanity, treatment of, by electricity, 200–201, 210–211, 236–237
Itinerancy, 12, 17, 19, 72, 74, 75, 92–94, 166, 201–202, 206–207, 229, 239, 240, 257–258, 262, 263, 270, 277

Jefferson, Thomas/Jeffersonianism, 15, 20, 48, 75, 135, 202, 230–232, 236, 245, 246, 283, 316n48, 346n15
Jeffrey, Francis, 145
Johnson, Samuel, 218–220
Johnson, William, 112, 312n30
Jokes. *See* Play; Pleasure
Jones, John, 77–78
Jones, W. and S., 340n34
Jones, William, of Nayland, 152
Junto (Philadelphia), 28, 68–69

Kalm, Per, 123
Kant, Immanuel, 3, 6, 278
Kayashuta, 125
Kelpius, Johannes, 236, 237
King, Rufus, 139, 352n35
King, William, 206–207, 262
Kinnersley, Ebenezer, 8, 11, 31, 60, 72, 74, 75, 87–90, 93–109, 114–115, 120, 122, 123, 126–128, 129–130, 153, 154, 189, 190, 192, 193, 202, 206, 213, 215, 225, 281, 296n34, 309n13, 335n7
Kircher, Athanasius, 25, 91
Kite, Charles, 223
Kite experiments, 1–4, 55–58, 60–64, 100, *161, 163*, 181

Laboratories, 6, 14, 46, 53–54, 56, 60, 73, 77, 85, 87, 100, 106, 162, 167, 176–177, 180–183, 189, 190, 192, 199, 297n4
Lafayette, Marquis de, 139, 346n15
Landriani, Marsilio, 81
Langworthy, Charles, 254, 261, 264, 266, 267, 268, 270
Lathrop, John, 77–78, 80, 81, 222, 224, 303n27
Latrobe, Benjamin, 246
Launy, David, 258
Laurens, Henry, 312n30
Lavater, Johann Caspar, 157–158
Law, Thomas, 336n13
Laws of nature, 24, 43–44, 66, 69–70, 85–86, 91, 100, 215, 216, 260, 282
Lee, Ann, 133

Lee, Arthur, 81
Lee, Charles, 265
Lee, Elias, 217
Leiper, Thomas, 81–82
Léry, Jean de, 178
Leyden jar, 14–16, 33–38, 42–45, 53–54, 73, 98–99, 112, 123, 165, 178–179, 197, 205, 209, 228, 259, 329n16
Library Company of Philadelphia, 28, 29, 30, 31, 34, 92, 297n4
Lightning, 3–4, 5, 50–86, 101, 107, 123, 126, 207, 223, 232, 234, 303n27, 305n34, 306n38
Lightning rods, 5, 11, 50–86, 99–100, 107, 111, 112, 114, 146, 147–148, 156–157, 158, 231, 260, 276, 302n25, 303n26, 304nn31–32, 305nn33–34, 341n41
Ligon, Richard, 16–17
Lining, John, 71
Linnaeus, Carolus, 17, 22, 29, 165, 171–172, 173, 195, 325n2
Locke, John, 25, 219
Logan, James, 205
London Coffee House (Philadelphia), 94, 154
Lorenzini, Stefano, 178
Lott, Frans van der, 179, 188, 189, 329n18, 329n22
Lovett, Richard, 209, 227
Loyalists, 156–164, 171, 325n38
Lucretius, 116

Macclesfield, Lord, 59, 147
Macrery, Joseph, 257
Magic picture (demonstration), 129–132
Manigault, Ann Ashby, 111–112
Manigault, Gabriel, 111–112
Marshall, John, 267, 351n35
Martin, Benjamin, 91, 111
Martinet, Johannes, 71
Materialism, 25, 77, 106, 181, 203, 219
Mather, Cotton, 21, 23, 64–66, 76
Mather, Increase, 21, 64–65, 76
Matthias, the Prophet, 217
Mayr, Otto, 139–141

Mazéas, Abbé Guillaume, 58, 298n9
M'Cabe, John, 48
Mechanical philosophy, 42, 178
Medical electricity, 12, 117, 188–189, 198, 200–232, 233–234, 237, 239–243, 249–277, 288n7, 334n5, 335nn7–8, 339n28, 348n20, 350n31
Meigs, Josiah, 254, 267
Melville, Herman, 305n33
Mercer, John Francis, 135
Merry, Robert, 139
Mesmer, Franz Anton/Mesmerism, 205, 206, 240, 242, 248, 251, 252, 264–265, 270, 272, 276, 346n15
Miles, Henry, 233
Millenarianism, 12, 202, 213, 215–217, 222, 231–232, 237
Miller, Samuel, 114, 258, 277
Miller, William, 217
Mitchill, Samuel Latham, 246
Morse, Jedediah, 224, 267
Morton, Charles, 22–23
Moyes, Henry, 93
Muhlenberg, Henry Melchior, 109
Munson, Eneas, 254, 260
Museums, 111–112
Musschenbroek, Pieter van, 14, 58, 98

Native Americans, 21, 125, 167, 172, 176, 183–189, 199, 281, 316n48, 330nn23,24,26
Natural history/botany, 16–18, 29, 116, 167–176, 186, 195, 245, 327n8, 330n23
Natural philosophy, 39, 85–86, 107, 112, 114, 124–125, 150–153, 173–174, 219, 229, 234, 245, 276
Natural theology. See Physico-theology
Neale, John, 91
Nervous disorders, treatment of, by electricity, 205, 207–208, 210–211, 235
Nervous system and electricity, 204, 208, 235, 256
Networks, 13, 15, 17, 18, 22, 28, 29, 96, 167, 172, 240, 266, 280
Newton, Isaac/Newtonianism, 20, 21, 23–24, 32, 39, 41, 43, 64, 92, 97, 98, 131, 144,

145, 151, 178, 202, 213, 217, 219, 245, 256, 293n18, 323n29
Niderburg, Sigismund, 207, 257–258, 262–263
Nollet, Abbé Jean-Antoine, 26, 32, 40, 41, 42, 58, 59, 82, 96, 98, 180, 189, 205

Obeah, 126–127
Odell, Jonathan, 154, 156
Oliver, Peter, 154, 156, 157, 159, 160
Otto, Louis, 246

Paine, Tom, 247–248, 264
Parker, Benjamin, 240
Patents, 74, 258–259, 264–265, 305n33, 350n31
Patterson, Robert, 82, 112
Peale's Museum, 112
Penn, Thomas, 34, 47, 293n20
Penn, William, 102
Perkinean Society, 240, 255, 260, 267, 268
Perkinism, 239–243, 249–277
Perkins, Benjamin Douglas, 240, 250, 252, 253, 255, 258, 260, 261, 262, 265, 266, 268, 269, 271, 272, 274, 275, 277
Perkins, Elisha, 12, 239, 249–250, 251, 252, 253, 258, 259, 262, 264, 265, 266, 268, 269, 279–280, 281, 283, 347n18, 351n35
Philosophical modesty, 39–40, 44–46, 59, 145, 151, 282–283
Physico-theology, 64–65, 70, 103–104, 175–176, 215
Physiocrats, 42–43
Pickering, Timothy, 265
Pitt the Elder, William, 134
Placebo trials, 270–273
Play, 27, 91, 119, 120, 122, 129–130, 132, 263, 271–272, 274–275, 314n38, 315n42
Pleasure, 28, 31, 56, 59, 73, 76, 90–91, 101, 102–103, 109–110, 112, 113, 117, 119–120, 122, 315n42
Pliny the Elder, 177
Pluche, Abbé Noel-Antoine, 219, 300n14
Poe, Edgar Allan, 224
Politeness, culture of, 26, 76, 90, 110–128

Politics and electricity, 129–164, 278–279, 280–283
Poor Richard's Almanack, 74
Porta, Giambattista della, 91
Porter, Andrew, 47
Premature burial, 224
Priestley, Joseph, 46–47, 55–56, 60, 89, 106–107, 122, 124, 127, 146–147, 158, 282, 295n32, 335n7
Prince, John, 229
Prince, Thomas, 66–72, 83, 107
Provincialism, 20–22, 58, 90, 96–97, 101, 143, 218, 230, 238, 266, 276

Quackery, 240, 242, 262–277

Rabiqueau, Charles, 117
Race and electricity, 124–128, 187–189, 281, 316n48, 317n49
Rafn, J. D., 261
Ralegh, Walter, 167
Randolph, John, 306n38
Rathbun, Valentine, 159
Rational recreations, 90–91
Ray, John, 65, 173, 300n14
Reanimation, 203, 220–225, 257, 339n28
Reason vs. superstition, 3–4, 51–53, 68–69, 114, 127, 219, 281
Réaumur, René-Antoine de, 14, 180, 181, 188, 329n18
Reiche, Charles, 71
Reid, Thomas, 248–249
Religious views of electricity, 45, 51–52, 60, 64–72, 73, 77, 80, 83–86, 90, 96, 102–109, 131, 133–134, 141–142, 159, 202–203, 209, 212–220, 223, 225, 231, 234, 278, 279–280, 282
Republic of letters, 20, 22, 59
Republicanism, 134–143, 150, 153, 156, 160, 161, 164, 206, 213, 221, 222–223, 225–226, 236, 237, 242, 244–245, 247–248, 263–264, 283
Revealed knowledge, 11, 68, 141, 142, 159, 202, 266, 280
Rich, Caleb, 133

Richer, Jean, 178
Richmann, George, 60, 123, 149
Rittenhouse, David, 77–78, 81, 82, 144, 192, 193
Rivers, Lord, 267
Robinson, Ebenezer, 269
Rousseau, Jean-Jacques/Rousseauvianism, 176, 184
Royal Africa Company, 17–18
Royal Institution, 244
Royal Society, 17, 18, 22, 29, 34, 40, 56, 58–59, 60, 64, 122, 158, 165, 171, 172, 268, 281
Rush, Benjamin, 208, 225–226, 228, 251, 257, 259, 281

Sade, Marquis de, 196
Satire, 269–275
Scribonius Largus, 177
Seven Years' War, 172–173
Sewall, Susanna, 233
Sexuality and electricity, 116–119, 154, 155, 195–196. *See also* Venus electrificata; Women, gender, and electricity
Shaftesbury, Third Earl of (Anthony Ashley Cooper), 246–248, 249
Shebbeare, John, 117
Sheffield, Lord, 135
Shippen, William, 81
Sigaud de la Fond, Joseph Aignan, 117
Silliman, Benjamin, 275
Slavery / Slaves, 17, 75, 84–85, 126–128, 133, 167, 168–171, 172, 173, 183–189, 199, 208, 220, 281, 316n48
Sloane, Hans, 18, 173, 175
Smith, Adam, 221, 247
Smith, William, 124–125, 144–145
Solander, Daniel, 195
Spencer, Archibald, 28, 93, 97
Spontaneous combustion, 12, 203, 232–238, 261, 342n43
Stedman, John Gabriel, 182
Stiles, Ezra, 47, 59, 72
Swedenborg, Immanuel, 217
Sympathy, 137–138, 221, 222, 225, 246–247

Syng, Philip, 31, 34, 48
System-building, 40, 59, 151, 248

Tennent, Gilbert, 66
Terror, 14, 70, 76, 108, 114, 123, 127, 131, 211–212, 237, 238, 245, 272
Thacher, Thomas, 221, 222
Theatricality, 87–128, 129–130, 243, 262–277
Theories vs. observations (and facts), 39–42, 59, 151, 249–255, 262, 283
Thistlewood, Thomas, 75, 316n48
Thompson, Benjamin (Count Rumford), 244
Thunder houses, 72–73, 100
Tilton, James, 252, 259, 266, 267
Tocqueville, Alexis de, 245, 248, 280
Tode, Professor, 268–269
Toderini, Giambattista, 147
"Tom Telescope," 112
Tractors, Perkins' metallic, 12, 239–243, 249–277, 350n31
Tucker, Josiah, 159
Tucker, St. George, 306n38
Turgot, Anne-Robert Jacques, 3, 149
Turner, Robert, 329n16

Useful knowledge, 13, 24, 28, 31, 92, 145–146, 203, 212–213, 242, 243–246, 249, 262, 263–264, 272, 276–277, 280

Valli, Eusebio, 261
Vancouver, Charles, 233, 234
Vaughan, John, 239, 252, 259–260, 261, 267, 349n25
Venus electrificata (electric kiss), 30, 88, 115–119, 195, 211–212. *See also* Sexuality and electricity; Women, gender, and electricity
Volney, Constantin, 50–51, 238
Volta, Alessandro, 198–199, 256–257, 261, 333n39
Voltaic pile. *See* Electric battery
Voltaire, 243

Wakefield, Priscilla, 111
Walker, James, 118
Walpole, Horace, 8
Walsh, John, 197–198, 256, 329n16, 333n38
Warren, James, 135
Warren, Joseph, 228
Washington, George, 80, 139, 265, 267
Watson, William, 60, 96, 122
Webster, Noah, 235, 238
Wedderburn, Alexander, 153, 154, 156, 157, 158, 324n35
Wentworth, Paul, 171, 326n6
Wesley, John, 104, 209, 226–227, 282
West, Benjamin, 148–149
Wheler, Granville, 26, 29
Whitefield, George, 94, 96, 102, 107, 108
Whitman, Walt, 141
Wiatt, Susanna, 233
Wieland, or, The Transformation (Brockden Brown, 1798), 235–238
Wilkes, John, *161*, 162
Wilkinson, Charles, 264
Willard, Moses, 228, 257
Williams, Samuel, 103–104, 109
Williams, Thomas, 170–171
Williamson, Hugh, 83, 192–193, 194, 197, 305n34, 332n35

Wilson, Benjamin, 82, 156
Winchester, Elhanan, 253
Winthrop, John, IV, 1, 44, 59, 68–72, 83, 103, 114, 304n32
Wistar, Caspar/Wistarburg Glassworks, 30, 293n16
Wollaston, William, 300n14
Wollstonecraft, Mary, 116, 118
Women, gender, and electricity, 110–119, *155*, 187, 195–196, 211–212, 299n13, 336n13. *See also* Sexuality and electricity; Venus electrificata
Wonder/Wonders, 4, 8–10, 44–46, 62–63, 65, 89–90, 97–99, 101, 103–105, 114, 119–120, 125–126, 141–142, 164, 176, 206, 215, 237, 242–243, 262, 263, 270–276, 280, 282–283, 314n38, 316n48
Woodmason, Charles, 72, 305n36

Yatman, Matthew, 260, 261
Yeldal, Anthony, 263
Yellow fever, 208, 235, 251, 353n40

Zangari, Cornelia, 233, 234